Dynamics of Astrophysical Discs

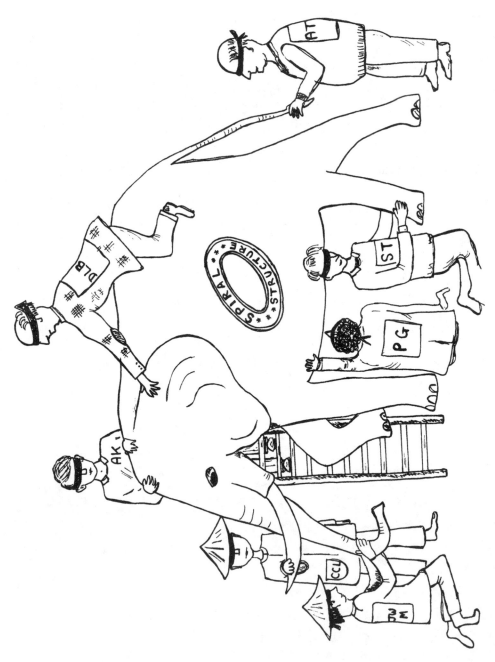

Cartoon, drawn by Janet Sellwood in 1984, based on Toomre's assessment of the state of spiral structure theory in 1980. Apart from a few extra blindfolded individuals, this still seems appropriate today.

Dynamics of Astrophysical Discs

The proceedings of a conference
at the Department of Astronomy,
University of Manchester,
13 - 16 December 1988.

Edited by

J. A. SELLWOOD

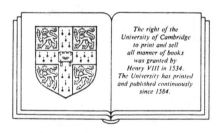

The right of the
University of Cambridge
to print and sell
all manner of books
was granted by
Henry VIII in 1534.
The University has printed
and published continuously
since 1584.

CAMBRIDGE UNIVERSITY PRESS

Cambridge

New York Port Chester

Melbourne Sydney

PUBLISHED BY THE PRESS SYNDICATE OF THE UNIVERSITY OF CAMBRIDGE
The Pitt Building, Trumpington Street, Cambridge, United Kingdom

CAMBRIDGE UNIVERSITY PRESS
The Edinburgh Building, Cambridge CB2 2RU, UK
40 West 20th Street, New York NY 10011–4211, USA
477 Williamstown Road, Port Melbourne, VIC 3207, Australia
Ruiz de Alarcón 13, 28014 Madrid, Spain
Dock House, The Waterfront, Cape Town 8001, South Africa

http://www.cambridge.org

First published 1989
First paperback edition 2003

A catalogue record for this book is available from the British Library

ISBN 0 521 37485 5 hardback
ISBN 0 521 54814 4 paperback

Table of contents

Titles of invited papers are **highlighted.** *Names of co-authors who did not attend the conference are given* in smaller type.

Preface

Discs occurring in a wide diversity of astronomical objects prompt similar questions about their dynamical behaviour. Astrophysicists working on problems related to just one type of disc may find that a similar problem has already been addressed in a different context. This is especially true of the dynamical behaviour: the dispersion relation for spiral density waves was originally derived for galaxies but has been applied, with more success even, to Saturn's rings, the formalism of a global mode treatment for gaseous accretions discs has some similarity to that for the collisionless stellar discs of galaxies, aspects of the dynamics of planet formation around a young stellar object are reflected in the response of a galaxy disc to a co-orbiting giant molecular cloud complex, *etc.*

In order to encourage thinking along these lines, the Department of Astronomy in the University of Manchester organised a four day conference in December 1988 to bring together experts on discs in a number of contexts. In rough order of increasing physical size, these are: planetary ring systems, accretion discs in cataclysmic binary stars and active galactic nuclei, protoplanetary and protostellar discs and disc galaxies.

The aim of the conference, and of these proceedings, was to present those aspects of the behaviour of each type of disc that could be of relevance to other types. To emphasise this theme, the sequence of talks was deliberately arranged so as ensure that many different disc types were discussed on any one day. Though this worked very successfully at the meeting, it is less appropriate for a reference volume; the sequence of papers in these proceedings has therefore been rearranged so that those concerning each particular disc type are grouped together.

Of the papers presented at the meeting, this volume contains articles by all ten review speakers, reports of 15 out of the 17 contributed papers, 33 out of the 34 poster papers and summary talks of both the observational data and theoretical work. The editor is particularly grateful to the overwhelming majority of these authors for transmitting their manuscripts to him in machine-readable form and for not exceeding the requested lengths. Two of the papers by authors from the Soviet Academy of Sciences were given considerably more space than other contributed papers as their work is less widely available in the West.

Full postal addresses of all the participants are listed at the front of the book – the affiliations only of the authors are given in each paper. The abbreviations used for the main astronomical journals are those recommended in *Astronomy and Astrophysics Abstracts*. I have included a citation index of all papers referenced in the book which is followed by an index of authors. I hope that together these will provide more useful information than the customary names index, which has no obvious advantage (aside from boosting the egos of those most frequently cited!).

The enthusiasm of the ninety-five participants, who came from sixteen different countries, and the generally high standard of their presentations ensured the success of the meeting. The conference could not have taken place, however, without the willing help of a very able group of Manchester locals.

The invaluable and constant guidance of Professor Franz Kahn, in matters both scientific and organisational, is greatly appreciated. The local and financial arrangements for the meeting were almost exclusively, and ably handled by Mrs Ellen Carling, assisted by Miss Melanie Thomas. Mrs Carling also typed most of the manuscripts not received in machine readable form. Mrs Yvonne Cain printed the conference programmes *etc.* and an especially well presented book of *Abstracts* for participants. Mr John Turner assisted with slides, videos and other aids during the sessions. The conference photograph was taken and printed by Mr Ian Callaghan.

The text of this volume was prepared using TeX on the Manchester Starlink facilities. I hope it will serve both as an accurate record of the meeting and a useful reference work in years to come.

JERRY SELLWOOD
June 1989

Names and addresses of participants

Peter Allan   Department of Astronomy, The University, Manchester M13 9PL

Moray Anderson   Department of Astronomy, The University, Glasgow G12 8QW

Pawel Artymowicz   Space Telescope Science Institute, 3700 San Martin Drive,
                       Baltimore MD 21218, USA

Lia Athanassoula   Observatoire de Marseille, 2 Place le Verrier, 13248 Marseille Cedex 04,
                       France

Mark Bailey   Department of Astronomy, The University, Manchester M13 9PL

Joss Bland   Department of Astronomy & Space Physics, Rice University,
                       Houston TX 77251, USA

Mike Bode   School of Physics & Astronomy, Lancashire Polytechnic,
                       Corporation Street, Preston PR1 2TQ

Nicole Borderies   Jet Propulsion Laboratory 301-150, 4800 Oak Grove Drive,
                       Pasadena CA 91109, USA

Genevieve Brebner   School of Physics & Astronomy, Lancashire Polytechnic,
                       Corporation Street, Preston PR1 2TQ

Adrian Brooks   Department of Astronomy, The University, Manchester M13 9PL

Gene Byrd   Department of Physics & Astronomy, University of Alabama, Box 1921,
                       Tuscaloosa AL 35487, USA

Massimo Calvani   International School for Advanced Studies, Strada Costiera 11,
                       34014 Trieste, Italy

Claude Carignan   Départment de Physique, Université de Montréal, CP 6128 Succ "A",
                       Montréal, Quebec, Canada

Jim Cohen   Nuffield Radio Astronomy Laboratories, Jodrell Bank, Macclesfield,
                       SK11 9DL

Neil Comins   Department of Physics and Astronomy, University of Maine,
                       Bennett Hall, Orono ME 04469, USA

Michal Czerny   Astronomical Observatory, Warsaw University, Al. Ujazdowskie 4, 00-
                       478 Warszawa, Poland

J Robert Davies   Department of Physics, University of Wales, P O Box 913,
                       Cardiff CF1 3TH

Fernando de Felice   Istituto di Fisica Matematica 'J.-Louis Lagrange', Via Carlo Alberto 10,
                       10123 Torino, Italy

Herwig Dejonghe   Astronomical Observatory, Krijgslaan 281, B-9000 Gent, Belgium

Marc de Vries   Sterrenkundig Instituut, Postbus 80000, Utrecht, The Netherlands

Luke Dones   Mail Stop 245-3, NASA Ames Research Center, Moffet Field CA 94035,
                         USA

Karl Donner   Observatory and Astrophysics Laboratory, Tähtiforninmäki, SF-
                         00130 Helsinki, Finland

Mark Dubal   Department of Astronomy, University of Leicester, University Road,
                         Leicester LE1 7RH

Bérengère Dubrulle   Observatoire du Pic du Midi et de Toulouse, 14 Avenue Edouard Belin,
                         31400 Toulouse, France

Wolfgang Duschl   Institut für Theoretische Astrophysik,
                         Universität Heidelberg, Im Neuenheimer Feld 561, D-6900 Heidelberg 1,
                         Federal Republic of Germany

John Dyson   Department of Astronomy, The University, Manchester M13 9PL

George Efstathiou   Department of Astrophysics, 1 Keble Road, Oxford OX1 3NP

Stefan Engström   Department of Astronomy/Astrophysics,
                         Chalmers University of Technology, S-41296 Göteborg, Sweden

N Wyn Evans   Astronomy Unit, School of Mathematics, Queen Mary College,
                         Mile End Road, London E1 4NS

Juhan Frank   Max-Planck-Institut für Astrophysik, Karl-Schwarzschild Straße 1,
                         8046 Garching bei München, Federal Republic Germany

Alexei Fridman   Astronomical Council, USSR Academy of Sciences, Pyatnitskaya St 48,
                         109017 Moscow, USSR

Michael Friedjung   Institut D'Astrophysique, 98 bis, Boulevard Arago, 75014 Paris, France

Daniel Friedli   Observatoire de Genève, Chemin des Maillettes 51, CH-1290 Sauverny,
                         Switzerland

Maryvonne Gerin   Radioastronomie Millimetrique, Laboratoire de Physique de l'ENS,
                         24 Rue Lhomond, F75231 Paris Cedex 05, France

Vasilios Geroyannis   Astronomy Laboratory, Department of Physics, University of Patras,
                         26110 Patras, Greece

Jamshid Ghanbari   Department of Astronomy, The University, Manchester M13 9PL

Wolfgang Glatzel   Max-Planck-Institut für Astrophysik, Karl-Schwarzschild Straße 1,
                         8046 Garching bei München, Federal Republic Germany

Richard James   Department of Astronomy, The University, Manchester M13 9PL

Jelle Kaastra	Laboratory for Space Research, P O Box 9504, 2300 RA Leiden, The Netherlands
Franz Kahn	Department of Astronomy, The University, Manchester M13 9PL
Simon Kemp	Department of Astronomy, The University, Manchester M13 9PL
Andrew King	Department of Astronomy, University of Leicester, University Road, Leicester LE1 7RH
Christopher Kitchin	The Hatfield Polytechnic Observatory, Bayfordbury, Near Hertford SG13 8LD
Basil Laspias	Department of Astronomy, The University, Manchester M13 9PL
Umin Lee	Department of Astronomy, University of Tokyo, Bunkyo-ku, Tokyo 113, Japan
Doug Lin	Lick Observatory, University of California, Santa Cruz CA 95064, USA
Peter te Lintel Hekkert	Sterrewacht, Postbus 9513, 2300 RA Leiden, The Netherlands
Jack Lissauer	Earth and Space Science Department, State University of New York, Stony Brook NY 11794, USA
Simon Litchfield	Department of Astronomy, University of Leicester, University Road, Leicester LE1 7RH
Huw Lloyd	Department of Astronomy, The University, Manchester M13 9PL
Stephen Lubow	Space Telescope Science Institute, 3700 San Martin Drive, Baltimore MD 21218, USA
Matt Malkan	Department of Astronomy, University of California, Math-Science Building, Los Angeles CA 90024, USA
Leon Mestel	Astronomy Centre, University of Sussex, Falmer, Brighton BN1 9QN
Shin Mineshige	Astronomy Department, University of Texas, RLM 15.308, Austin, TX 78712-1083, USA
Gerard Muratorio	Observatoire de Marseille, 2 Place le Verrier, 13248 Marseille Cedex 04, France
Carl Murray	Astronomy Unit, School of Mathematical Sciences, Queen Mary College, Mile End Road, London E1 4NS
Luciano Nobili	Dipartimento di Fisica, Universita di Padova, Via F Marzolo 8, 35131 Padova, Italy
Masafumi Noguchi	National Astronomical Observatory, Mitaka, Tokyo 181, Japan

Mark O'Reilly	Department of Astronomy, University of Leicester, University Road, Leicester LE1 7RH
Rachael Padman	Mullard Radio Astronomy Observatory, Cavendish Laboratory, Madingley Road, Cambridge CB3 0HE
Philip Palmer	Astronomy Unit, School of Mathematical Sciences, Queen Mary College, Mile End Road, London E1 4NS
John Papaloizou	Astronomy Unit, Department of Mathematics, Queen Mary College, Mile End Road, London E1 4NS
Alan Pedlar	Nuffield Radio Astronomy Laboratories, Jodrell Bank, Macclesfield SK11 9DL
Daniel Pfenniger	Observatoire de Genève, Chemin des Maillettes 51, CH-1290 Sauverny, Switzerland
Valerij Polyachenko	Astronomical Council, USSR Academy of Sciences, Pyatnitskaya St 48, 109017 Moscow, USSR
Neil Raha	Department of Astronomy, The University, Manchester M13 9PL
N Ramamani	181A Huntingdon Road, Cambridge CB3 0DJ
John Richer	Mullard Radio Astronomy Observatory, Cavendish Laboratory, Madingley Road, Cambridge CB3 0HE
Keith Robinson	School of Physics & Astronomy, Lancashire Polytechnic, Corporation Street, Preston PR1 2TQ
Alessandro Romeo	Scuola Internazionale Superiore di Studi Avanzati, Strada Costiera 11, 34014 Trieste, Italy
Steven Ruden	Astronomy Department, 601 Campbell Hall, University of California, Berkeley CA 94720, USA
Renzo Sancisi	Kapteyn Laboratory, Postbus 800, 9700 AV Groningen, The Netherlands
Gertjan Savonije	Sterrenkundig Instituut, Universiteit van Amsterdam, Roeters Straat 15, 1018 WB Amsterdam, The Netherlands
Mike Scarrott	Department of Physics, University of Durham, South Road, Durham DH1 3LE
Jerry Sellwood	Department of Astronomy, The University, Manchester M13 9PL
Ron Snell	Five College Radio Observatory, University of Massachusetts, Graduate Research Center, Tower B, Amherst MA 01003, USA
Linda Sparke	Washburn Observatory, University of Wisconsin, 475 N Charter Street, Madison WI 53706, USA

Ralph Spencer   Nuffield Radio Astronomy Laboratories, Jodrell Bank,
                         Macclesfield SK11 9DL

Tom Statler   Joint Institute for Laboratory Astrophysics, University of Colorado,
                         Boulder CO 80309-0440, USA

Tom Steiman-Cameron   Mail Stop 245-3, NASA Ames Research Center, Moffett Field CA 94035,
                         USA

Björn Sundelius   Department of Astronomy/Astrophysics, Chalmers
                         University of Technology, S-41296 Göteborg, Sweden

Maria Sundin   Department of Astronomy/Astrophysics, Chalmers University
                         of Technology, S-41296 Göteborg, Sweden

Jean Sygnet   Institut d'Astrophysique de Paris, 98 bis, Boulevard Arago, 75014 Paris,
                         France

Magnus Thomasson   Onsala Space Observatory, S-43900 Onsala, Sweden

Robert Thompson   Astronomy Unit, School of Mathematics, Queen Mary College,
                         Mile End Road, London E1 4NS

Alar Toomre   Department of Mathematics, Room 2-371, MIT, Cambridge MA 02139,
                         USA

José Torelles   Instituto de Astrofísica de Andalucía, Ap Correos 2144, Granada 18080,
                         Spain

Scott Tremaine   Director CITA, McLennan Labs, University of Toronto,
                         Toronto ON M5S 1A1, Canada

Roberto Turolla   Dipartimento di Fisica, Universita di Padova, Via F Marzolo 8,
                         35131 Padova, Italy

Esko Valtaoja   Department of Physical Sciences, University of Turku, SF-20500 Turku,
                         Finland

Leena Valtaoja   Turku University Observatory, Tuorla, SF-21500 Piikkio, Finland

Tony Weeks   Department of Astronomy, The University, Manchester M13 9PL

Althea Wilkinson   Department of Astronomy, The University, Manchester M13 9PL

Ray Wolstencroft   Royal Observatory, Blackford Hill, Edinburgh EH9 3HJ

Basil Zafiropoulos   Department of Astronomy, The University, Manchester M13 9PL

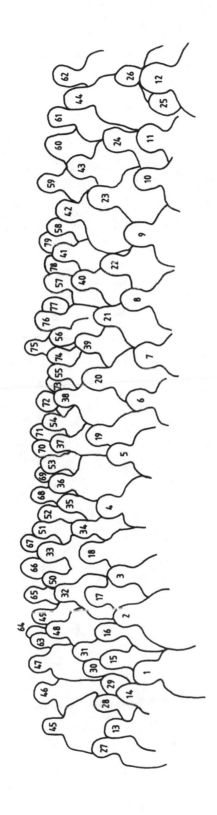

1 Ramamani	11 Bailey	21 Cohen	31 Malkan	41 te Lintel	51 Statler	61 Friedli	71 E Valtaoja
2 Steiman-Cameron	12 Sellwood	22 Dyson	32 Carignan	42 Lubow	52 Sygnet	62 Polyachenko	72 Dones
3 Ward-Thompson	13 Sparke	23 Duschl	33 Lee	43 Mestel	53 Glatzel	63 Snell	73 Kaastra
4 Byrd	14 L Valtao₋a	24 de Vries	34 Athanassoula	44 Pfenniger	54 Turolla	64 Evans	74 Muratorio
5 Palmer	15 Torrelles	25 Borderies	35 Friedjung	45 Mr Turner	55 Dejongie	65 Wolstencroft	75 Thomasson
6 Gerin	16 Engström	26 Brebner	36 Savonije	46 Papaloizou	56 Czerny	66 Toomre	76 Litchfield
7 Dubrulle	17 Efstathiou	27 Mrs Carling	37 Romeo	47 Tremaine	57 Comins	67 Sundelius	77 Lin
8 King	18 Robinson	28 Miss Thomas	38 de Felice	48 Noguchi	58 Richer	68 Donner	78 Frank
9 Kahn	19 Sancisi	29 Sundin	39 Calvani	49 Davies	59 James	69 Dubal	79 Scarrott
10 Lissauer	20 Pedlar	30 Artymowicz	40 Nobili	50 Mineshige	60 Padman	70 Ruden	

Spiral waves in Saturn's rings

Jack J. Lissauer*

State University of New York, Stony Brook, USA

Abstract Spiral density waves and spiral bending waves have been observed at dozens of locations within Saturn's rings. These waves are excited by resonant gravitational perturbations from moons orbiting outside the ring system. Modelling of spiral waves yields the best available estimates for the mass and the thickness of Saturn's ring system. Angular momentum transport due to spiral density waves may cause significant orbital evolution of Saturn's rings and inner moons. Similar angular momentum transfer may occur in other astrophysical systems such as protoplanetary discs, binary star systems with discs and spiral galaxies with satellites.

1 Introduction

Saturn's ring system was the first astrophysical disc to be discovered. When Galileo observed the rings in 1610, he believed them to be two giant moons in orbit about the planet. However, these "moons" appeared fixed in position, unlike the four satellites of Jupiter which he had previously observed. Moreover, Saturn's "moons" had disappeared completely by the time Galileo resumed his observations of the planet in 1612. Many explanations were put forth to explain Saturn's "strange appendages", which grew, shrank and disappeared every 15 years. In 1655, Huygens finally deduced the correct explanation, that Saturn's strange appendages are a flattened disc of material in Saturn's equatorial plane, which appear to vanish when the Earth passes through the plane of the disc (Figure 1). The length of time between Galileo's first observations of Saturn's rings and Huygens' correct explanation was due in part to the poor resolution of early telescopes. However, a greater difficulty was recognition of the possibility and plausibility of astrophysical disc systems. Contrast this to the situation today, when almost any flattened object observed in the heavens is initially suspected of being a disc.

The understanding of Saturn's ring system progressed slowly in the three centuries following Huygens [see Alexander (1962) for a historical review]. During this period, attention gradually shifted away from Saturn's rings as other astrophysical discs were observed (*e.g.* spiral galaxies) or proposed based on theoretical considerations (*e.g.* the protoplanetary disc, Kant 1745, Laplace 1796).

Similarities between Saturn's rings and spiral galaxies were first remarked upon by Maxwell (1859) "I am not aware that any practical use has been made of Saturn's Rings ... But when we contemplate the Rings from a purely scientific point of view, they become the most remarkable bodies in the heavens, except, perhaps, those still less useful bodies – the spiral nebulae". Today we know that the type of waves responsible for the grand design spiral structure observed in many disc galaxies (Lin & Shu 1964) are also present on much smaller scale within Saturn's rings (Cuzzi *et al.* 1981). The presence of density waves within Saturn's rings was predicted on theoretical grounds by Goldreich & Tremaine (1978a). Spiral bending waves, first proposed to explain galactic warps (Hunter & Toomre 1969), have also been observed within the rings of Saturn

---

* Alfred P. Sloan Research Fellow

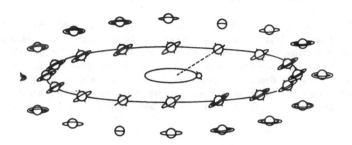

Figure 1. Schematic of Saturn and its rings as viewed
from Earth at various longitudes of Saturn's orbit.

(Shu, Cuzzi & Lissauer 1983, henceforth SCL). Angular momentum transport by spiral
density waves may be important to Saturn's rings, the proto-solar nebula, galaxies and
other astrophysical accretion discs.

Knowledge of planetary rings increased manifold during the decade from 1977 to
1986. Within this period ring systems were discovered about Uranus (Elliot *et al.* 1977),
Jupiter (Smith *et al.* 1979) and Neptune (Hubbard *et al.* 1986). Spacecraft visited all
of the ringed planets except Neptune, including three flybys of Saturn. Theoretical
developments advanced almost as rapidly as observations.

Each of the planetary ring systems has its own distinctive character. Good general
reviews are available for each ring system (Jupiter: Burns *et al.* 1984; Saturn: Cuzzi *et
al.* 1984; Uranus: Cuzzi & Esposito 1987; Neptune: Nicholson *et al.* 1989). This review
will focus on spiral density waves and spiral bending waves in planetary rings. Spiral
waves generated by gravitational perturbations of external moons have been observed at
several dozen locations within Saturn's rings and have tentatively been detected within
the rings of Uranus (Horn *et al.* 1988). They represent one of the best understood
forms of structure within planetary rings, and have been very useful as diagnostics of
ring properties such as surface mass density and local thickness. However, the angular
momentum transfer associated with the excitation of density waves within Saturn's
rings leads to characteristic orbital evolution time-scales of Saturn's A ring and inner
moons which are much shorter than the age of the solar system.

The theory of spiral waves in Saturn's rings is reviewed briefly in §2. §3 discusses
the observations and the ring properties derived therefrom. The short timescale of
ring evolution predicted by density wave torques and other outstanding questions are
discussed in §4. Conclusions are summarized in §5.

2 Theory

Spiral density waves are horizontal density oscillations which result from the bunching
of streamlines of particles on eccentric orbits (Figure 2). Spiral bending waves, in
contrast, are vertical corrugations of the ring plane resulting from the inclinations of

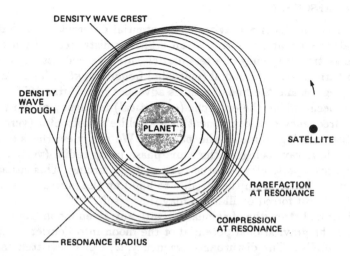

DENSITY WAVE CREST

DENSITY
WAVE
TROUGH

PLANET

SATELLITE

RAREFACTION
AT RESONANCE

COMPRESSION
AT RESONANCE

RESONANCE RADIUS

Figure 2. Schematic of particle streamlines within a resonantly excited two-armed spiral density wave. Density waves observed in Saturn's rings are much more tightly wound. (From Lissauer & Cuzzi 1985)

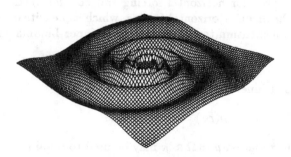

Figure 3. Schematic of a spiral bending wave showing variation of vertical displacement with angle and radius for a two-armed spiral. Bending waves observed in Saturn's rings are much more tightly wound. (From SCL)

particle orbits (Figure 3). In Saturn's rings, both types of spiral waves are excited at resonances with Saturn's moons, and propagate due to the collective self-gravity of the particles within the ring disc.

Ring particles move along paths which are very nearly Keplerian ellipses with one focus at the centre of Saturn. However, small perturbations due to the wave force a coherent relationship between particle eccentricities (in the case of density waves) or inclinations (in the case of bending waves) that produces the observed spiral pattern. The theory of excitation and propagation of linear spiral waves within planetary rings has been reviewed by Shu (1984). An abbreviated summary of aspects of the waves which may be observed and analyzed to determine ring properties is presented below.

2.1 ORBITS AND RESONANCES

Saturn's oblateness produces a gravitational potential different from that of a point mass. Near Saturn's equatorial plane, at an arbitrary distance, r, from the centre of the planet, the vertical frequency of a test particle, $\mu(r)$, exceeds its angular frequency, $\Omega(r)$, which in turn exceeds the radial (epicyclic) frequency, $\kappa(r)$. Thus the nodes of a particle's orbit regress and the line of apsides advances (*e.g.* Burns 1976).

Resonances occur where the epicyclic (or vertical) frequency of the ring particles is equal to the frequency of a component of a satellite's horizontal (vertical) forcing, as sensed in the rotating frame of the particle. We can view the situation as the resonating particle always being at the same phase in its radial (vertical) oscillation when it experiences a particular phase of the satellite's forcing. This situation enables continued coherent "kicks" from the satellite to build up the particle's radial (vertical) motion, and significant forced oscillations may thus result.

The locations and strengths of resonances with any given moon can be determined by decomposing the gravitational potential of the moon into Fourier components [see Shu (1984) for details]. The disturbance frequency, ω, can be written as the sum of integer multiples of the satellite's angular, vertical and radial frequencies:

$$\omega = m\Omega_{\rm M} + n\mu_{\rm M} + p\kappa_{\rm M}, \tag{1}$$

where the azimuthal symmetry number, m, is a non-negative integer, and n and p are integers, with n being even for horizontal forcing and odd for vertical forcing. The subscript M refers to the moon. Horizontal forcing, which can excite density waves and open gaps by angular momentum transport, occurs at inner Lindblad resonances, $r_{\rm L}$, where

$$\kappa(r_{\rm L}) = m\Omega(r_{\rm L}) - \omega. \tag{2}$$

Vertical forcing occurs at inner vertical resonances, $r_{\rm V}$, where

$$\mu(r_{\rm V}) = m\Omega(r_{\rm V}) - \omega. \tag{3}$$

When $m \neq 1$, the approximation $\mu \approx \Omega \approx \kappa$ may be used to obtain

$$\frac{\Omega(r_{\rm L,V})}{\Omega_{\rm M}} = \frac{m + n + p}{m - 1}. \tag{4}$$

The $(m+n+p)/(m-1)$ or $(m+n+p):(m-1)$ notation is commonly used to identify a given resonance.

Lindblad resonances with $m = 1$ depend on apsidal precession due to the difference between $\Omega(r)$ and $\kappa(r)$ and are referred to as apsidal resonances. Vertical resonances with $m = 1$ depend on the regression of the nodes of the ring particles upon the ring plane caused by the difference between $\Omega(r)$ and $\mu(r)$ and are called nodal resonances.

The strength of the forcing by the satellite depends, to lowest order, on the satellite's eccentricity, e, and inclination, i, as $e^{|p|}(\sin i)^{|n|}$. The strongest horizontal resonances have $n = p = 0$, and are of the form $m : (m - 1)$. The strongest vertical resonances have $n = 1$, $p = 0$, and are of the form $(m + 1) : (m - 1)$. The locations and strengths of such orbital resonances are easily calculated from known satellite masses and orbital

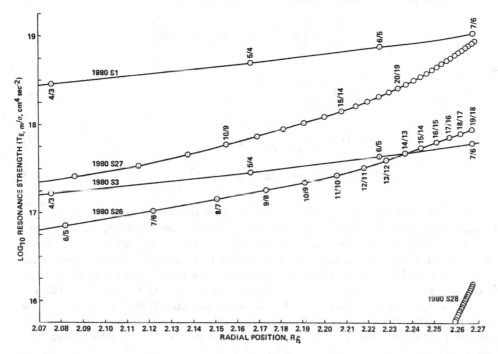

Figure 4. Locations and strengths of major Lindblad resonances in Saturn's A ring for Saturn's five closest "ring moons", Janus (1980S1), Epimetheus (1980S3), Pandora (1980S26), Prometheus (1980S27) and Atlas (1980S28). The closer moons have more closely spaced resonances with strength increasing outward more rapidly. (From Lissauer & Cuzzi 1982)

parameters and the gravitational moments of Saturn (Lissauer & Cuzzi 1982, SCL). By far the lion's share of the strong resonances lie within the outer A Ring (Figure 4).

2.2 BENDING WAVES

The vertical component of the gravitational force exerted by a satellite on an orbit inclined with respect to the plane of the rings excites motion of the ring particles in a direction perpendicular to the ring plane. The vertical excursions of the particles are generally quite small (Burns *et al.* 1979). However, at vertical resonances in Saturn's rings, the natural vertical oscillation frequency of a particle, $\mu(r)$, is equal to the frequency at which the vertical force due to one Fourier component of the satellite's gravitational potential is applied. Such coherent vertical perturbations can produce significant out-of-plane motions (Cook 1976). Self-gravity of the ring disc supplies a restoring force that enables bending waves to propagate away from resonance creating a corrugated spiral pattern. For $m > 1$, bending waves propagate toward Saturn (SCL); nodal bending waves ($m = 1$) propagate away from the planet (Rosen & Lissauer 1988).

In the inviscid linear theory, where viscous damping is ignored and the slope of the bent ring mid-plane is assumed to be small, the height of the local ring mid-plane relative to the Laplace (invariant) plane (Figure 5) is given by

$$Z(r) = \frac{|A_V|}{\sqrt{\pi}} \Re \left(e^{i(\Phi_V + \frac{\pi}{4})} e^{iq\xi^2} \int_{-\infty}^{\xi} e^{-iq\eta^2} d\eta \right) \qquad (5)$$

where

$$|A_V| = \mathcal{F} \sqrt{\frac{r_V}{G\sigma|\mathcal{D}|}}, \qquad (6)$$

$$\xi \equiv q(r_V - r)\sqrt{\frac{|\mathcal{D}|}{4\pi G\sigma r_V}}, \qquad (7)$$

$$\mathcal{D} \equiv \left[r\frac{d}{dr}(\mu^2(r) - m^2(\Omega_p - \Omega(r))^2) \right]_{r-r_V}, \qquad (8)$$

r is the distance from Saturn, r_V is the location of the vertical resonance and $\Omega_p = \omega/m$ is the angular frequency of the reference frame in which the wave pattern remains fixed (Gresh *et al.* 1986). The number of spiral arms in the wave is equal to the azimuthal symmetry number of the resonance, m. The surface mass density of the ring material is denoted by σ and G is the gravitational constant. The forcing strength, $\mathcal{F}$, and the phase of the wave, Φ_V, depend on the satellite in a manner given by equations (45-47) of SCL for $m > 1$ and equations (11-13) of Rosen & Lissauer (1988) for $m = 1$. The operator $\Re$ signifies the real part of the quantity. The sign of $\mathcal{D}$ is given by q, which is equal to $+1$ for inner vertical resonances, at which all observed bending waves are excited.

The oscillations of bending waves are governed by the Fresnel integral in equation (5). In the asymptotic far-field approximation, the oscillations remote from resonance have a wavelength

$$\lambda = \frac{4\pi^2 G\sigma}{m^2 \left[\Omega_p - \Omega(r)\right]^2 - \mu^2(r)}. \qquad (9)$$

Equation (9) can be simplified by approximating the orbits of the ring particles as Keplerian, $\mu(r) = (GM_S/r^3)^{1/2} = \Omega(r)$, for the $m > 1$ case and approximating the departure from Keplerian behaviour to be due to the quadrupole term of Saturn's gravitational potential, $\frac{3}{2}J_2\frac{R_S^2}{r^2}GM_S$, $J_2 = 0.0163$, for the $m = 1$ case. (The symbols M_S and R_S refer to the mass and equatorial radius of Saturn, respectively.) The resulting formulae are (Rosen 1989)

$$\lambda(r) \simeq 3.08 \left(\frac{r_V}{R_S}\right)^4 \frac{\sigma}{m-1} \frac{1}{r_V - r} \qquad (m > 1) \qquad (10a)$$

$$\lambda(r) \simeq 54.1 \left(\frac{r_V}{R_S}\right)^6 \sigma \frac{1}{r_V - r} \qquad (m = 1) \qquad (10b)$$

where λ, r and r_V are measured in kilometers, and σ is in g cm^{-2}. Equations (10) afford a means of deducing the surface density from measured wavelengths.

Inelastic collisions between ring particles act to damp bending waves. A rigorous kinetic theory of the damping of bending waves has not yet been developed. SCL used

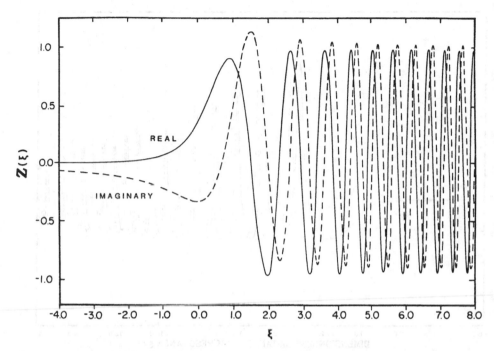

Figure 5. Theoretical height profile of an undamped linear spiral bending wave as a function of distance from resonance. The length and height scales are in arbitrary units. The solid and dashed lines represent two profiles of the same wave plotted at different azimuths. (Adapted from SCL)

---

a fluid approximation and the assumption that damping is weak to derive an amplitude variation of in terms of the viscosity of the ring material, ν. The collisional model of Goldreich & Tremaine (1978a) can be used to calculate the local scale height, H, of the ring from the measured viscosity:

$$H = \sqrt{\frac{2\nu}{\Omega} \frac{(1 + \tau^2)}{\tau}} \qquad (11)$$

where τ is the optical depth of the ring.

2.3 DENSITY WAVES

The gravity of a moon on an arbitrary orbit about Saturn has a component which produces epicyclic (radial and azimuthal) motions of ring particles. However, as in the case of vertical excursions induced by moons on inclined orbits, the epicyclic excursions are generally extremely small. An exception occurs near Lindblad (horizontal) resonances (equation 2), where coherent perturbations are able to excite significant epicyclic motions. In a manner analogous to the situation at vertical resonances, self-gravity of the ring disc supplies a restoring force that enables density waves to propagate away from Lindblad resonance (Goldreich & Tremaine 1978a). All density waves identified within

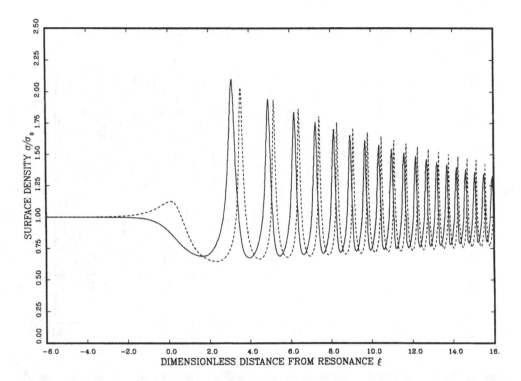

Figure 6. Theoretical surface density profile of a damped non-linear spiral density wave. The solid and dashed lines represent two profiles of the same wave plotted at different azimuths. (Adapted from Shu *et al.* 1985b)

Saturn's rings are excited at Inner Lindblad Resonances, and propagate outward, away from the planet.

The theory of spiral density waves is analogous to that of spiral bending waves, with the fractional perturbation in surface mass density, $\Delta\sigma/\sigma$, replacing the slope of the disc, dZ/dr. The relationship in the linear theory ($\Delta\sigma/\sigma \ll 1$) which is analogous to equation (5) is (Rosen 1989)

$$\frac{\Delta\sigma}{\sigma} = \Re\left\{ A_{\mathrm{L}}\left[i + \xi_0 e^{-i\xi_0^2/2} \int_{-\infty}^{\xi_0} e^{i\eta^2/2}d\eta\right] e^{-im\theta_0}\right\} \qquad (12)$$

where i denotes $\sqrt{-1}$ and the amplitude is given by

$$A_{\mathrm{L}} = \frac{-\psi_{\mathrm{M}}}{2\pi Gr_{\mathrm{L}}} \qquad (13)$$

The forcing function, ψ_{M}, is given for various cases by Shu (1984) and Rosen (1989). The phase of the wave is given by

$$\Phi_{\mathrm{L}} = \arg(A_{\mathrm{L}}) + \frac{\pi}{2} - m\theta_0 \qquad (14)$$

Far from resonance, equation (12) predicts a spacing between wave crests given by

$$\lambda = \frac{4\pi G\sigma}{m^2[\Omega_p - \Omega(r)]^2 - \kappa^2(r)} \tag{15}$$

Equations (10), with R_V and r_V replaced by R_L and r_L, are valid for density waves under the same approximations as for bending waves. The theory of damping of linear density waves is given by Goldreich & Tremaine (1978a).

Most density waves observed in Saturn's rings have $\Delta\sigma/\sigma \sim 1$. At such large amplitudes the linear theory breaks down. The theory of non-linear density waves has been developed by Shu *et al.* (1985a, b) and Borderies *et al.* (1986). The principal results of the non-linear theory are as follows: non-linear density waves depart from the smooth sinusoidal pattern predicted by equation (12), and become highly peaked (Figure 6). The theoretical wave profiles have broad, shallow troughs with surface density never dropping below half of the ambient value, and are qualitatively similar to observed waves (compare Figures 6 and 10). (A more detailed correspondence may be achieved if the background surface density is assumed to vary within the wave region, Longaretti & Borderies 1986.) Shocks do not occur, *i.e.* neighbouring streamlines never cross, even though they become arbitrarily close at wave crests as the wave propagates outward if viscous damping is not included (Shu *et al.* 1985a). The non-linear torques exerted by Saturn's moons are similar to those calculated using linear theory (Shu *et al.* 1985b).

3 Observations

Spiral waves in planetary rings are extremely tightly wound, with typical winding angles being 10^{-4} to 10^{-3} degrees. Such waves have very short wavelengths, of order 10 km. Earth-based photographs and Pioneer spacecraft images have resolution inadequate to detect such small scale features, so all available observations of spiral waves in planetary rings are from the Voyager spacecraft. Stellar occultations by Saturn's rings visible from Earth and HST may provide additional data during the next few years.

Spiral waves in Saturn's rings were detected by four instruments on the Voyager spacecraft. Voyager images have been used to detect density waves due to brightness contrasts between crests and troughs both in reflected light on the sunlit face of the rings (Figure 7; Dones 1987) and in diffuse transmission of sunlight to the dark side of the ring plane (Cuzzi *et al.* 1981). Bending waves are visible on Voyager images of the lit face of the rings (Figure 7) due to the dependence of brightness on local solar elevation angle (SCL). Bending waves appear on images of the unlit face of the rings due to the dependence of the slant optical depth, through which sunlight diffused, on local ring slope (Lissauer 1985).

Both density waves and bending waves were detected in the data received when the Voyager radio signal was attenuated by the occulting rings on its way to Earth (Figure 8; Marouf *et al.* 1986, Gresh *et al.* 1986). Density waves are observable by occultation experiments because bunching of particle streamlines increases the optical depth, τ, of crests; bending waves, although they leave optical depth normal to the ring plane unchanged, can be detected because the tilt of the ring plane causes oscillations in the observed slant optical depth (Figure 9). Similarly, the Voyager Photopolarimeter

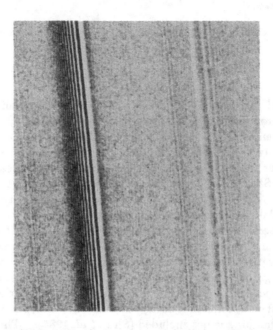

Figure 7. Voyager 2 image (FDS 43999.19) of a portion of the lit face of Saturn's A ring showing two prominent wave patterns. The feature on the left is the Mimas 5:3 bending wave; its contrast is high because the tilt of the local ring plane due to the wave was comparable to the solar elevation angle ($\sim 8°$) when the image was taken. The Mimas 5:3 density wave is seen on the right. The separation between the locations of the two waves is due to the non-closure of orbits caused by Saturn's oblateness. The other linear features in the images are unresolved density waves excited by the moons Pandora and Prometheus. Saturn is off to the left.

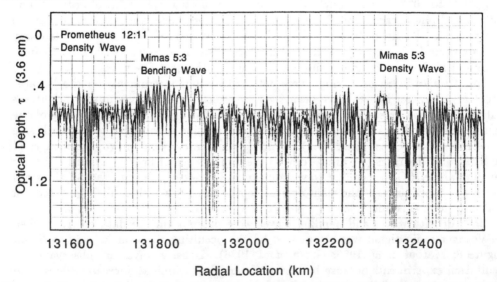

Figure 8. Examples of wave features observed in the radio occultation data of Saturn's A ring. The solid curved is measured normal optical depth, τ, plotted to increase downward. The gray shaded region represents the 70% confidence bounds on the measurement. (From Rosen 1989)

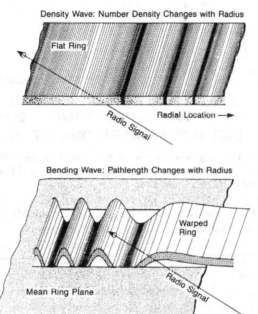

Radio Occultation as a Probe of
Density & Bending Waves

Density Wave: Number Density Changes with Radius

Flat Ring

Radio Signal

Radial Location ⟶

Bending Wave: Pathlength Changes with Radius

Warped Ring

Radio Signal

Mean Ring Plane

Radial Location ⟶

Figure 9. Schematic of a radio occultation of spiral waves. (From Rosen 1989)

(PPS) and Ultraviolet Spectrometer observed the diminution of light as a star passed behind the rings, thereby detecting changes in the slant optical depth of the rings caused by spiral waves (Lane *et al.* 1982, Holberg *et al.* 1982; Figure 10).

Five bending waves and several dozen density waves in Saturn's rings have thus far been identified with exciting resonances and analyzed to determine the local surface mass density of the rings. Results of the first analyses are tabulated by Esposito *et al.* (1984). More recent studies have been performed by Cuzzi *et al.* (1984), Lissauer (1985), Gresh *et al.* (1986), Longaretti & Borderies (1986), Rosen & Lissauer (1988) and Rosen (1989). The surface density, σ, at most wave locations in the optically thick A and B rings is of order 50 g cm^{-2}. Measured values in the optically thin C ring are $\sigma \approx 1$ g cm^{-2}; an intermediate value of 10 g cm^{-2} has been estimated for Cassini's Division.

The damping behaviour of three bending waves have also been analyzed to place upper bounds on the viscosity and local thickness of the rings (Lissauer *et al.* 1984, Esposito *et al.* 1983, Gresh *et al.* 1986, Rosen & Lissauer 1988). The A ring appears to have a local thickness of a few tens of meters; the thickness of the C ring is < 5 m. Viscosity measurements from the damping of density waves are less reliable (Lissauer *et al.* 1984, Shu *et al.* 1985b).

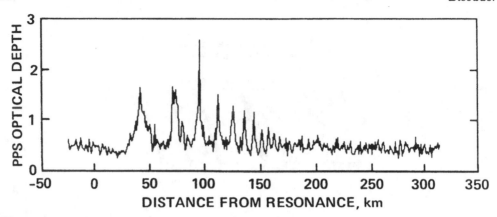

Figure 10. The Mimas 5:3 density wave as viewed from the Voyager 2 PPS stellar occultation experiment, plotted so that τ increases upwards. Note the sharp peaks and broad flat troughs caused by non-linearities (*cf.* Figure 6). (From Cuzzi *et al.* 1984)

The outer edges of the B and A rings are maintained by the Mimas 2:1 and Janus 7:6 resonances, which are the strongest resonances within the ring system (Smith *et al.* 1981, Holberg *et al.* 1982, Porco *et al.* 1984a, Lissauer & Cuzzi 1982, Borderies *et al.* 1982). Nearly empty gaps with embedded optically thick ringlets have been observed at strong resonances located in optically thin regions of the rings (Holberg *et al.* 1982). These features are probably caused by a resonance-related process; however, no explanation for the embedded ringlets currently exists. Nearly empty gaps with embedded ringlets have also been observed at non-resonant locations (Porco *et al.* 1984b).

4 Unresolved issues

Although spiral waves in Saturn's rings are well understood by astrophysical standards, there remain several major outstanding issues. These problems will be summarized in this section in a sequence beginning with those of an observational nature and ending with theoretical questions concerning angular momentum transport and the age of Saturn's rings. The latter issues are probably more relevant to the understanding of spiral waves in other disc systems.

The first mystery about density waves is why they are seen at all in Voyager images (Dones 1987). The optical depths of the A and B rings are sufficiently large and the solar elevation angles at the Voyager 1 and 2 encounters so low (4° and 8°, respectively) that very little sunlight could diffuse through the rings. This means that the brightness of the lit face of the rings should not be very sensitive to ring optical depth, as very little light would be able to diffuse down to the "extra" material near the unlit face, be reflected, and then diffuse back out to reach the Voyager cameras. The brightness contrast predicted by the wave amplitudes directly observed via stellar and radio occultations (which are generally similar to the theoretical amplitudes) is far less than the contrast observed in the images. All density waves imaged on the lit face of the rings are non-linear. The strong particle perturbations in such regions cause high-velocity collisions that can produce a physical thickening of the rings at density wave crests. The "top surface" of the ring plane could therefore vary in elevation, and local slopes could lead to brightness variations in a manner analogous to bending waves (Dones 1987). Alternatively, the ring particles could become close-packed near the peaks, which would enhance ring brightness by reducing the effects of shadowing. Another possibility is that the observed waves may be visible due to particle size variations: small particles could be produced by high velocity inter-particle collisions near the crests, and then be absorbed by lower velocity impacts at other wave phases. A quantitative study of these processes is currently being undertaken (L. Dones, private communication 1989).

The damping behaviour of most observed spiral waves differs significantly from that predicted by the simple fluid approximation in a constant viscosity disc (Goldreich & Tremaine 1978a, SCL). This has led to severe problems with attempts to estimate the viscosity (and thereby the thickness, see equation 11) of the rings using observations of wave damping. Some of the factors which can influence the damping behaviour of waves are variations in the background surface density and velocity dispersion, interference with other waves and wave non-linearities (Lissauer *et al.* 1984). The damping of non-linear density waves has been studied in detail by Shu *et al.* (1985b), who find that damping rates can be very sensitive to particle collision properties and to optical depth. Thus, the anomalous damping behaviour of many spiral waves presents both a challenge and an opportunity for researchers attempting to deduce ring properties other than surface mass density from the study of non-linear waves (Longaretti & Borderies 1986).

The cause of the observed enhancement of material in regions where strong waves propagate is poorly understood. Density waves excited at inner Lindblad resonances carry negative angular momentum, *i.e.* the angular momentum of the ring particles is temporarily reduced by the passage of these waves. When such waves damp, particles drift inwards. Resonant removal of angular momentum causes sharp outer ring edges and gaps to be produced at the strongest resonances within the ring system (Borderies *et al.* 1982). Waves are observed to be excited at the strongest resonances which do not produce gaps. However, waves do not appear to deplete material from the regions in which they propagate; on the contrary, a surface density enhancement is often observed in such regions (Holberg *et al.* 1982, Longaretti & Borderies 1986, Rosen & Lissauer 1988, Rosen 1989; see however, Esposito 1986). The problem of "dredging" of ring material due to wave damping has never been solved in a self-consistent manner, in which the evolution of a region due to wave propagation, wave damping and general ring viscosity is following until a quasi-steady state is attained. (The most detailed study of this problem thus far attempted is presented by Borderies *et al.* 1986.) Damping of the wave in the outer portion of the region in which it propagates could bring material towards resonance, but what stops (or at least slows) this material from further inward drift? The answer to this question may be relevant to two of the major questions facing ring theorists today: what maintains inner edges of the major rings of Saturn? and do the short time-scales for ring evolution due to density wave torques imply Saturn's rings are much younger than the planet itself?

Before Voyager arrived at Saturn, Goldreich & Tremaine (1978a) predicted that torques due to density waves excited by the moon Mimas at its 2:1 resonance had removed sufficient angular momentum from ring material to have cleared out Cassini's Division, a 4000 km wide region of depressed surface mass density located between the broad high density A and B rings. Although the hypothesis of density waves clearing Cassini's Division remains unverified, Voyager found a multitude of density waves excited by small newly-discovered satellites orbiting near the rings. Although the torque at these individual resonances is less than that at Mimas's 2:1 resonance, the sum of their torques is much greater. Moreover, the waves are observed and amplitudes agree with theory to within a factor of order unity.

Goldreich & Tremaine (1982) pointed out that the back torque the rings exert on the inner moons causes them to recede on a timescale short compared to the age of the solar system; current estimates suggest that all of the small moons orbiting inside the orbit of Mimas should have been at the outer edge of the A ring within the past 10^9 years, with Prometheus' journey outward occurring on a timescale of only a few million years. Resonance locking to outer more massive moons could slow the outward recession of the small inner moons; however, angular momentum removed from the ring particles should force the entire A ring into the B ring in a few times 10^8 yr (Lissauer *et al.* 1984, Borderies *et al.* 1984). If the calculations of torques are correct, and if no currently unknown force counterbalances them, then small inner moons and/or the rings must be "new", *i.e.* much younger than the age of the solar system. Both of these possibilities appear *a priori* to be highly unlikely. Rings could be remnants of Saturn's proto-satellite disc which never accreted into moon-sized bodies due to the strong tidal forces of Saturn inside Roche's limit, in which case they would be 4.5×10^9 yr old. Alternatively, Saturn's ring particles could be part of the debris from a moon that was collisionally disrupted, in which case they would most likely date from the first $\sim 10^9$ years of the solar system, when much more debris large enough to cause such a disruption was available than is today (Lissauer *et al.* 1988). Recent accretion of the inner moons within the ring system may be possible (Borderies *et al.* 1984), but why did such accretion only occur during the past $\sim 10^9$ yr? For these reasons, the issue of short time-scales due to density wave torques is a major outstanding problem in the field of planetary rings.

The issue of angular momentum transport via resonant excitation of density waves is of great interest in studies of other astrophysical disc systems. For example, Goldreich & Tremaine (1980) showed that density wave torques could have led to significant orbital evolution of Jupiter within the protoplanetary disc on a timescale of a few thousand years.

5 Conclusions

Resonantly excited density waves and bending waves are among the best understood features in planetary rings, and planetary rings are certainly the location in which spiral waves have best been observed and analyzed. Waves are seen to propagate from all of the strong satellite resonances, except those resonances which produce gaps. The locations of these waves agree with predicted values to within the observational uncertainties, which are often less than 1 part in 10^4. The wavelength behaviour also agrees well with the theory, and wavelength analysis has been used to obtain the best available estimates of the surface mass density of the rings; the total ring mass determined in this manner is roughly equivalent to that of an ice moon between 150 km and 200 km in radius. The damping behaviour of spiral waves appears to be very complex, and theoretical studies suggest that it may be very sensitive to particle collision properties.

The major outstanding issue in planetary rings is the origin and age of ring systems. Angular momentum transport via resonantly excited density waves suggest the rings must be much younger than the age of the solar system. However, various arguments suggest that such a recent origin of Saturn's rings is extremely unlikely. Resolution of this problem is important if angular momentum transport due to density waves is to be applied to models of other, less well observed, astrophysical disc systems, such as protoplanetary discs and accretion discs in binary star systems.

I thank Luke Dones, Stan Peale and Paul Rosen for their valuable comments on the manuscript. This work was supported in part by the NASA Planetary Geology and Geophysics Program under grant NAGW-1107.

References

Alexander, A. F., 1962. *The Planet Saturn*, MacMillan Co, New York, NY.

Borderies, N., Goldreich, P. & Tremaine, S., 1982. *Nature*, **299**, 209.

Borderies, N., Goldreich, P. & Tremaine, S., 1986. *Icarus*, **68**, 522.

Burns, J. A., 1976. *Amer. J. Phys.*, **44**, 944.

Burns, J. A., Showalter, M. R. & Morfill, G. E., 1984. In *Planetary Rings*, p. 200, eds. Greenberg, R. & Brahic, A., University of Arizona Press, Tuscon, AZ.

Cook, A. F., II, 1976. *Gravitational resonances in Saturn's rings. II: Perturbations due to Janus, Mimas and Enceladus: The initial profile in Saturn's equatorial plane and warping of the rings*, Preprint No. 588, Center for Astrophysics, Cambridge, Mass.

Cuzzi, J. N. & Esposito, L. W., 1987. *Sci. Amer.*, **237**, 52.

Cuzzi, J. N., Lissauer, J. J., Esposito, L. W., Holberg, J. B., Marouf, E. A., Tyler, G. L. & Boischot, A., 1984. In *Planetary Rings*, p. 73, eds. Greenberg R. & Brahic, A., University of Arizona Press, Tuscon, AZ.

Cuzzi, J. N., Lissauer, J. J. & Shu, F. H., 1981. *Nature*, **292**, 703.

Dones, L., 1987. *PhD thesis*, University of California, Berkeley.

Elliot, J. L., Dunham, E. W. & Mink, D. J., 1977. *Nature*, **267**, 328.

Esposito, L. W., 1986. *Icarus*, **67**, 345.

Esposito, L. W., Cuzzi, J. N., Holberg, J. B., Marouf, E. A., Tyler, G. L. & Porco, C. C., 1984. In *Saturn*, p. 463, eds. Gehrels, T. & Matthews, M. S., University of Arizona Press, Tuscon, AZ.

Esposito, L. W., O'Callahan, M. & West, R. A., 1983. *Icarus*, **56**, 439.

Goldreich, P. & Tremaine, S., 1978a. *Icarus*, **34**, 227.

Goldreich, P. & Tremaine, S., 1978a. *Icarus*, **34**, 240.

Goldreich, P. & Tremaine, S., 1980. *Astrophys. J.*, **241**, 425.

Gresh, D. L., Rosen, P. A., Tyler, G. L. & Lissauer, J. J., 1986. *Icarus*, **68**, 502.

Holberg, J. B., Forrester, W. T. & Lissauer, J. J., 1982. *Nature*, **297**, 115.

Horn, L. J., Yanamandra-Fisher, P. A., Esposito, L. W. & Lane, A. L., 1988. *Icarus*, **76**, 485.

Hubbard, W. B., Brahic, A., Sicardy, B., Elicer, L-R., Roques, F. & Vilas, F., 1986. *Nature*, **319**, 636.

Kant, I., 1755. *Allegmeine Naturgeschichte und Theorie des Himmels*.

Lane, A. L., *et al.*, 1982. *Science*, **215**, 537.

Laplace, P., 1796. *Exposition du systeme du monde*, Paris.

Lin, C. C. & Shu, F. H., 1964. *Astrophys. J.*, **140**, 646.

Lissauer, J. J., 1985. *Icarus*, **62**, 425.

Lissauer, J. J. & Cuzzi, J. N., 1982. *Astron. J.*, **87**, 1051.

Lissauer, J. J. & Cuzzi, J. N., 1985. In *Protostars and Planets II*, p. 920, eds. Black, D. C. & Matthews, M. S., University of Arizona Press, Tuscon, AZ.

Lissauer, J. J., Shu, F. H. & Cuzzi, J. N., 1984. In *Planetary Rings*, Proc. IAU Colloq. 75, p. 385, ed. Brahic, A., Cepadues, Toulouse.

Lissauer, J. J., Squyres, S. W. & Hartmann, W. K., 1988. *J. Geophys. Res.*, **93**, 13776.

Longaretti, P. Y. & Borderies, N., 1986. *Icarus,* **67**, 211.

Maxwell, J. C., 1859. *On the stability of the motions of Saturn's Rings*, Cambridge and London: MacMillan and Co. Also reprinted in *Maxwell on Saturn's Rings*, 1983. p. 73, eds. Brush, S. G., Everitt, C. W. F. & Barber, E., The MIT Press, Cambridge, MA.

Nicholson, P.D., Cooke, M. L., Matthews, K., Elias, J. H. & Gilmore, G., 1989. *Icarus,* , submitted.

Porco, C., Danielson, G. E., Goldreich, P., Holberg, J. B. & Lane, A. L., 1984a. *Icarus,* **60**, 17.

Porco, C., *et al.*, 1984b. *Icarus,* **60**, 1.

Rosen, P. A., 1989. *PhD thesis,* Stanford University.

Rosen, P. A. & Lissauer, J. J., 1988. *Science,* **241**, 690.

Shu, F. H., 1984. In *Planetary Rings*, p513, eds. Greenberg, R. & Brahic, A., University of Arizona Press, Tuscon, AZ.

Shu, F. H., Cuzzi, J. N. & Lissauer, J. J., 1983. *Icarus,* **53**, 185 (SCL).

Shu, F. H., Dones, L., Lissauer, J. J., Yuan, C. & Cuzzi, J. N., 1985b. *Astrophys. J.,* **299**, 542.

Shu, F. H., Yuan, C. & Lissauer, J. J., 1985a. *Astrophys. J.,* **291**, 356.

Smith, B. A. *et al.*, 1979. *Science,* **204**, 951.

Smith, B. A. *et al.*, 1981. *Science,* **212**, 163.

Structure of the Uranian rings

C.D. Murray and **R.P. Thompson**

Queen Mary College, London

1 Introduction

Nine rings of Uranus were discovered by Earth-based stellar occultations in 1977. In order of increasing semi-major axis from the planet they have been denoted 6, 5, 4, α, β, η, γ, δ and ϵ. The rings are very dark, very narrow (typically < 5 km) and have extremely sharp edges. The standard theory for the confinement of narrow rings against the spreading effects of Poynting-Robertson drag and collisions (Goldreich & Tremaine 1977) proposes that each ring is bounded by a pair of Lindblad resonances from shepherding satellites. Saturn's F-ring is now known to be shepherded by the two satellites Pandora and Prometheus, and in January 1986 Voyager 2 images of the Uranian system in back scattered light showed the presence of two satellites, Cordelia and Ophelia, on either side of the ϵ ring. Eight other small satellites, all exterior to the main rings, were discovered by Voyager 2.

Porco & Goldreich (1987) showed that the inner edge of the ϵ ring is within a kilometre of the 24:25 outer Lindblad resonance with Cordelia while the outer edge is within 300 m of the 14:13 inner Lindblad resonance with Ophelia. They identified two other possible resonances between ring features and these two satellites. A series of narrow angle Voyager 2 images of the rings were taken at fixed, non-rotating positions to search for small satellites orbiting between the rings. No further satellites were found down to a detection limit of 10 km (Smith *et al.* 1986).

2 The rings in forward scattered light

We have analysed a unique image of the rings taken in forward scattered light from behind the main ring system. The image reveals the presence of a number of narrow rings and diffuse features which were not detectable in occultations or back scattered

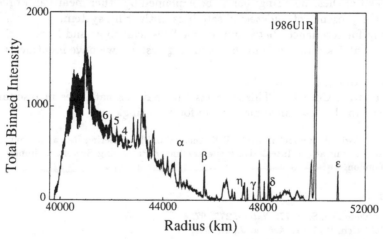

Figure 1. Total binned intensity as a function of orbital radius.

images. We have attempted to (i) provide a catalogue of all the new ring features, (ii) search for correlations between the new ring features and resonances with the newly discovered satellites, and (iii) provide constraints on the possible locations of small (radius < 10 km) satellites orbiting in the main ring system.

The 96 s exposure time of the image and the relative motion between the spacecraft and the rings meant that portions of the image were affected by radial smear. However, we were able to select a region (encompassing $\sim 130\,000$ pixels) which had been targeted to be relatively smear-free. Using the known geometry of the eccentric rings, we navigated the image to calculate the radius, longitude and intensity at each pixel.

Figure 1 shows the total binned intensity as a function of orbital radius for the selected region of the forward scattered image. We have identified the nine 'classical' rings and the new ring, 1986U1R. With the exception of the faint ring just outside the β ring, all the additional fine structure shown in Figure 1 has not been detected in occultations or in the back scattered images. The brightness of 1986U1R in forward scattered light is the most striking feature in the scan plot. It has extremely sharp edges and a profile that is wider than that suggested by the Voyager occultation data.

By calculating the locations of all the first order Lindblad resonances within the ring system due to the known satellites, we found no further resonance correlations with the forward scattered features other than those previously reported by Porco & Goldreich for the 'classical' rings – the known satellites cannot account for the locations of all the observed rings. In order to search for possible locations of undiscovered satellites we calculated the resonances associated with hypothetical satellites at 1 km intervals in the rings, looking for correlations with a number of different ring features simultaneously. Various locations were found where satellites could have Lindblad resonances with three or more 'classical' ring features. There appears to be a significant correlation between the locations of these hypothetical satellites and first order resonances with Cordelia.

After cataloguing the new ring features we have concluded that:

1. The new features seen in the forward scattered image are not correlated with resonances with either the five 'classical' satellites or the ten new small satellites. The confinement of these new rings has to be explained by other means and presumably requires the presence of undetected satellites in the ring system.

2. By looking for resonances between the nine 'classical' rings and hypothetical satellites placed at 1 km intervals in the main ring system, we have identified a number of locations where such a satellite has Lindblad resonances with three or more rings. There is a striking correlation between these locations and the location of first order resonances with Cordelia. This suggests that pairs of satellites in resonance may be involved in the resonance mechanism for each ring.

We gratefully acknowledge support from the U.K. Science and Engineering Research Council. We are also grateful to the staff at the Interactive Planetary Image Processing System at Imperial College, University of London.

References

Goldreich, P. & Tremaine, S., 1977. *Nature,* **277**, 97.
Porco, C. & Goldreich, P., 1987. *Astron. J.,* **93**, 724.
Smith, B. A. *et al.* 1986. *Science,* **233**, 43.

Planetary rings: theory

Nicole Borderies

Observatoire du Pic du Midi et du Toulouse, France
and Jet Propulsion Laboratory, Pasadena, USA

Abstract The dynamics of planetary rings is described in terms of streamlines and how these streamlines interact. The reversal of the viscous flux of angular momentum plays an important role in this dynamics. This paper reports the current status on questions related to broad rings, narrow rings and arcs of rings.

1 Introduction

The recent ground-based and spacecraft observations of planetary rings have revealed how complex these astrophysical discs are. In spite of this complexity, a number of unifying elements exist in the theory. I begin this review by a presentation of the concept of streamlines, which provides a general description of the kinematics. Then I describe a physical process which plays an important part in the dynamics of planetary rings: it is the reversal of viscous angular momentum flux from its usual outward direction to the inward direction. The next three sections are devoted to broad rings, narrow rings and arcs of rings.

2 Streamlines

For non-axisymmetric perturbations, the shape of the streamlines is described in polar coordinates r, ϕ in a frame rotating with the pattern speed Ω_p by the equations

$$r = a\left[1 - e(a, \bar{\phi}, t)\cos\left(m\bar{\phi} + m\Delta(a, \bar{\phi}, t)\right)\right], \tag{1}$$

$$\phi = \bar{\phi} + 2e(a, \bar{\phi}, t)\sin\left(m\bar{\phi} + m\Delta(a, \bar{\phi}, t)\right), \tag{2}$$

$$\bar{\phi} = \bar{\phi}_0 + \Omega(a)t. \tag{3}$$

The streamline description given by equations (1), (2) and (3) was suggested by Borderies, Goldreich & Tremaine (*e.g.* 1982, 1983a, 1983b). It represents a class of solutions which correspond to small perturbations ($e \ll 1$) such that, to zero order in the eccentricity e,

$$m\bar{\phi} + m\Delta = E, \tag{4}$$

where E is the eccentric anomaly and Δ does not have any secular variations.

The separation between neighbouring streamlines can be expressed in terms of

$$J = \frac{\partial r}{\partial a} = 1 - q\cos(m\bar{\phi} + m\Delta + \gamma), \tag{5}$$

with

$$q\cos\gamma = \frac{\partial(ae)}{\partial a}, \tag{6}$$

$$q\sin\gamma = ae\frac{\partial\left(m\bar{\phi} + m\Delta\right)}{\partial a}. \tag{7}$$

We must have $0 \leq q < 1$. The streamlines are parallel to each other for $q = 0$, and they intersect for $q = 1$. An unperturbed ring corresponds to the case where the streamlines are circular.

3 Angular momentum transport

Collisions between particles redistribute angular momentum. In a fluid dynamical approach, the effects of collisions are modelled through a coefficient of kinematic viscosity (Goldreich & Tremaine 1978a)

$$\nu \sim \frac{v^2 \tau}{\Omega \left(1 + \tau^2\right)}. \tag{8}$$

where v is the velocity dispersion and τ is the optical depth. In planetary rings typical values are $v \simeq 0.2$ cm/s, $\Omega \simeq 2 \ 10^{-4}$ rd/s, $\tau \simeq 1$, and $\nu \simeq 50$ cm^2/s.

The viscous flux of angular momentum, that is to say the rate at which angular momentum flows across a unit length of a streamline per unit of time, is (Borderies, Goldreich & Tremaine 1982)

$$F_H = 2\Sigma_0 \nu \Omega a \left(\frac{1}{J} - \frac{1}{4J^2}\right) \tag{9}$$

where Σ_0 is the unperturbed surface density. The luminosity of angular momentum, or the integral of the flux is (Borderies, Goldreich & Tremaine 1982)

$$L_H = \pi \Sigma_0 \nu \Omega a^2 \frac{(3 - 4q^2)}{(1 - q^2)}. \tag{10}$$

In a unperturbed ring we have

$$L_H = 3\pi \Sigma_0 \nu \Omega a^2. \tag{11}$$

In a unperturbed ring, because $L_H > 0$, angular momentum flows outward. The time-scale for the spreading of an unperturbed ring of width Δa is $t_\nu = (\Delta a)^2/\nu$. With a viscosity of 50 cm^2/s, the characteristic spreading time for an unperturbed ring of width 25,000 km, like Saturn's B ring, is equal to the age of the solar system.

In a perturbed ring, the viscous flux of angular momentum is reversed at some longitude for $q > 3/4$ and the luminosity of angular momentum is reversed for $q > \sqrt{3/4}$. A ring such that $q > \sqrt{3/4}$ contracts instead of spreading radially. Borderies, Goldreich & Tremaine (1983a) found the same qualitative results from kinetic theory. The fluid dynamical approach and the kinetic theory converge for large optical depth, but at small optical depth, the kinetic theory gives more precise results. One finds that, for $\tau < 1$, the reversal of angular momentum flux occurs at smaller q than predicted by the fluid dynamical approach.

4 Broad rings

One of the most conspicuous morphological features of Saturn's rings is the outer edge of the B ring, associated with a 2:1 resonance with Mimas. Goldreich & Tremaine

(1978b) presented the first model for confinement of rings. The idea is that the non-axisymmetric distribution of mass associated with the wave excited at the 2:1 resonance with Mimas exerts a positive gravitational torque on Mimas, while the ring is subjected to a negative torque. The viscous luminosity of angular momentum which flows toward the ring's edge is transferred to the satellite by the wave. This model has been modified by Borderies, Goldreich & Tremaine (1982) by the introduction of the effect of viscous torque reversal which explains why the edge is so sharp.

There is an unresolved time-scale problem associated with the confinement of the rings of Saturn. The satellites which border these rings excite a number of density waves in the A ring. The resulting gravitational torques exerted on the satellites make them recede from the rings. The time-scale of evolution is of the order of 10 million years, very short compared to the age of the solar system.

Most density waves observed in Saturn's rings are non-linear. A non-linear representation of the surface density of a density wave is given by

$$\Sigma = \frac{\Sigma_0}{1 - q\cos(m\bar{\phi} + m\Delta + \gamma)}. \tag{12}$$

The non-linear representation allows one to reproduce the large troughs and narrow peaks which are typical of the optical depth profiles of non-linear density waves. Analysis of data (Longaretti & Borderies 1986), theoretical arguments (Borderies, Goldreich & Tremaine 1984a) and developments of the non-linear density wave theory (Shu, Yuan & Lissauer 1985) show that the non-linear torques are not very different from those calculated by the linear theory. Hence, non-linearity cannot solve the short time-scale problem. Other developments of the non-linear theory have been made by Shu *et al.* (1985) and Borderies, Goldreich & Tremaine (1986).

Besides the type of gap represented by that opened by Mimas at the outer edge of the B ring, another type of gap can be produced by satellites embedded in the rings. Cuzzi & Scargle (1985) have inferred the existence of a satellite orbiting in the centre of the Encke gap from the wavy edges seen in Voyager images. Showalter *et al.* (1986) have analyzed Voyager observations of the gap and used a kinematic model to determine the moonlet orbit and mass while Borderies, Goldreich & Tremaine (1989) have studied the wake dynamics in the presence of interactions between the particles.

Saturn constitutes the best example of a planet surrounded by broad rings. However, it has been found that highly inclined rings are theoretically possible around Neptune (Dobrovolskis 1980, Borderies 1987, 1989). Highly inclined rings around Neptune would be similar to polar rings of galaxies (Athanassoula & Bosma 1985) with the triaxial potential of the latter being replaced by the Neptune-Triton gravity field. Dobrovolskis, Steiman-Cameron & Borderies (1988) have shown that thin polar rings are stable against collisions between particles by using a twofold approach, based on numerical and analytical computations.

5 Narrow rings

Narrow rings must be confined, unless they are very young. Goldreich & Tremaine (1979b) proposed that the rings of Uranus are confined by shepherd satellites orbiting on each side of them. Voyager 2 has discovered the two shepherd satellites of the ϵ

ring. Porco & Goldreich (1987) have identified the Lindblad resonances associated with the edges of the ϵ ring and found other resonances associated with the rings δ, γ and 1986U1R. The shepherds of the other rings have not been found, presumably because they are too small and too dark.

French *et al.* (1988) have fitted kinematic models to the data of the Uranian rings. They found that the rings 6, 5, 4 α, β and ϵ have shapes which correspond to the mode $m = 1$, that is to say these rings have an elliptical shape with one focus coinciding with the centre of mass of Uranus. Each ring rotates around the planet with an angular speed approximatively equal to the apsidal precession rate induced by the gravitational field of the planet. The shape of the ring δ corresponds to the mode $m = 2$, which means that it is an ellipse whose centre coincides with the centre of mass of the planet. The ring γ contains two modes: the mode $m = 1$ and the mode $m = 0$, which corresponds to axisymmetric oscillations with the epicyclic frequency.

The origin of these modes could be the excitation by satellites as proposed by Goldreich & Tremaine (1981) or a viscous instability found by Borderies, Goldreich & Tremaine (1985). Borderies, Goldreich & Tremaine (1984b) showed that ring inclinations can also be excited by satellites.

Even if we understand that satellites or a gravitational instability can excite the eccentricities of narrow rings, we need to explain how these rings can maintain a uniform precession in spite of the tendency towards differential precession associated with the oblateness of the planet. Goldreich & Tremaine (1979a) and Borderies, Goldreich & Tremaine (1983b) have developed a model in which the self-gravity of the ring balances the differential precession due to the oblateness of the planet. Borderies, Goldreich & Tremaine (1983c) showed also that the rigid precession of inclined rings can be enforced by their self-gravity.

A recent analysis of the ϵ ring, by Borderies *et al.* (1988), showed that, in several perturbed portions of the ϵ ring, the streamlines are not aligned ellipses but have a more complex shape coming from the superposition of another mode. The testing of the self-gravity model for the ϵ ring will require taking into account the complex shape of the streamlines.

On the other hand the self-gravity model encounters serious difficulties with the rings α and β. Theoretical arguments (Goldreich & Porco 1987) indicate that the mass deduced from the self-gravity model for these rings is too small.

6 Ring arcs

Ground based observations of stellar occultations by Neptune have revealed the presence of incomplete rings around this planet (Hubbard *et al.* 1986). This was a remarkable discovery: the rings of Jupiter, Saturn and Uranus are complete circles, but the rings of Neptune are short arcs. The occultation data suggest that there are a number of short arcs at several radii with lengths between 1000 and 10,000 km. Moreover, these arcs are narrow.

The presence of arcs strongly suggest that they are made of particles orbiting around a co-rotation resonance.

Lissauer (1985) published a model for ring arcs in which the incomplete rings of Neptune consist of particles orbiting around a triangular Lagrange point of a hypothetical satellite. Furthermore, each arc is radially confined by at least one shepherding moon on a nearby orbit. Lissauer estimated the radius of these satellites to be at least of the order 50 to 100 km. Since arcs have been observed at different distances from the planet, Lissauer's model requires several big satellites.

Goldreich, Tremaine & Borderies (1986) have published another model in which the arcs are located at more complex co-rotation resonances involving a satellite on an inclined orbit. The highly inclined ($\simeq 160°$) and circular orbit of Triton indicates that some event, maybe an encounter with an object of the size of the Earth, has produced the high inclination. The same event could have produced the inclination of our hypothetical satellite. Tidal dissipation in the satellite would have damped its orbital eccentricity but not its inclination if the satellite is oblate.

In our model a single satellite gives rise to a set of co-rotation resonances. The satellite gives rise also to a Lindblad resonance near each of these co-rotation resonances. The effect of the Lindblad resonances is to excite the particle orbital eccentricities, providing in this way the energy necessary to confine the arcs. Thus, in our model a single satellite is responsible for the azimuthal and radial confinement of several arcs.

Lin, Papaloizou & Ruden (1987) have studied in more detail the conditions for the confinement of rings. In particular, they proved that confinement requires the Lindblad resonance to be on the same side as the satellite with respect to the co-rotation resonance.

This review was written while the author held a National Research Council – Jet Propulsion Laboratory Research Associateship.

References

Athanassoula, E. & Bosma, A., 1985. *Annu. Rev. Astron. Astrophys.*, **23**, 147.
Borderies, N., 1987. *Bull. Am. Astron. Soc.*, **19**, 891.
Borderies, N., 1989. *Icarus*, in press.
Borderies, N., Goldreich, P. & Tremaine, S., 1982. *Nature*, **299**, 209.
Borderies, N., Goldreich, P. & Tremaine, S., 1983a. *Icarus*, **55**, 124.
Borderies, N., Goldreich, P. & Tremaine, S., 1983b. *Astron. J.*, **88**, 1560.
Borderies, N., Goldreich, P. & Tremaine, S., 1983c. *Astron. J.*, **88**, 226.
Borderies, N., Goldreich, P. & Tremaine, S., 1984a. In *Planetary Rings*, p. 713, eds. Greenberg, R. & Brahic, A., University of Arizona Press, Tuscon, AZ.
Borderies, N., Goldreich, P. & Tremaine, S., 1984b. *Astrophys. J.*, **284**, 429.
Borderies, N., Goldreich, P. & Tremaine, S., 1985. *Icarus*, **63**, 406.
Borderies, N., Goldreich, P. & Tremaine, S., 1986. *Icarus*, **68**, 522.
Borderies, N., Goldreich, P. & Tremaine, S., 1989. *Icarus*, in press.
Borderies, N., Gresh, D. N., Longaretti, P-Y. & Marouf, E. A., 1988. *Bull. Am. Astron. Soc.*, **20**, 844.
Cuzzi, J. N. & Scargle, J. D., 1985. *Astrophys. J.*, **292**, 276.
Dobrovolskis, A. R., 1980. *Icarus*, **43**, 222.
Dobrovolskis, A. R., Steiman-Cameron, T. Y. & Borderies, N. J., 1988. *Bull. Am. Astron. Soc.*, **20**, 861.
French, R. G., Elliot, J. L., French, L. M., Kangas, J. A., Meech, K. J., Ressler, M. E., Buie, M. W., Frogel, J. A., Holberg, J. B., Fuensalida, J. J., Joy, M., 1988. *Icarus*, **73**, 349.
Goldreich, P. & Porco, C. C., 1987. *Astron. J.*, **93**, 730.
Goldreich, P. & Tremaine, S., 1978a. *Icarus*, **34**, 227.
Goldreich, P. & Tremaine, S., 1978b. *Icarus*, **34**, 240.

Goldreich, P. & Tremaine, S., 1979a. *Astron. J., 84*, 1638.
Goldreich, P. & Tremaine, S., 1979b. *Nature, 277*, 97.
Goldreich, P. & Tremaine, S., 1981. *Astrophys. J., 243*, 1062.
Goldreich, P., Tremaine, S. & Borderies, N., 1986. *Astron. J., 92*, 490.
Hubbard, W. B., Brahic, A., Sicardy, B., Elicer, L-R., Roques, F. & Vilas, F., 1986. *Nature, 319*, 636.
Lin, D. N. C., Papaloizou, J. C. B. & Ruden, S. P., 1987. *Mon. Not. R. Astron. Soc., 227*, 75.
Lissauer, J. J., 1985. *Nature, 318*, 544.
Longaretti, P. Y. & Borderies, N., 1986. *Icarus, 67*, 211.
Porco, C. C. & Goldreich, P., 1987. *Astron. J., 93*, 724.
Showalter, M. R., Cuzzi, J. N., Marouf, E. A. & Esposito, L. W., 1986. *Icarus, 66*, 297.
Shu, F. H., Dones, L., Lissauer, J. L., Yuan, C. & Cuzzi, J. N., 1985. *Astrophys. J., 299*, 542.
Shu, F. H., Yuan, C. & Lissauer, J. L., 1985. *Astrophys. J., 291*, 356.

Simulations of light scattering in planetary rings

L. Dones[1], M. R. Showalter[2] and J. N. Cuzzi[1]
[1]*NASA-Ames Research Center, California, USA*
[2]*Stanford University, USA*

1 Statement of the problem

Most studies of light scattering in planetary rings have assumed layers which are many particles thick, plane parallel, and homogeneous. However, real rings may be thin, vertically warped, and clumpy. We have developed a ray tracing code which calculates the light scattered by an arbitrary distribution of particles. This approach promises to clarify a number of puzzling observations of the Saturnian and Uranian rings.

(1) Many studies have concluded that Saturn's rings are many particles thick (*e.g.* Lumme *et al.* 1983), whereas dynamical calculations predict that optically thick rings should be physically thin (Wisdom & Tremaine 1988 and references therein). Lumme *et al.* argue that the particles in Saturn's B Ring fill only 2% of the volume of the ring, while Wisdom and Tremaine predict a filling factor of 20% or more.

The claim that Saturn's rings are thick is based on their observed opposition surge, a rapid brightening (0.3 mag in the V band) which occurs at phase angles below about 1.5°. The surge is attributed to particles covering their own shadows near opposition. Shadowing can occur either *between* discrete particles, or *within* the surface structure of a particle. The range in phase angle over which the brightening takes place is proportional to the volume filling factor of the ring or surface. Thus the very narrow opposition effect of Saturn's rings implies a very porous ring, unless individual particles backscatter extremely strongly. Such a backscattering phase function has been generally considered unlikely, since no satellite was known to have such a narrow surge (Cuzzi *et al.* 1984). However, the Uranian satellites Titania and Oberon may have an even steeper opposition surge than Saturn's rings (Goguen *et al.* 1989 and references therein). It is therefore possible that much of the observed surge for Saturn's rings occurs *within* individual ring particles, so that observations place little constraint on the thickness of the rings. A similar conclusion holds for the Uranian rings (Herbst *et al.* 1987).

(2) The B Ring and inner A Ring are darker in forward scatter than multiple scattering calculations predict, even if no diffracting, submicron particles are assumed present (Doyle 1987, Dones 1987). For a classical layer of particles, the ring brightness in this geometry is dominated by multiple scattering; the observed low reflectivity may be due to reduced multiple scattering in a thin layer. By contrast, the outermost part of the A Ring is fitted very well by classical models, suggesting that the energy input due to numerous density waves in this region puffs up the ring. This scenario is also consistent with the short damping length of waves in the outer A Ring (Shu *et al.* 1985).

(3) A number of other observations of the Saturnian and Uranian rings are hard to understand with classical, many-particle-thick models. (a) The existence of the quadrupole brightness variations ("azimuthal asymmetry") in the A Ring is not predicted by classical models (Thompson *et al.* 1981). (b) The A and B Rings show a strong increase of reflectivity I/F with normal optical depth τ; classical models predict only a very weak increase. Similarly, the ϵ Ring of Uranus shows only a weak variation

of radially integrated reflectivity with longitude, which is puzzling for an optically thick ring of variable width (Svitek & Danielson 1987). (c) The brightness of the B Ring increases as the solar and observer elevation angles increase at constant phase angle (Hämeen-Anttila & Pyykkö 1973). This is unlikely to be due to multiple scattering, because the ring particles have only a moderate albedo and are highly backscattering. (d) In images of the inner Cassini Division taken in diffusely transmitted light, I/F varies systematically with τ, such that the brightest bands have $\tau \approx 0.2$; classical models predict that the ring brightness should peak at $\tau \approx 0.1$ (Flynn & Cuzzi 1989).

2 Technique and results

We use our ray tracing code to calculate the singly-scattered light from a layer which can be one or many particles thick. The layer has periodic boundary conditions to simulate a ring of infinite horizontal extent. We typically consider 1000 identical, spherical particles, but our approach easily allows a size distribution. The Sun illuminates the layer; we divide the layer into many pixels and determine the brightness of each, using an assumed surface phase function for the particles. We then calculate the singly-scattered light which finally reaches the observer, allowing particles to shadow and block each other from view. The main output of the code is an "image" of the layer.

Preliminary results include the following: (1) We find excellent agreement with the theory of the opposition effect formulated by Irvine (1966). (2) Because shadowing is reduced in a thin layer, ring brightness is typically greater in backscatter than for classical rings. In typical Voyager 2 inbound images of the main rings, a monolayer would be about 65% brighter than a thick ring of equal optical depth. (3) In low-phase images, ring brightness is more sensitive to optical depth in the range $0.1 < \tau < 0.3$ for a thin ring than for a thick ring. (4) Azimuthal brightness variations as seen by Voyager can be produced if the particle distribution is very non-uniform (C. C. Porco, 1989, personal communication).

In the future we plan to study the opposition surges of rings and satellites in more detail. We hope to determine whether a broad particle size distribution can produce a narrow opposition surge even for a fairly dense layer, as claimed by Hapke (1986). We also intend to incorporate multiple scattering, which is important at high phase angles.

References

Cuzzi, J. N. *et al.*, 1984. In *Planetary Rings*, p. 73, eds. Greenberg, R. & Brahic, A., University of Arizona Press, Tuscon, AZ.
Dones, L., 1987. *PhD thesis*, University of California, Berkeley.
Doyle, L. R., 1987. *PhD thesis*, University of Heidelberg.
Flynn, B. C. & Cuzzi, J. N., 1989. *Icarus*, in press.
Goguen, J. D., Hammel, H. B. & Brown, R. H., 1989. *Icarus*, **77**, 239.
Hämeen-Anttila, K. A. & Pyykkö, S., 1973. *Astron. Astrophys.*, **19**, 235.
Hapke, B., 1986. *Icarus*, **67**, 264.
Herbst, T. M., Skrutskie, M. F. & Nicholson, P. D., 1987. *Icarus*, **71**, 103.
Irvine, W. M., 1966. *J. Geophys. Res.*, **71**, 2931.
Lumme, K., Irvine, W. M. & Esposito, L. W., 1983. *Icarus*, **53**, 174.
Shu, F. H., Dones, L., Lissauer, J. J., Yuan, C. & Cuzzi, J. N., 1985. *Astrophys. J.*, **299**, 542.
Svitek, T. & Danielson, G. E., 1987. *J. Geophys. Res.*, **92**, 14,979.
Thompson, W. T., Irvine, W. M., Baum, W. A., Lumme, K. & Esposito, L. W., 1981. *Icarus*, **46**, 187.
Wisdom, J. & Tremaine, S., 1988. *Astron. J.*, **95**, 925.

Accretion discs around young stellar objects and the proto-Sun

D. N. C. Lin

Lick Observatory, University of California, USA

Abstract   Observed infrared and ultraviolet excesses have widely been interpreted as signatures for accretion discs around young stellar objects. Analyses of the observed properties of these discs are important for the investigation of star formation as well as the dynamics of the protoplanetary disc out of which the solar system was formed. Accretion disc theories suggest that evolution of protoplanetary discs is determined by the efficiency of angular momentum transport. During the formation stages, the disc dynamics are regulated by mixing of infalling material and disc gas. In the outermost regions of the disc, self-gravity may promote the growth of non-axisymmetric perturbations which can transfer angular momentum outwards. After infall has ceased, convectively driven turbulence can redistribute angular momentum with an evolutionary time-scale of the order 10^{5-6} yr. Convection in protoplanetary discs may eventually be stabilized by surface heating as the disc material is depleted. Once the grains in the disc have settled to the midplane region, the disc can neither generate its own energy through viscous dissipation nor reflect radiation from the central star. Consequently, the infrared excess vanishes and the young stellar objects become "naked T Tauri stars." Protoplanetary formation modifies the structure and evolution of the disc when giant protoplanets acquire sufficient mass to truncate the disc. In this case, a protoplanet's tidal torque opens up a gap in disc. Gap formation also leads to the termination of protoplanetary growth by accretion. The condition for a proto-Jupiter to acquire its present mass implies that the viscous evolution time-scale for the disc is of the order 10^{5-6} yr which is comparable to the age of typical T Tauri stars with circumstellar protoplanetary discs.

1 Introduction

Excess infrared radiation detected in young stellar objects is often used as an indicator of the presence of circumstellar discs (Adams, Lada & Shu 1987, Shu, Adams & Lizano 1987, Kenyon & Hartmann 1987). Though unresolved directly, roughly half of all premain-sequence stars are deduced to have discs with masses in the range 0.01–0.1 $M_\odot$ and sizes from 10 to 100 AU (Strom, Edwards & Strom 1989). In the case of HL Tau where the image is resolved, a disc with a size 2000 AU and mass 0.1–1 $M_\odot$ is deduced from radio velocity maps (Sargent & Beckwith 1987).

Theoretical investigation of disc dynamics in this context is important not only for the analysis of these observed properties but also for the theory of formation of rotating and multiple stars. For example, discs around T Tauri stars either reflect central stellar radiation or in some cases produce their own infrared radiation. In some cases, UV excess is also observed and it has been interpreted as radiation from a boundary layer which separates the stellar surface from the disc (Herbig & Goodrich 1986, Bertout, Basri & Bouvier 1988).

Accretion disc theory is based on the conjecture that the source of energy is viscous dissipation of differentially rotating gas in the disc. Associated with energy dissipation is angular momentum transfer and mass diffusion. The rate of disc evolution is determined by the magnitude of an effective viscosity (Lynden-Bell & Pringle 1974). Molecular viscosity is generally too small to make a significant contribution. Disc models provide estimates on the time-scale and physical mechanism for the viscous evolution. If the infrared excess is due to viscous dissipation or reprocessing of stellar radiation, the disappearance of discs, as young stellar objects evolve towards a "naked" T Tauri phase, would probably occur on a viscous evolution time-scale.

Another important motivation for constructing protoplanetary disc models around young stellar objects is to apply these models to study the origin of the solar system. In this case, the disc provides the environment out of which protoplanets are formed. A model for planet formation involves the consideration of additional physical processes such as chemical and dynamical evolution of grains (Morfill & Völk 1984, Weidenschilling 1984). condensation and coagulation of planetesimals (Safronov 1969, Greenberg *et al.* 1978, Hayashi, Nakazawa & Nakagawa 1985), the formation and growth of terrestrial protoplanets and giant protoplanetary cores (Wetherill & Stewart 1988, Lecar & Aarseth 1985) and gas accretion and orbital evolution of giant protoplanets (Bodenheimer & Pollack 1986, Lin & Papaloizou 1985). Most of these processes are regulated, to various degrees, by the structure of the disc and detailed analyses may yield important clues to the dynamical properties of the protoplanetary disc. For example, giant planets are composed mostly of gas and therefore must have been formed in a gas-rich environment. But there is little of it left between the planets today. The determination of a characteristic time-scale for gas depletion in the disc provides a constraint on the formation epoch of giant protoplanets. This constraint also applies to terrestrial protoplanet formation since giant protoplanet formation proceeds through the formation of solid cores, which have masses comparable to or greater than those of the terrestrial planets.

Close comparison between the conditions which led to the formation of the solar system and the properties of discs around young stellar objects can provide observational tests for theories of cosmogony. At the same time, we can deduce signatures of protoplanetary formation in accretion discs around young stellar objects. Although we have no intention of discussing the detailed scenarios for protoplanetary formation in this context, we consider some effects of planet formation on the evolution of the disc. For example, in §2, we show that the depletion of grains can cause the disc to become optically thin and attain an isothermal structure in the direction normal to the plane of the disc. This process may be the reason T Tauri stars become "naked." Another protoplanetary formation process which may influence the dynamical evolution of the disc is protoplanet-disc tidal interaction (Lin & Papaloizou 1980, 1985). In §4, we show that a detailed analysis of the protoplanet-disc interaction can provide an important constraint on the magnitude of viscosity and the evolutionary time-scale of protoplanetary discs.

2 Thermal convection: source of heat and angular momentum transport

Mass flow in a differentially rotating, geometrically thin, disc requires both angular momentum transport and energy dissipation (Lynden-Bell & Pringle 1974). These two processes can operate simultaneously through an effective viscous stress. Molecular viscosity in typical accretion discs is too small to be of astrophysical interest. In a variety of accretion discs, turbulent viscosity is often assumed to be responsible for both angular momentum transport and viscous dissipation – despite the lack of rigorous proof that turbulence may occur intrinsically (Shakura & Sunyaev 1973, Lynden-Bell &

Pringle 1974). Protoplanetary discs, however, are unstable against thermal convection, in the direction normal to the plane of the disc which makes the flow turbulent (Lin & Papaloizou 1980, 1985, Lin 1981).

2.1 INTRINSIC NATURE OF CONVECTIVE INSTABILITY

In order to show that the protoplanetary disc is intrinsically unstable against thermal convection, we analyze the structure of a turbulent-free protoplanetary disc in which rotation prevents gas from migrating in the radial direction. Gas can, however, contract in the vertical direction. If the disc is not in hydrostatic equilibrium initially, it would rapidly evolve towards such a state. Using a one-dimensional numerical hydrodynamic scheme, Ruden (1987) showed that a disc of cold gas contracts towards the midplane. After the disc has settled into a quasi-hydrostatic equilibrium, slow contraction continues as thermal energy is lost from the disc's surface. In the absence of any source of energy, surface radiation causes heat to diffuse from the midplane to the surface region. The associated reduction in the pressure support leads to a readjustment towards a new hydrostatic equilibrium. In the typical temperature range of ten to a few thousand degrees, such adjustment leads to a superadiabatic structure in the vertical direction because the opacity, primarily due to dust grains, is an increasing function of temperature. In this case, the cooler surface region has a lower opacity and cools more efficiently (Lin & Papaloizou 1980, 1985, Lin 1981). This condition is not generally satisfied in other accretion discs. According to the standard Schwarzschild criterion, the superadiabatic gradient induces the disc to become convectively unstable in the vertical direction.

2.2 GLOBAL CONVECTIVE PATTERN AND ANGULAR MOMENTUM TRANSPORT

The role of convection in a disc is not limited to heat transport in the vertical direction. Convective eddies induce mixing over a radial extent comparable to their own size and therefore transfer angular momentum. Convection also generates turbulence which causes dissipation of energy stored in differential rotation. Perhaps the simplest treatment for convection is to use the mixing length prescription in which the eddy viscosity is assumed to be the product of the convective speed and an effective mixing length which is comparable to the size of the eddies (Lin & Papaloizou 1980). From such a treatment, we can build self-consistent models in which convection is responsible for: (1) energy dissipation, (2) heat transport in the vertical direction, (3) angular momentum and mass transport in the radial direction.

The mixing length model, though informative and easy to use, is based on an *ad hoc* prescription of eddy viscosity. In a convective disc, eddies with a variety of scales are generated. Similar to typical turbulent shear flows, the largest eddies often dominate the momentum transfer whereas the smallest eddies provide most of the energy dissipation in the disc. The scale of the largest convective eddies is comparable to the vertical scale of the entire convectively unstable zone which itself extends over a significant fraction of the thickness of the protoplanetary disc. On these large scales, global effects such as rotation and radiative losses are important. Thus, convection must be examined with a global analysis. In an attempt to carry out a global analysis, Cabot *et al.* (1987a,b) computed a vertically-averaged effective viscosity which is derived from integrating the linear growth-rate through various distances above the midplane of the disc. This growth-rate varies greatly in the vertical direction and therefore cannot be attributed to a given eddy.

A more appropriate global treatment is to determine a unique growth-rate, and its associated eigenfunction, for each characteristic convective mode (Ruden, Papaloizou & Lin 1988). These eigenfunctions extend over finite radial distances. In the thin-disc limit, the WKB approximation may be used to describe the radial dependence of temperature and density. These global linear stability analyses of axisymmetric perturbation indicate that:

(1) rotation and compressibility tend to reduce the growth-rate of the disturbances;

(2) the growth-rate is proportional to the square root of the radial wave number and is bounded by the maximum values of the Brunt-Väisälä frequency;

(3) the maximum radial size of eddies scales as the square root of the superadiabaticity times the size of the convective region;

(4) due to radiative losses, the short wavelength modes become overstable and only the fundamental and the first harmonic modes, where the wavelength is comparable to the thickness of the disc, can grow effectively; and

(5) both even and odd modes exist in which a single eddy may either be confined to one side of or thread through the midplane and have a characteristic scale comparable to the thickness of the entire disc.

Convective eddies effectively couple different parts of the disc, dissipate heat and transport angular momentum. The magnitude of the effective viscosity can be derived under the assumption that gas, within a radial wavelength, mixes efficiently over the characteristic growth time-scale. This estimate generally agrees well with that derived on the basis of the mixing length model, *i.e.* it yields an effective $\alpha \sim 0.01$. A more rigorous next step would be to carry out a global analysis of the initial linear growth of non-axisymmetric disturbances and its growth into the non-linear region where dissipative processes become important. The determination of the torque associated with the growing non-axisymmetric disturbance will provide a more rigorous estimate of the efficiency of angular momentum transport.

2.3 EFFECTS OF INFALL AND SURFACE HEATING

The superadiabatic gradient in the vertical direction can be significantly modified by the boundary condition at the surface of the disc. For example, during the formation stage of the disc, shock dissipation near the region where the infalling material joins the disc induces an isothermal structure for the disc. If the infall rate onto the disc remains constant, shock dissipation near the surface dominates energy loss due to contraction towards the midplane so that convection is suppressed. However, when the infall rate is reduced, the superadiabatic gradient may be established as the surface heating becomes less important (Ruden 1987).

Recently, Nagagawa, Watanabe & Nakazawa (1989) have shown that surface heating due to radiation from the central star can also reduce the temperature gradient in the vertical direction in a manner similar to shock dissipation associated with infall. In their calculation, Nagagawa *et al.* show that convection may be stabilized if there is a relatively large stellar radiation flux incident onto the surface of a disc with a relatively low surface density. Their choices of parameters are biased toward reducing the temperature gradient in the disc. A more general analysis (Bell, Lin & Ruden 1989) indicates that convection is suppressed only when the surface heating is sufficiently large to induce a black body temperature comparable to temperature near the midplane. This critical flux increases with the surface density of the disc since opacity and consequently temperature, near the midplane, also increases with the surface density.

In the limit where surface heating exceeds the critical flux required to stabilize against convection, it remains unclear whether the disc may be subject to axisymmetric instabilities (Lin 1989). Consider an axisymmetric perturbation in which one region of the disc has a slightly higher temperature. The disc thickness would increase there and expose that region of the disc to additional solar radiation. Although the disc may be stabilized against convection, this relatively thick region of the disc would cast shadows and reduce surface heating for the exterior regions. A decrease in the incident solar radiation in the shielded region would cause a reduction in the disc temperature there. Consequently, the disc opacity is reduced and the cooling efficiency increased. Although this process may not be thermally unstable, it may generate a large temperature gradient in the radial direction across the interface between the exposed inner hot region and the shielded outer cool regions. For perturbations with wavelengths comparable to or shorter than the thickness of the disc, the disc may become convectively unstable in the radial direction. Mixing in the radial direction would limit the magnitude of the radial temperature gradient. The consequence of surface heating may be the formation of ripples on the disc surface.

2.4 OTHER MECHANISMS OF ANGULAR MOMENTUM TRANSPORT

The discovery of a massive disc around HL Tau (Sargent & Beckwith 1987) reveals the existence of discs whose self-gravity, in the vertical direction, is comparable to that due to the central star. The relative importance of self-gravity is measured by a dimensionless parameter Q (Toomre 1964). When Q is of order unity, self-gravity of the disc may cause unstable growth of axisymmetric and non-axisymmetric perturbations. In the context of angular momentum transfer non-axisymmetric instabilities are more interesting and relevant not only because they induce non-axisymmetric torques (Savonije, this volume) but also because, under certain circumstances, they can persist even when the disc is stable against axisymmetric perturbations (Papaloizou, this volume, Sellwood & Lin 1989).

The condition for gravitational instability against axisymmetric perturbations can be obtained through a local analysis and it is $Q < 1$ (Safronov 1960, Toomre 1964). The stability analyses of non-axisymmetric perturbations generally require global analyses because non-axisymmetric perturbations induce torques which have strong effects over extended regions of the disc. Recently, we carried out a linear normal-mode analysis (Papaloizou & Lin 1989) in which we derived the growth-rate for unstable non-axisymmetric normal modes in several different disc models. We found that if the disc contains surface density variations with a characteristic scale-length comparable to, or shorter than, the distance separating the Lindblad and co-rotation resonances of the perturbations, non-axisymmetric perturbations may grow for $Q > 1$. Similar instability can be found in particle discs around a massive central star (Sellwood, this volume,

Sellwood & Lin 1989). Based on these analyses, it is tempting to derive an approximate formula for the effective torque generated by non-axisymmetric instabilities in self-gravitating discs (Lin & Pringle 1987). This approximate formula, though very crude, is easily applied to the calculations of the global evolution of a self-gravitating disc. Note that the effective torque induced by gravitational instability is due to global transfer and it would not necessarily lead to a turbulent flow pattern.

An alternative scheme for angular momentum transport has been proposed by Donner (1979) and Spruit (1987). In this scheme, self-similar shock waves may induce effective angular momentum transport in accretion discs (Larson 1989). The self-similarity of these shock waves implies that they have zero pattern speed and are therefore stationary. Wave-like disturbances induced by the perturbation at very large disc radii may steepen into shock waves as they propagate inward. This requires that there is little or no dissipative damping of wave action as the waves propagate inwards. If these shock waves can be induced, the radial spacing between successive shock fronts continues to reduce. When the spacing between shock fronts is reduced to less than the vertical scale height of the disc, wave propagation in the vertical direction becomes important.

Although non-linear wave propagation in a disc remains to be analyzed, a three-dimensional linear analysis has been carried out (Lin, Papaloizou & Savonije 1989). Due to differential rotation, the wavelength of linear waves is reduced as they propagate through the disc. When the wavelength becomes comparable to the vertical scale height, wave propagation becomes influenced by the vertical structure of the disc. In the optically thick region of a protoplanetary disc, the vertical structure is thermally stratified such that the temperature and sound speed decrease with distance from the midplane. The magnitude of the pressure gradient in the vertical direction is also larger than that in the radial direction. Consequently, initially radially propagating waves are continually deflected in the vertical direction. For moderate temperature contrast between the disc's midplane and its surface, waves with wavelength comparable to or shorter than the vertical scale height, can be transmitted into and dissipated at the tenuous upper atmosphere. Thus, wave propagation is not an effective mechanism for angular momentum transport over extended regions in the disc. Nevertheless on a local scale, wave propagation can lead to effective dissipation. If the energy in the shear can be continually and effectively transferred into waves through some unstable growing modes, wave energy dissipation may provide a significant source of local viscosity.

These results indicate that dissipative damping may prevent linear waves from steepening into shock waves. Even if shock waves are induced, refraction effects may cause the shock waves to bend and propagate in the vertical direction. Thus, we remain pessimistic that self-similar propagation of shock waves can be an effective mechanism for angular momentum transport.

3 Dynamical evolution

The evolution of protostellar discs can be analyzed with a time-dependent diffusion equation (Lüst 1952, Lynden-Bell & Pringle 1974, Lin & Papaloizou 1985)

$$\frac{\partial \Sigma}{\partial t} - \frac{3}{r}\frac{\partial}{\partial r}\left(r^{1/2}\frac{\partial}{\partial r}(\Sigma\nu r^{1/2}) - \frac{2S_\Sigma(r,t)J(r,t)}{\Omega}\right) - S_\Sigma(r,t) = 0 \qquad (1)$$

where Ω, Σ and ν are the angular frequency, surface density, and viscosity of the disc material respectively, $S_\Sigma(r,t)$ and $J(r,t)$ are the mass flux and excess angular momentum respectively, of the infalling material. Clearly, the evolutionary time-scale and disc structure depend on both the magnitude of viscosity and the properties of the infalling material. During the initial formation of the disc, $S_\Sigma(r,t)$ may be particularly important in determining the structure of the disc, whereas in the post infall stage when $S_\Sigma(r,t) = 0$, viscous diffusion dominates the evolution of the disc. We now discuss the three epochs of protostellar disc evolution.

3.1 FORMATION OF THE DISC

The formation stage of a protoplanetary disc is closely associated with the initial collapse of the protostellar cloud and the formation of the central star itself. The specific angular momentum of a typical protostellar cloud is 10^{20-21} cm^2 s^{-1} (Goldsmith & Arquilla 1985). At the lower end of this range, the specific angular momentum of the cloud is comparable to that in the solar system today, whereas at the upper end, rotational effects would inhibit the collapse of the cloud at a size $\sim 10^{16}$ cm. Recently, numerical calculation has been carried out for both limits.

In an attempt to investigate the collapse of cloud with relatively low angular momentum, Morfill, Tscharnuter & Völk (1985) carried out a series of two-dimensional numerical hydrodynamical calculations in which they adopted an *ad hoc* α prescription (Shakura & Sunyaev 1973) for angular momentum transport, with $\alpha = 0.3$. Since the magnitude of viscosity is a function of temperature, the detailed treatment of radiation transfer is important in determining the structure and evolution of the disc. Morfill *et al.* modelled the radiation transfer process using an Eddington approximation. The outcome of collapse was the formation of a central condensation surrounded by a disc. The energy dissipation associated with the formation and subsequent infall keeps the disc moderately hot. In a more comprehensive analysis, Bodenheimer *et al.* (1989) adopted a different numerical hydrodynamic scheme and modelled the radiative transport with a diffusion equation. Their results indicate the formation of a rapidly rotating central object which is surrounded by a relatively hot and geometrically thick disc spreading from 1 to 60 AU. The disc is essentially stable against axisymmetric gravitational instability.

The collapse of high-angular momentum clouds would lead to the formation of more extended discs, which are likely to be optically thin. In these regions, the cooling time-scale may be short compared with the dynamical time-scale so that the disc is likely to be relatively cool. The combination of low temperature and a large distance from the central condensation provides a favourable condition for the disc to become gravitationally unstable. Using a simple prescription for the effective torque induced by growing non-axisymmetric perturbations (Lin & Pringle 1987), we compute the formation of an extended disc (Lin & Pringle, in preparation). In these computations, we also include turbulent transport of angular momentum, with an α prescription, but we assume that the turbulent viscosity is relatively small on the basis that convection is probably stabilized by the shock dissipation near the disc surface (see §2.2). Radiative transfer for both optically thin and thick limits are included. Finally, the difference in the specific angular momentum carried by the infalling and disc material is taken into account. In general, we find that non-axisymmetric instability can induce efficient angular momentum transfer in the outer regions of the disc and regulate mass transfer, through the disc, at a rate $\sim 10^{-6} - 10^{-4}$ $M_\odot$ yr^{-1}. Thus, the typical time-scale for disc evolution is $\sim 10^{5-6}$ yr which may be long compared with the time-scale for the infall phase. Consequently there could be a substantial residual disc after the infall is effectively switched off.

3.2 MAIN EVOLUTIONARY PHASE

We now turn our attention to the main evolutionary phase after infall onto the disc has finished. The dynamics of the disc during this stage is essentially determined by the viscous diffusion process and can be analyzed by applying a simplified prescription for convectively induced turbulent viscosity (see §2.2) in the diffusion equation. In principle, the surface density distribution at the end of the infall stage may be used as the initial condition for the main evolution phases. The evolution of the disc can also be computed with arbitrary initial conditions.

In the context of solar system formation, an interesting initial condition is the "minimum mass" nebula model (Cameron 1973, Hayashi 1981). In this model, the surface density distribution is derived by augmenting gas to the present mass distribution in the solar system based on the assumption that the protoplanetary disc had a solar composition and planetary formation is totally efficient at retaining all the heavy elements in the disc. Applying the effective viscosity associated with convective viscosity into equation (1), we find the evolution of the minimum-mass nebula with a physical dimension comparable to that of the present day solar nebula takes place on the time-scale of $\sim 10^6$ yr (Lin & Papaloizou 1985). The temperature distribution resembles that deduced from the condensation temperature for various terrestrial planets and satellites (Lewis 1972). At radii interior to the orbit of Mercury, the midplane temperature exceeds 2 000 K and grains would be mostly evaporated.

The disc becomes optically thin exterior to the orbit of Neptune and attains an isothermal vertical structure. Consequently, convection would not occur and dust would settle towards the midplane. Even if there are some other local instabilities, we do not expect turbulence to be sustained for an extended period. In the marginally optically thin region, the characteristic time-scale for heat loss is comparable to, or shorter than, the orbital time-scale of the disc. Energy transfer between eddies with different sizes occurs on the eddy turnover time-scale which is comparable to, or longer than, the orbital time-scale. Since thermal energy loss is faster than the kinetic energy of the eddies, the eddy motion becomes supersonic. Shocks during eddy mixing could cause turbulence to decay.

However, the absence of local turbulence may not imply the termination of disc evolution. Accompanying the decay of turbulence is the decline in mass diffusion and energy dissipation so that the disc temperature and optical depth decrease with the surface density. The effect of self-gravity also increases with time. Eventually, the Q value decreases to order unity and gravitational instability becomes important in promoting the growth of non-axisymmetric disturbances and in inducing angular momentum transfer (Lin & Pringle 1989). In this region, the evolutionary time-scale is determined by the cooling time-scale since cooling is essential in maintaining a sufficiently low value for Q. In a relatively extended massive disc, when the temperature decreases to ~ 10 K, the cooling time-scale at 100 AU exceeds 10^{6-7} yr. The disc temperature can not decrease below 10 K, which is the temperature of the protostellar cloud. Thus, when the disc surface density decreases to a sufficiently small value, the effect of self-gravity can no longer be important and the evolution of the solar nebula must stop at that region. A natural outcome for this process is the development of an extended, circumstellar, optically-thin disc.

For the minimum-mass solar nebula model, the surface density of the disc is so low that self-gravity becomes at best marginally important beyond the orbit of Neptune even when the disc temperature is 10 K. If the outer region is both optically thin and non-self-gravitating, the disc terminates its evolution there. This may be the reason why there is no major planet beyond the orbit of Neptune. Inside the outermost region, the disc remains opaque and viscous evolution continues. A fraction of disc material would be deposited into the outermost region to carry the excess angular momentum. While gas may eventually be evaporated by the photo-dissociation process, dust particles would be left behind. In the absence of turbulence, dust particles descend towards the midplane. When the dust layer becomes sufficiently thin, gravitational instability would cause the dust to clump and form 10 km size objects which may be the progenitors of planetesimals or comets (Goldreich & Ward 1973).

3.3 CAUSES FOR UNSTEADY FLOW AND OUTBURST PHENOMENA

In addition to the usual infrared signatures of discs, the prototype of young stellar objects, T Tau, has a variable light curve. Large magnitude luminosity variations, on the time-scale of months to years, are found in other young stellar objects such as FU Ori and V1057 Cyg (Herbig 1977). It was suggested that these outbursts are associated with variation in mass transfer rate in accretion discs (Lin & Papaloizou 1985). There is observational evidence which supports this scenario (Hartmann & Kenyon 1987a,b). For example, the pre-outburst spectrum of V1057 Cyg is that of a typical T Tau star. The accretion disc around a typical T Tau star has a mass transfer rate of 10^{-7} to 10^{-6} $M_\odot$ yr^{-1}. In outbursts, however, the estimated mass accretion rate ranges from $\sim 10^{-4}$ $M_\odot$ yr^{-1} for FU Ori to $\sim 10^{-3}$ $M_\odot$ yr^{-1} (Kenyon, Hartmann & Hewett 1988).

The short rise time-scale for the FU Orionis events implies that the variation in accretion rate occurred in the inner regions of the disc close to the accreting star. If this variation is communicated through the disc on a viscous diffusion time-scale, the radius of sudden increase in mass transfer rate must be confined to within 10^{12} cm which is only slightly larger than the dimension of the accreting protostar. However, during the rise, the inferred disc mass being accreted by the star is $\sim 10^{-3}$ $M_\odot$. From standard disc models, we deduce that such a large mass concentration in the inner regions of the disc requires perturbations which can sweep all the disc material from 10^{14} cm into the region within 10^{12} cm (Clarke, Lin & Pringle 1989).

Mass transfer variation may originate from a more extended region but it would require a more rapid propagation of disc response than that which can be provided by viscous diffusion. One possible mechanism for inducing a rapid rise is thermal instability in the disc analogous to that attributed to dwarf nova outbursts (Lin & Papaloizou 1985). For accretion rates exceeding $\sim 10^{-5}$ $M_\odot$ yr^{-1}, the associated energy dissipation would partially ionize the gas in the inner regions of a protoplanetary disc. According to standard α models, partially ionized regions of accretion discs are thermally unstable because the opacity increases rapidly with temperature (Faulkner, Lin & Papaloizou 1983) – a small temperature increase would cause a large increase in opacity and a decrease in the heat diffusion rate. Consequently, the disc undergoes an upward thermal transition. If the magnitude of viscosity is a function of temperature, as is the case in the α prescription, thermal instabilities would lead to changes in the mass and angular momentum transfer rate. Thermal transition at one radius would cause a large local temperature gradient in the radial direction. The sudden increase in viscous stress across the transition front would cause the front to propagate throughout the disc such that the entire disc enters an outburst state. If the mass infall rate onto the disc is insufficient to sustain the rate of mass transfer in the disc during the outburst, the disc material would be depleted and the disc would return to quiescence. This type of thermal relaxation limit cycle has been examined in detail for dwarf nova outbursts (Lin, Papaloizou & Faulkner 1985, Papaloizou, Faulkner & Lin 1983). In the case of protoplanetary discs, in order to cause a T Tauri type disc to become thermally unstable, large amplitude perturbations are needed (Clarke, Lin & Pringle 1989). Unlike the situation in dwarf novae, such a large amplitude perturbation is unlikely to be caused by thermal instability alone.

Also, unlike dwarf novae where the outbursts are sustained for a few days, FU Orionis has remained in a post outburst high state for more than thirty years. In V1057 Cyg, the decline rate is very slow. An important theoretical issue is how to maintain the high state. One difference is that protoplanetary discs are much larger than dwarf-nova discs so that the critical mass transfer rate for the onset of thermal instability is also larger (Clarke, Lin & Papaloizou 1989). After an upward transition, the thickness is a large fraction of the radius of the disc. The advective transport of heat is much more important in the FU Orionis discs than in dwarf-nova discs. Over the temperature range near the transition front, advective heat transport can stabilize against thermal runaway and therefore prevent the propagation of the transition front throughout the disc. It is of interest to note that this stabilizing effect is only effective for discs with a mass transfer rate comparable to or larger than 10^{-4} $M_{\odot}$ yr^{-1}. Therefore, it may be very useful for the observers to determine whether there is any correlation between the light curves and mass accretion rate in the FU Orionis stars.

Spectroscopic data for several FU Ori stars are available (Kenyon, Hartmann & Hewett 1988). These spectra can be decomposed into stellar and steady state disc components. Conspicuously absent is any break in the continuum which is often used as an indicator of boundary layer radiation. Since the energy released in the boundary layer is comparable to that from the entire disc, it would be detectable unless it is redistributed over an area larger than the accreting star's surface. In an outburst state, significant radiation transfer in the radial direction can occur since (1) the opacity near the midplane is small compared to that near the surface of the disc and (2) the thickness is comparable to the radius of the disc. Detailed radiative transfer calculations are underway (Bell, Clarke & Lin 1989).

Another interesting implication of these models is that if all T Tauri stars have gone through an FU Orionis stage, the disc accretion rate at one stage of their evolution would have been so large that the disc temperature at a few AU could reach several thousand degrees. This speculative deduction may have interesting implications for cosmochemists.

There are several possible mechanisms which may induce large amplitude perturbations to trigger thermal instability in the disc. For example passage of hypothetical companions through the inner regions of the disc may cause significant perturbations to the surface density and mass transfer rate in the disc (Clarke & Pringle 1989). Alternatively, rapidly accreting discs dissipate energy at a high rate. If the surface of the disc is exposed to the radiation from the star or boundary layer, convection may be stabilized. If convectively driven turbulence is the only source of viscosity, accretion flow may be quenched. The outer regions of a disc with such a large accretion rate are probably self-gravitating. Large surface heating may sustain a relatively high disc temperature so that the disc may be stabilized against gravitational instability. In either case, a potential feedback mechanism may be induced if the surface heating flux is proportional to the mass transfer rate into the inner regions of the disc. Preliminary investigation indicates that such a feedback mechanism may lead to regular or chaotic limit cycles. Large amplitude variations in the mass transfer rate in these cycles provide a good model for FU Ori outbursts.

3.4 CLEARING PROCESSES

On the evolutionary time-scale of typical T Tauri stars, $\sim 10^6$ yr, approximately half of these young stellar objects become naked T Tauri, *i.e.* they lose any indication of the presence of a circumstellar disc. If the convectively driven turbulent viscosity is applied to equation (1), we deduce the time-scale for the disc to evolve to an optically thin system to be $\sim 10^{8-9}$ yr. Even with the assumption that some hypothetical transonic turbulence may be responsible for angular momentum transfer, the time-scale for the disc to evolve into an optically thin state is still larger than 10^7 yr. These arguments imply that the disappearance of IR and UV excess is not entirely due to viscous diffusion.

One possible mechanism for eliminating the UV and IR excess from the disc is through dust settling. In §2.3, we indicate that the surface heating effect can stabilize against convection (Nagagawa, Watanabe & Makazawa 1989) provided the surface density of the disc is sufficiently low (Bell, Lin & Ruden 1989). During the evolution of the disc, the surface density of the disc continually decreases. For a minimum mass nebula model, the surface density of the outer regions of the disc is reduced to a sufficiently low value such that these regions are stabilized after a few million years.

When the disc is stabilized, turbulence decays unless there are other instabilities. In the absence of turbulence, dust can settle toward the midplane on the relatively short time-scale of 10^{3-4} yr (Hayashi, Nakazawa & Nakagawa 1985). Once settled to the midplane, dust layers can either undergo gravitational instability to form kilometre size planetesimals (Goldreich & Ward 1973) or coagulate rapidly (Weidenschilling 1984). These processes would cause the protoplanetary disc to become transparent so that it would no longer reprocess radiation from the central star. Furthermore, the lack of turbulent viscosity implies that there would not be any viscous dissipation to heat the disc or to supply disc mass to be accreted by the central star. Through this mechanism, radiative flux from the disc is significantly reduced while most of the disc gas is retained. The disc gas may be eventually eliminated on a somewhat longer time-scale by (1) stellar wind ablation, (2) wind generated by the dissociation or ionization of disc gas by the incident solar radiation, and (3) disc-protoplanet tidal interaction.

It is also possible that convective instability is not the only source of turbulence. In this case, although the surface heating effect may induce an isothermal vertical structure and eliminate convection, it could also provide a more favourable condition for sound waves to propagate a large distance. Note that in the absence of vertical thermal stratification, refraction of waves is eliminated. Consequently, when the surface density has dropped below the critical value, the efficiency of angular momentum transport may actually increase despite the lack of convection. We are currently examining the possibility that the disc could evolve to a state where it becomes completely optically thin on a time-scale of a few million years.

4 Protoplanet-disc tidal interaction

We now examine the effect of giant protoplanet formation on the structure and evolution of the disc. Through its tidal torque on the protoplanetary disc, a giant protoplanet can excite waves which carry angular momentum and energy. The dissipation of wave energy in the disc induces angular momentum transport and can have significant effects on the evolution of the disc.

4.1 PROPAGATION OF TIDAL DISTURBANCE

Tidally induced waves are directly observed in planetary rings (Cuzzi *et al.* 1984). An important issue in protoplanet-disc interaction is how far can the tidally induced wave propagate before it is dissipated. In the regions of the disc interior to the protoplanet's orbit, these waves carry negative angular momentum flux with respect to the local fluid. When the wave is dissipated, it deposits negative angular momentum and thereby causes the disc material to drift inwards. If these waves can propagate deep into the interior region of the disc, the tidal influence of the protoplanet is distributed over an extended region, whereas if these waves are dissipated in the close vicinity of the protoplanet's orbit, the tidal effect is localized. Because the protoplanetary disc has a thermally stratified structure in the vertical direction, *i.e.* the temperature decreases with distance from the midplane, the propagation of the waves in the radial and vertical directions is closely linked. In §2.4 we indicated that thermal stratification causes refraction such that initially radially propagating waves are deflected in the vertical direction. For moderate temperature contrast between the disc's midplane and its surface, wave transmission into the tenuous upper atmosphere is allowed. For protoplanetary discs, these two effects inhibit wave propagation through large distances in the radial direction. Consequently, waves excited by tidal disturbances due to a protoplanet are not expected to reach the interior regions of the disc with a significant amplitude (Lin, Papaloizou & Savonije 1989). Based on these results, the rate of energy dissipation can be evaluated as a function of the distance from where the waves are launched.

4.2 TIDAL TRUNCATION OF THE DISC

The rate of angular momentum transfer between the protoplanet and the disc is independent of the nature of dissipation provided the waves are dissipated (Goldreich

& Tremaine 1982). However, the structure of the disc in the neighbourhood of the protoplanet does depend on the propagation speed and dissipation rate of the waves in the disc (Papaloizou & Lin 1985). Using a prescription for the effective viscosity, the structure of the disc in the neighbourhood of the protoplanet can be computed with a two dimensional numerical hydrodynamic model in the limit that the disc is relatively thin (Lin & Papaloizou 1986 a,b). The results of these computations indicate that when the mass of the protoplanet is sufficiently large, its tidal torque may induce the formation of a gap in the vicinity of its orbit.

There are two conditions for gap formation. One criterion is a necessary condition: the Roche radius of the protoplanet must be comparable to or larger than the scale height of the protoplanetary disc. If the necessary condition is not satisfied, the moment a protoplanet tries to open up a gap, the pressure gradient, at the boundary of the disc near the protoplanet, would become sufficiently large to force the epicyclic frequency to become negative and therefore cause dynamical instability. The other criterion is that the rate of angular momentum transfer between the protoplanet and the disc, due to the tidal torque of the protoplanet, must exceed that within the disc due to viscous stress. In typical protoplanetary disc models, the second condition is automatically satisfied when the first criterion is satisfied provided the effective α is somewhat less than unity.

After a protoplanet opens up a gap in vicinity of its orbit, mass growth of the protoplanet is essentially terminated. Thus, we can deduce some useful constraints on the dynamics of the primordial solar nebula from the present mass of Jupiter. For example, let us consider the scenario that the disc is not turbulent and viscous dissipation is ineffective. In this case, only the first criterion for gap formation applies. In order to avoid gap formation before Jupiter acquired its present mass, the nebula must have a relatively large vertical scale height. The nebular temperature necessary for such a large scale height is several times that which can be provided by solar radiation. Thus, an additional heat source is required. Viscous dissipation in the disc can provide sufficiently high temperature but it requires an effective viscosity considerably larger than the molecular viscosity which is inconsistent with the assumption that the disc is quiescent.

Let us consider the other extreme scenario that the primordial solar nebula is massive and its evolution is regulated by torque induced by gravitational instabilities. In this case, the disc would be relatively hot and the viscosity relatively large, so that the disc truncation would be unlikely to occur until Jupiter had acquired a mass substantially larger than its present mass. Unless Jupiter could have lost most of its initial mass, this scenario also seems unlikely. From the above arguments, we deduce that in order for both criteria to be satisfied for the present mass of Jupiter, the viscous evolution time of the disc must be of the order $10^5 - 10^6$ yr. It is of interest to note that this is comparable to the typical age of the T Tauri stars.

4.3 PROTOPLANETARY ORBITAL EVOLUTION

After the protoplanet opens up a gap, it continues to tidally interact with the disc. Such interaction causes the orbit of the protoplanet to evolve and the protoplanet continues to have a strong influence on the structure and evolution of the disc (Lin &

Papaloizou 1986b). When gap formation first occurs, the amount of material cleared to either side of the gap is determined by the motion as well as the mass of the embedded protoplanet. The disc responds by modifying the local surface density gradient such that the angular momentum transfer rate between the protoplanet and the region interior to it is delicately balanced by that between the protoplanet and the disc exterior to it. If the mass of the protoplanet is relatively large compared with that of the disc, the protoplanet would not undergo significant orbital evolution. The disc interior to the protoplanet would be depleted on a relatively short time-scale. The disc mass exterior to the protoplanet would be conserved; the surface density decreases slowly as it expands viscously. This process would lead to the formation of a hole in the central region of the disc which may be observed directly. In the limit that the protoplanet's mass is small compared with that of the disc, the protoplanet's orbital evolution would proceed with the disc on the viscous diffusion time-scale of the disc. In this limit, if the protoplanet were formed at a relatively small radius, it would migrate inward on a relatively short time-scale, whereas if the protoplanet were formed at a region near the outer edge of the disc, it would migrate outward until the disc material interior to its orbit was somewhat depleted.

Based on these results, one can deduce the mass distribution of the protoplanetary disc from Jupiter's orbit. The disc mass exterior to the proto-Jupiter could not have exceeded that interior to its orbit by more than 0.1 $M_{\odot}$ otherwise proto-Jupiter would have migrated significantly toward the Sun. Similarly, it can also be argued that the mass of the disc interior to Jupiter could not exceeded that exterior to its orbit by more than 0.1 $M_{\odot}$ otherwise the resonant asteroids would have escaped from their commensurable orbit with Jupiter.

5 Discussion

We have here reviewed some important aspects of various dynamical processes in the protoplanetary disc. Mass redistribution is determined by angular momentum transfer. The rate of angular momentum transfer is regulated by the magnitude of the effective viscosity. During the formation of the protoplanetary disc, mixing of infalling material with the disc gas can lead to significant mass transfer. Rapidly rotating clouds can lead to the formation of extended discs. In the outer regions of the disc, self-gravity of the disc can promote growth of non-axisymmetric disturbances. The tidal torque associated with these growing unstable modes induces angular momentum transport and regulates mass transfer. During the main phase of disc evolution, convectively driven turbulence provides an effective viscosity such that the typical evolutionary time-scale of the disc is $\sim 10^6$ yr. This time-scale is consistent with the typical age of young stellar objects with signatures of circumstellar discs. It is also consistent with the time-scale derived from the condition for tidal truncation of protoplanetary discs by proto-Jupiter.

Aspects of the work reported here are due to P. H. Bodenheimer, R. Bell, C. Clarke, L. Hartmann, J. C. B. Papaloizou, J. E. Pringle, S. Ruden, G. Savonije, J. A.Sellwood and F. Shu. Their contributions are greatly appreciated. We thank I. Roxburgh for hospitality while part of this work was carried out at the School for Mathematical Sciences, Queen Mary College, London. This work is supported in part by grants AST-85-21636 and NAGW 1211 from NSF and NASA respectively. Part of this work has been conducted under the auspices of a special NASA astrophysics theory program that supports a Joint Center for Star Formation Studies at NASA-Ames Research Center, University of California, Berkeley and University of California, Santa Cruz.

References

Adams, F. C., Lada, C. & Shu, F. H., 1987. *Astrophys. J.*, **312**, 788.

Bell, R., Clarke, C. & Lin, D. N. C., 1989. in preparation.

Bell, R., Lin, D. N. C. & Ruden, S. P., 1989. in preparation.

Bertout, C., Basri, G. & Bouvier, J., 1988. *Astrophys. J.*, **330**, 350.

Bodenheimer, P. H. & Pollack, J., 1986. *Icarus*, **67**, 391.

Bodenheimer, P., Yorke, H. W., Rozyczka, M. & Tohline, J. E., 1989. In *Formation and Evolution of Low-Mass Stars*, eds. Dupress, A. & Lago, M. T. V. T., Reidel, Dordrecht, in press.

Cabot, W., Canuto, V. M., Hubickyj, O. & Pallock, J. B., 1987a. *Icarus*, **69**, 387.

Cabot, W., Canuto, V. M., Hubickyj, O. & Pallock, J. B., 1987b. *Icarus*, **69**, 423.

Cameron, A. G. W., 1973. *Icarus*, **18**, 407.

Clarke, C., Lin, D. N. C. & Papaloizou, J. C. B., 1989. *Mon. Not. R. Astron. Soc.*, in press.

Clarke, C., Lin, D. N. C. & Pringle, J. E., 1989. in preparation.

Clarke, C. & Pringle, J. E., 1989. in preparation.

Cuzzi, J., Lissauer, J. J., Esposito, L. W. Holberg, J. B., Marouf, E. A., Tyler, G. L. & Boischot, A., 1984. In *Planetary Rings*, p. 73, eds. Greenberg, R. & Brahic, A., University of Arizona Press, Tuscon, AZ.

Donner, K., 1979. *PhD thesis*, Cambridge University.

Faulkner, J., Lin, D. N. C. & Papaloizou, J. C. B., 1983. *Mon. Not. R. Astron. Soc.*, **205**, 359.

Goldreich, P. & Tremaine, S., 1982, *Annu. Rev. Astron. Astrophys.*, **20**, 249.

Goldreich, P. & Ward, W. R., 1973. *Astrophys. J.*, **183**, 1051.

Goldsmith, P. F. & Arquilla, R., 1985. In *Protostars and Planets II*, p. 137, eds. Black, D. & Matthews, M. S., University of Arizona Press, Tuscon, AZ.

Greenberg, R., Wacker, J. F., Hartmann, W. K. & Chapman, C. R., 1978. *Icarus*, **35**, 1.

Hartmann, L. & Kenyon, S. J., 1987a. *Astrophys. J.*, **312**, 243.

Hartmann, L. & Kenyon, S. J., 1987b. *Astrophys. J.*, **322**, 393.

Hayashi, C., 1981. *Prog. Theor. Phys., Suppl.*, **70**, 35.

Hayashi, C., Nakazawa, K. & Nakagawa, Y., 1985. In *Protostars and Planets II*, p. 1100, eds. Black, D. & Matthews, M. S., University of Arizona Press, Tuscon, AZ.

Herbig, G. H., 1977. *Astrophys. J.*, **217**, 693.

Herbig, G. H. & Goodrich, R. W., 1986. *Astrophys. J.*, **309**, 294.

Kenyon, S. J. & Hartmann, L., 1987. *Astrophys. J.*, **323**, 714.

Kenyon, S. J., Hartmann, L. & Hewett, R., 1988. *Astrophys. J.*, **325**, 231.

Larson, R., 1989. In *The Formation and Evolution of Planetary Systems*, eds Weaver, H. A., Patesce, F. & Danly, L., Cambridge University Press, Cambridge, in press.

Lecar, M. & Aarseth, S. J., 1985. *Astrophys. J.*, **305**, 564.

Lewis, J. S., 1972. *Icarus*, **16**, 241.

Lin, D. N. C., 1981. *Astrophys. J.*, **242**, 780.

Lin, D. N. C., 1989. In *The Formation and Evolution of Planetary Systems*, eds Weaver, H. A., Patesce, F. & Danly, L., Cambridge University Press, Cambridge, in press.

Lin, D. N. C. & Papaloizou, J. C. B., 1980. *Mon. Not. R. Astron. Soc.*, **191**, 37.

Lin, D. N. C. & Papaloizou, J. C. B., 1985. In *Protostars and Planets II*, p. 981, eds. Black, D. & Matthews, M. S., University of Arizona Press, Tuscon, AZ.

Lin, D. N. C. & Papaloizou, J. C. B., 1986a. *Astrophys. J.*, **307**, 395.

Lin, D. N. C. & Papaloizou, J. C. B., 1986b. *Astrophys. J.*, **309**, 846.

Lin, D. N. C., Papaloizou, J. C. B. & Faulkner, J., 1985. *Mon. Not. R. Astron. Soc.*, **212**, 105.

Lin, D. N. C., Papaloizou, J. C. B. & Savonije, G. J., 1989. in preparation.

Lin, D. N. C. & Pringle, J. E., 1987. *Astrophys. J.*, **225**, 607.

Lin, D. N. C. & Pringle, J. E., 1989. in preparation.

Lüst, R., 1952. *Z. Naturforsch., A,* **7a**, 87.

Lynden-Bell, D. & Pringle, J. E., 1974. *Mon. Not. R. Astron. Soc.,* **168**, 603.

Morfill, G. E. & Völk, H. J., 1984. *Astrophys. J.,* **287**, 371.

Morfill, G. E., Tscharnuter, W. & Völk, H. J., 1985. In *Protostars and Planets II,* p. 493, eds. Black, D. & Matthews, M. S., University of Arizona Press, Tuscon, AZ.

Nagagawa, Y., Watanabe, S. & Nakazawa, K., 1989. In *The Formation and Evolution of Planetary Systems,* eds. Weaver, H. A., Patesce, F. & Danly, L., Cambridge University Press, Cambridge, in press.

Papaloizou, J. C. B. & Lin, D. N. C., 1985. *Astrophys. J.,* **285**, 818.

Papaloizou, J. C. B. & Lin, D. N. C., 1989. *Astrophys. J.,* in press.

Papaloizou, J. C. B., Faulkner, J. & Lin, D. N. C., 1983. *Mon. Not. R. Astron. Soc.,* **205**, 487.

Ruden, S. P., 1987. *PhD thesis,* University of California, Santa Cruz.

Ruden, S. P., Papaloizou, J. C. B. & Lin, D. N. C., 1988. *Astrophys. J.,* **329**, 739.

Safronov, V. S., 1960. *Sov. Phys. Dokl.,* **5**, 13.

Safronov, V. S., 1969. *Evolution of the Protoplanetary Cloud and the Formation of the Earth and Planets,* (Nauka, Moscow), NASA TT F-677, 1972.

Sargent, A. I. & Beckwith, S., 1987. *Astrophys. J.,* **323**, 294.

Sellwood, J. A. & Lin, D. N. C., 1989. *Mon. Not. R. Astron. Soc.,* in press.

Shakura, N. I. & Sunyaev, R. A., 1973. *Astron. Astrophys.,* **24**, 337.

Shu, F.H., Adams, F.C. & Lizano, S., 1987. *Annu. Rev. Astron. Astrophys.,* **25**, 23.

Spruit, H. C., 1987. *Astron. Astrophys.,* **184**, 173.

Strom, S. E., Edwards, S. & Strom, K. M., 1989. preprint.

Toomre, A., 1964. *Astrophys. J.,* **139**, 1217.

Weidenschilling, S. J., 1984. *Icarus,* **60**, 555.

Wetherill G. W. & Stewart, G. R., 1988. *Icarus,* **74**, 543.

The Beta Pictoris disc: a planetary rather than a protoplanetary one

Pawel Artymowicz[1]

Space Telescope Science Institute, Baltimore, USA

1 Physical processes

We distinguish between the two possibilities indicated in the title by analysing the physical process operating in the β Pic system. Based on recent models of the disc (Artymowicz *et al.* 1989) and the information on gaseous constituents of the disc (Vidal-Madjar *et al.* 1986, Lagrange-Henri *et al.* 1988) we consider the following processes, which we expect to determine the size distribution of grains and influence the disc appearance:

1 Inter-particle collisions. In the densest parts of the disc ($\sim$ 20 to 50 AU from the star) grains collide typically once in several hundred orbits ($\sim 10^3$ yr). At 100 AU, the time-scale is 10^5 yr and at 1000 AU of order 10^8 yr. The outcome of a typical collision, which from our knowledge of the disc geometry occurs at impact speeds $\sim$ 0.1 times the local Keplerian velocity, is the erosional cratering of larger particles and the destructive shattering of smaller ones. No agglomeration through grain sticking is possible.

2 Poynting-Robertson (P-R) effect. In most previous work, the P-R drag was suggested to play a dominant role. This is not correct. The P-R time-scale for even the smallest (~ 2 μm-sized) particles is too long, $\sim 4 \times 10^6$ yr at 100 AU and increasing with the square of the radius. Whenever collisions act on shorter time-scales, the P-R drag effectively acts on the total mass of the disc, not just the smallest grains, hence the time-scales given are merely lower limits.

3 UV sputtering of grains. Photo-sputtering may affect grains if they are icy or ice-coated. In such a case, the time needed for a complete erosion of the smallest (2 μm) grains is almost 10 times shorter that the P-R drag time throughout the disc. Even so, collisions prevail at most disc radii. If there is a particulate stellar wind from β Pic, then the refractory grains may also be sputtered, but presumably at a much lower rate.

4 Gas Drag. There is a considerable uncertainty about the actual radial distribution of gas. However, from a known column density, and with a reasonable assumption as to the power-law index of the gas density profile (> 1.5), we estimate that gas drag may act on competitively short time-scales with respect to collisions only in the innermost (< 10 AU) disc region. Otherwise its characteristic times are closer to those of UV sputtering than to the dominant collisional time-scale.

5 Grain growth by accretion. Independent of the actual steepness of the gas profile, grain growth by adsorption of the infalling atoms of refractory elements is slower than the P-R drag time everywhere in the disc, and hence may be neglected.

---

[1] On leave from N. Copernicus Astronomical Centre, Warsaw, Poland

2 Discussion

We conclude that typical grains are being gradually eroded in the β Pic disc rather than agglomerating through "snowballing" to form larger bodies. Once smaller than $2 - 10$ microns they are removed from the system by radiation pressure (Artymowicz 1988). If the β Pic disc is a long lived structure then most of its mass must reside in large disc bodies, firstly to provide the mass reservoir, and secondly to heat it up dynamically to its observed thickness through gravitational scattering by embedded objects in the lunar mass range. The disc therefore appears to have already formed its large solid bodies, rather than having still to form them as would be the case in a protoplanetary system.

At the present rate of erosion, the system might have lost 10^{29} g of solid material during its 2×10^8 yr lifetime. This high erosion rate probably implies that the β Pic disc will evolve towards a less conspicuous phase in the next few $\times 10^8$ yr, in which it is "cleared" of much of the present solid circum-stellar material, possibly leaving a solar system-type remnant.

Based on the existing coronographic images (Artymowicz *et al.* 1989), we find that no large (*i.e.* Jupiter-sized) bodies are present in the observed regions (> 80 AU from β Pic). However, such bodies may be hidden behind the coronograph mask. We also find that embedded gap-opening bodies as small as 0.01 Earth masses may be detected with the Hubble Space Telescope.

References

Artymowicz, P., 1988. *Astrophys. J. Lett.,* **335**, L79.

Artymowicz, P., Burrows, C. & Paresce, F., 1989. *Astrophys. J.,* **337**, 494.

Lagrange-Henri, A. M., Vidal-Madjar, A. & Ferlet, R., 1988. *Astron. Astrophys.,* **190**, 275.

Vidal-Madjar, A., Hobbs, L. M., Ferlet, R. & Albert, C. E., 1986. *Astron. Astrophys.,* **167**, 325.

Optical polarimetry and thermal imaging
of the disc around Beta Pictoris

R. D. Wolstencroft[1], S. M. Scarrott[2] R. F. Warren-Smith[2]
C. M. Telesco[3], R. Decher[3] and E. E. Becklin[4]

[1] *Royal Observatory Edinburgh*
[2] *Department of Physics, University of Durham*
[3] *NASA Marshall Space Flight Center, Alabama, USA*
[4] *University of Hawaii, USA*

1 Introduction

The presence of a dust disc around the main sequence A5 star β Pic is now well established (Smith & Terrile 1984, 1987; Paresce & Burrows, 1987). Models based on the integrated thermal emission measured from IRAS and the ground (5 μm to 100 μm), as well as multi-aperture photometry and IRAS slow-scan data, have been constructed by Backman, Gillett & Witteborn (1989), who conclude that there is a dust-free zone around the star at a radius $\sim$ 20 AU, with a (face on) surface density of dust grains which decreases quite slowly with distance out to its outer edge at $\sim$ 1000 AU. However, models by Artymowicz, Burrows & Paresce (1988) based mainly on the optical images suggest that beyond 100 AU the surface density falls as r^{-2} or faster. A possible explanation of this discrepancy could be that there are two separate populations of grains responsible for the optical and infrared emission from the disc which have radically different spatial distributions.

2 Polarimetry

One valuable piece of information that could add significantly to our understanding of the disc is its optical polarization. By analogy with studies of the zodiacal light, the dependence of polarization on angular distance from the star can provide constraints on the radial dependence of grain number density, and the wavelength dependence of polarization sets limits on the size distribution. To see what is possible Wolstencroft, Scarrott & Warren-Smith (WSW) (1989) carried out an experiment using the Durham polarimeter on the SAAO 1.9 m telescope: the star was occulted by a N-S grid in the focal plane of the polarimeter and a sequence of 16 polarization images was obtained both for β Pic and for the nearby star α Pic which is believed to be a point source. In the absence of a Lyot stop, the diffracted light was appreciable, but nevertheless because the polarization of the diffracted light turned out to be small and the disc polarization "high" it was possible to obtain a useful result which is illustrated in Figure 1. Contours of the total brightness are shown, which reveal the disc at the correct position angle of 29°, with superposed vectors whose length is proportional to the polarized intensity. A preliminary analysis of these data indicates the following:

(1) the degree of polarization is about 15% which is comparable to that observed in the zodiacal light at a phase angle of 90°;

(2) the polarization vectors are normal to the scattering plane within the errors (20°) thus confirming that the optical feature is indeed a dust disc made visible by scat-

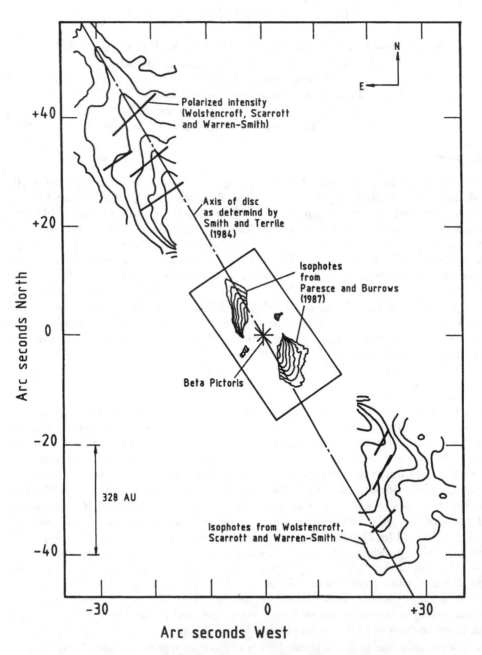

Figure 1. Optical isophotes for the inner (Paresce & Burrows 1987) and outer parts (WSW) of the β Pic disc, on which are superposed polarized intensity vectors (obtained by WSW) whose lengths are proportional to the polarized intensity. There is evidence that the disc may contain two branches separated in position angle by about 10°.

tering rather than say an emission feature such as a "jet"; and

(3) the polarized intensity, like the total intensity, is greatest on the NE as opposed to the SW side of the disc.

3 Thermal Imaging

Thermal imaging of the central $13''$ diameter region ($4.5''$ pixels) surrounding β Pic was carried out at 10 and 20 μm during August 1987 by Telesco *et al.* (1988) using the NASA Marshall Space Flight Center 20 pixel bolometer array. After correction for the contribution of the stellar photosphere the emission is found to be strongly peaked in the central pixel; however, extended emission is also seen which is distributed approximately along the edge-on circum-stellar disc apparent at optical wavelengths. Dust temperatures estimated from the observed flux ratios at 10 and 20 μm are not consistent with black body emission. The black body temperature for grains located at a projected distance of 80 AU from β Pic is close to 50 K whereas the observed temperature is about 130 K, implying a Planck averaged grain emissivity of about 2%. If the grains are either silicate or graphite spheres then using the emissivity calculations of Draine & Lee (1984), Telesco *et al.* (1988) deduced a mean grain size, a, in the range 0.1 to 0.3 μm. Starlight scattered by a power law grain size population with this relatively small value of a/λ is likely to be measurably blue whereas the observations of Paresce & Burrows (1987) and Gradie *et al.* (1987) indicate that the disc is either the same colour or redder than the central star. Evidence for small grains (~ 1 μm or smaller) also comes from the models of Backman, Gillett & Witteborn (1989) and from the recent detection of β Pic at 800 μm by Becklin & Zuckerman (1989) using the James Clerk Maxwell Telescope. Artymowicz (1988) has shown that spherical grains with radii between approximately 0.01 and 1 μm are likely to be removed by radiation pressure and that porous grains as large as 10 μm might be blown out. The observational evidence discussed above suggests that the upper limit is closer to 1 μm than to 10 μm.

4 How common are main sequence discs?

It is clearly important to find more discs like that surrounding β Pic which are both near enough and bright enough to permit detailed study and which in particular can be observed optically. A number of authors have searched for nearby main sequence stars with a far infrared excess comparable to that of β Pic and the other prototype stars Vega, Fomalhaut and ϵ Eridani. Using co-added IRAS survey data, Aumann (1985, 1988) and Backman & Gillett (1987) examined the far infrared excesses of stars within 25 pc (in the Gliese Catalogue): with slightly different criteria they listed 28 and 25 candidates respectively, with 11 stars including the prototypes being common to both lists. Prominent among the stars with high excess that are in both lists are ζ Leporis (A3V), β Leo (A3V) and τ^1 Eri (F6V).

A search for stars with comparable excesses, which also show evidence of extension in the IRAS Working Survey Database and are not restricted in distance, reveals a further 18 stars (Walker & Wolstencroft 1988). Early type stars tend to dominate these lists because of "luminosity bias", but when proper allowance is made for this selection effect it is evident that the incidence of dust shells around main sequence A, F and G stars is about equal, and that shells, and by inference discs, are the rule rather than the exception (Backman & Gillett 1987, Aumann 1988). Surprisingly, no candidates with optical discs have been detected apart from β Pic: Smith & Terrile (1987) have examined more than 100 nearby stars using their coronagraphic technique and have failed so far to detect 'any other circumstellar disc comparable in visible optical properties to those of the β Pictoris disc.' Just how unusual β Pic is cannot yet be assessed since not all the above candidates were known to Smith & Terrile, and the list of stars which they examined has not yet been published.

Future progress in our understanding of these discs can be expected soon not only from the ground based techniques described in this paper, but also from the application of the Hubble Space Telescope: the coronagraphic facility on the Faint Object Camera should soon provide fascinating ultraviolet images at high angular resolution to within about $1''$ of β Pic. These will allow direct confirmation of the existence of the "dust free" zone at ~ 20 AU from the star (Backman, Gillett & Witteborn 1989) and a search for the "gaps" that Artymowicz, Burrows & Paresce (1989) suggest could be present in the isophotes if orbiting planets sweep up dust along their orbits.

References

Artymowicz, P., 1988. *Astrophys. J. Lett.*, **335**, L79.
Artymowicz, P., Burrows, C. & Paresce, F., 1989. *Astrophys. J.*, **337**, 494.
Aumann, H. H., 1985. *Publ. Astron. Soc. Pac.*, **97**, 885.
Aumann, H. H., 1988. *Astron. J.*, **96**, 1415.
Backman, D. E. & Gillett, F. C., 1987. In *Cool Stars, Stellar Systems and The Sun*, p. 340, ed. Linsky, J. L. & Stencel, R. E., Springer-Verlag, New York.
Backman, D. E., Gillett, F. C. & Witteborn, F. C., 1989. *Astrophys. J.*, (in press).
Becklin, E. E. & Zuckerman, B. 1989. In *Millimetre and Submillimetre Astronomy*, proceedings of the URSI Symposium, ed. Philipps, T., Kluwer Academic Press, Dordrecht, in preparation.
Draine, B. T. & Lee, H. M., 1984. *Astrophys. J.*, **285**, 89.
Gradie, J., Hayashi, J., Zuckerman, B., Epps, H. & Howell, R., 1987. In *Proc. 18th Lunar and Planetary Conference* Vol. I, p. 351, Cambridge University Press and the Lunar and Planetary Institute.
Paresce, F. & Burrows, C., 1987. *Astrophys. J. Lett.*, **319**, L23.
Smith, B. A. & Terrile, R. J., 1984. *Science*, **226**, 1421.
Smith, B. A. & Terrile, R. J., 1987. *Bull. Am. Astron. Soc.*, **19**, 289.
Telesco, C. M., Becklin, E. E., Wolstencroft, R. D. & Decher, R., 1988. *Nature*, **335**, 51.
Walker, H. J. & Wolstencroft, R. D., 1988. *Publ. Astron. Soc. Pac.*, **100**, 1509.
Wolstencroft, R. D., Scarrott, S. M. & Warren-Smith, R. F., 1989. In preparation (WSW).

Observations of discs around protostars and young stars

Ronald L. Snell

University of Massachusetts, USA

Abstract The observational evidence for the presence of discs around protostars and young stars consists of spectral and polarimetric data from which the existence of circum-stellar discs are indirectly inferred and data in which discs are directly imaged. A review of both the direct and indirect evidence for discs is presented as well as summary of the properties of these discs and their relationship to bipolar outflows and stellar jets.

1 Introduction

This review of discs associated with protostars and young stars stars is limited to two types of young stellar objects (YSOs): infrared sources (objects that emit only at infrared wavelengths) and optically visible T Tauri and FU Orionis stars. Current star formation models suggest that discs should commonly be associated with protostars and young stars. In fact, the flattened nature of our Solar system provides strong circumstantial evidence that discs have played a role in the formation of at least one star and planetary system.

Interest in circum-stellar discs has been heightened by the recent discovery of bipolar molecular outflows and stellar jets associated with many YSOs. An attractive model for the collimation and generation of energetic outflows assumes that accretion of material onto a young star through a viscous disc ultimately powers this energetic phenomenon. Unfortunately, direct imaging of these discs has proven difficult and only a few circum-stellar discs have been unambiguously detected. Instead the efforts to detect circum-stellar discs have frequently uncovered evidence for much larger structures, often called "interstellar discs", surrounding YSOs. The most persuasive evidence that circum-stellar discs commonly surround YSOs is indirect and based on a variety of observational properties of young stars that are most readily explained by the presence of such discs.

The role discs may play in the star formation process has been summarized by Shu, Adams & Lizano (1987) in their recent review of current theories of star formation. Though many of the details of this star formation process are controversial, the general scenario outlined by them underlies most current star formation models. Shu, Adams & Lizano identified four phases in the star formation process:

- In the first phase, slowly rotating cloud cores form in molecular clouds due to the combined effects of gravity and ambipolar diffusion.

- Next, a central protostar forms in the core and accretes material via a circum-stellar disc; the viscous accretion of material through the disc and onto the star accounts for most of the star's luminosity.

- In the third phase, an energetic wind develops and is directed out of the poles of the star; this wind impedes the flow of material onto the star and slowly erodes material from the disc.

- In the final stage, the wind has removed most of the opaque accreting material and a visible star emerges surrounded by a spatially thin, inactive disc.

If this star formation scenario is pertinent, then the observed infrared sources must represent YSOs in an evolutionary phase in which they are still accreting material and are surrounded by dense circum-stellar discs. The frequent association of molecular outflows with these sources (Lada 1985, Snell 1987) would suggest that most have already evolved to the point where they have developed energetic stellar winds. The optically visible T Tauri stars and FU Orionis stars are likely to be in the last phase of the star formation process and may still be surrounded by discs, but these discs are likely to be small and relatively inactive.

As we will see, many of the searches for discs have been directed toward objects with bipolar molecular outflows and thus, objects presumed to have active accreting discs. The best example of such an object is the infrared source, IRS-5, in the nearby dark cloud L1551 and is frequently referred to in this review. Figure 1 shows an overview of the activity associated with this young star and illustrates how a highly collimated stellar wind (Bieging, Cohen & Schwartz 1982, Snell *et al.* 1985) originates within tens of AU of IRS-5 and is responsible for sweeping up the ambient molecular gas into a large bipolar molecular outflow (Snell, Loren & Plambeck 1980). The last frame in Figure 1 shows schematically the orientation of a circum-stellar disc that may play a major role in the generation and collimation of the stellar wind. The presence of bipolar outflows provides a means of estimating the orientation of the equatorial plane of the disc. Highly collimated bipolar outflows sources are likely to be systems seen nearly equator on, an ideal geometry for studying discs.

I divide this review into two parts, summarizing first the indirect evidence for the existence of discs around YSOs. Though these indirect measurements provide the most comprehensive evidence that discs are commonly found around most YSOs, they do not provide irrefutable proof that discs actually exist. I then go on to summarize the direct imaging of circum-stellar discs. Although these data are meagre, they provide the only direct evidence that discs are associated with YSOs and that they play any role in the formation of stars and planetary systems.

2 Indirect evidence for discs

During the past decade there has been a large body of data accumulated that is very suggestive that circum-stellar discs (< 1000 AU in size) are present around most young stars. These data include the following:

- infrared polarization measurements,
- the presence of infrared excesses,
- the detection of disc accretion luminosity,
- the presence of only blue-shifted emission lines.

A summary of each of these observations in turn and how they imply the existence of circum-stellar discs is presented below.

2.1 INFRARED POLARIZATION MEASUREMENTS

Two mechanisms have been proposed to explain the observed optical and infrared polarization of stars; these are preferential extinction by aligned grains (Davis & Greenstein 1951) and scattering by dust distributed non-isotropically (Elsasser & Staude 1978).

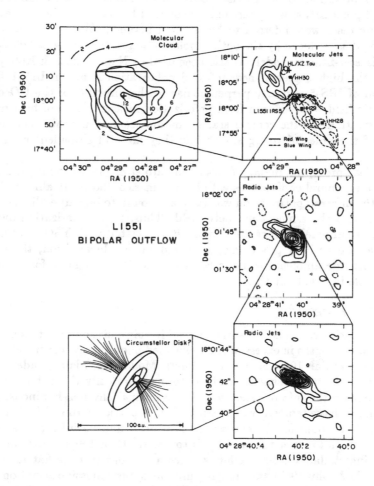

Figure 1. A composite figure from Snell *et al.* (1985) showing the relative sizes of the ambient cloud (upper left), the high-velocity molecular outflow (upper right), the radio emission from the partially ionized stellar wind (middle and lower right) and the proposed circumstellar disc (lower left).

In the latter model a central source is embedded in a circum-stellar disc. Observers viewing this disc edge-on will see highly polarized, scattered light from the poles of the star superimposed on the attenuated, unpolarized light coming directly from the star.

Observations of the 2 μm polarization of many YSOs have been carried out by Kobayashi *et al.* (1978), Dyck & Lonsdale (1979) and Heckert & Zeilik (1981) and compared with the optical polarization of background stars. In general, the YSOs could be divided into two groups; in the first, the position angle of the infrared polarization is the same as for the optical polarization observed for the background stars and in the second, the position angle of the infrared polarization is orthogonal to the optical polarization. It was concluded that in the first group both the optical and the infrared polarization were due to the Davis-Greenstein mechanism, but in the second group, the infrared polarization was presumably produced by scattering in sources containing discs. These discs were inferred to be oriented perpendicular to the direction of the magnetic field as deduced by the optical polarization of the background stars.

The Elsasser-Staude model has also been used by Nagata, Sato & Kobayashi (1983) to explain the high degree of near-infrared polarization found in L1551 IRS-5. The polarization of IRS-5 is nearly perpendicular to the direction of the CO outflow and the magnetic field – as deduced from polarization measurements of background stars. These observations fit the general star formation scenario in which collimated winds blow out of the poles of young stars while they are still shrouded by circum-stellar discs.

These data prompted Hodapp (1984) and Sato *et al.* (1985) to measure the infrared polarization of a number of sources with bipolar mass outflows. In almost all cases they observed, the infrared polarization was large and found to be perpendicular to both the outflow axis and the ambient magnetic field. Thus, these polarization measurements imply that circum-stellar discs are commonly associated with YSOs and in particular objects that are actively producing collimated winds. In addition, the alignment of the discs with the ambient magnetic fields suggest that magnetic forces may play an important role in disc formation.

2.2 INFRARED EXCESSES

A number of optically visible T Tauri stars and Herbig emission line stars have emissions at infrared wavelengths in excess of their expected photospheric contribution (Rucinski 1985, Rydgren & Zak 1986, Strom *et al.* 1989). Models by Adams, Lada & Shu (1987) have shown that reprocessing of starlight through optically thick, but spatially thin, discs can explain the infrared excesses. These intrinsically non-luminous, or so-called "passive discs", have temperature distributions proportional to $r^{-3/4}$ (Adams, Lada & Shu 1988). To account for the excesses seen at near- and far-infrared wavelengths, we require dust discs to extend from 0.1 AU to greater than 100 AU distance from their accompanying stellar sources. A few sources are found to have flat infrared spectra (Adams, Lada & Shu 1987) and may require much more massive accretion discs.

Strom *et al.* (1989) studied both T Tauri stars and "weak-line" T Tauri stars. They found that nearly 60% of the stars with ages younger than 3×10^6 years showed significant near-infrared excesses and that the fraction showing measurable excesses decreased with increasing stellar age. These observations are again consistent with identifying T Tauri stars as YSOs in the last phases of star formation in which winds are removing the last remnants of circum-stellar discs.

2.3 DISC ACCRETION LUMINOSITY

In addition to infrared excesses, some T Tauri stars show a significant ultraviolet excess over the predicted photospheric emission. Unlike the infrared excesses, the ultraviolet excess cannot be produced from reprocessed photospheric light. Bertout, Basri & Bouvier (1988) have modelled these T Tauri stars assuming that they are surrounded by "active" accretion discs. In this model the ultraviolet excess is produced at the boundary layer where the accretion disc meets the star. This model is not only successful in predicting the ultraviolet excess, but can also reproduce the observed infrared flux shortward of 10 μm. Accretion rates of 5×10^{-8} to 5×10^{-7} M$_\odot$ yr^{-1} are necessary to reproduce the strong ultraviolet excesses of some stars.

Accretion through a viscous disc can produce a substantial luminosity that may also contribute to the infrared excesses that are observed in some T Tauri stars. For example, an accretion rate of 10^{-7} M$_\odot$ yr^{-1} can produce an accretion luminosity of 1 L$_\odot$ (Lynden-Bell & Pringle 1974), a luminosity comparable to a typical T Tauri star's photospheric luminosity. Therefore, the infrared excesses observed in T Tauri stars could also be produced by active accretion discs. Unfortunately, the spectral index of an active Keplerian accretion disc is similar to that of a passive disc (Adams, Lada & Shu 1988), making it difficult to distinguish between these two possibilities based on the shape of their emergent spectra alone.

Photometrically eruptive YSOs, known as FU Orionis stars, have also been modelled by assuming that a substantial fraction of their emission during eruption arises in a self-luminous accretion disc (Hartmann & Kenyon 1985). A radial temperature gradient in the disc can account for the M-type spectrum observed in the near-infrared from these stars and the G-type spectrum observed at optical wavelengths. As pointed out by Hartmann & Kenyon (1985) a crucial prediction of a disc model is that weak absorption lines produced in the disc should be double-peaked due to the Keplerian motion of the gas within it. Such double-peaked lines have been detected in two FU Orionis stars, FU Ori and V1057 Cyg (Hartmann & Kenyon 1985, 1987a,b). In addition, high spectral resolution observations have shown that lines formed in the outer, cooler regions of the disc are narrower than lines formed in the inner, hotter, more rapidly rotating regions of the disc (Hartmann & Kenyon 1987a,b, Welty *et al.* 1989). Thus, these data reveal the differential rotation of the discs in these FU Orionis stars and provide additional confirmation of the disc models for these stars.

2.4 BLUE-SHIFTED EMISSION LINES

In T Tauri stars, the forbidden line emission of [OI] has been observed and modelled by Appenzeller, Jankovics & Ostriecher (1984) and Edwards *et al.* (1987). These emission lines presumably arise in the outer, lower density regions of T Tauri star winds that have typical velocities of 200 km s^{-1}. In almost all the T Tauri stars observed, only a blue-shifted forbidden line emission is seen. The absence of red-shifted emission can be understood if the young stars are surrounded by an opaque circum-stellar dust disc that obscures the emission from the red-shifted outflowing gas. To hide the red-shifted line emitting region it is estimated that the discs must have sizes of tens of AU and masses of at least 0.01 to 0.1 M$_\odot$. In fact, in many T Tauri stars optical jets are resolved (Mundt & Fried 1983, Strom *et al.* 1986) and are often one-sided, having a blue-shifted jet only. Thus, these discs must extend to much larger sizes to completely obscure the red-shifted optical jets.

Observations of the Hα and sodium D lines in FU Orionis stars have been made by Bastian & Mundt (1985) and Croswell, Hartmann & Avrett (1987). Like the forbidden lines in T Tauri stars, these absorption lines arise in the cooler, outer regions of stellar winds. The line profiles show predominately blue-shifted absorption. As in the T Tauri stars, the red-shifted outflow is presumably occulted by a circum-stellar disc.

3 Direct imaging of discs

With the construction of large millimetre and sub-millimetre telescopes and interferometers, and the development of infrared imaging techniques, more direct evidence for
discs has been obtained. However, even with these new instruments and techniques our
ability to image protoplanetary discs directly is limited. A number of problems exist,
which include:

- the large visual extinctions to regions of star formation,
- the problem of contrast between the disc emission and background and foreground
  cloud emission,
- the limited spatial resolution available for even the closest star forming sites.

The resolution problem is one of the most severe constraints on our ability to image
discs. Even using millimetre wavelength interferometers or infrared telescopes, the
spatial resolution at the distance of the nearest star forming regions is just barely
adequate to resolve circum-stellar or protoplanetary discs. In most star formation
regions, only size scales greater than several thousand AU can be probed. The data
that does exist on the direct imaging of discs is summarized in the following sections. A
summary of the infrared emission from discs is presented first, followed by a summary
of the observations of emission from molecular gas in discs.

3.1 INFRARED OBSERVATIONS

The resolution at near-infrared wavelengths is usually set by the "seeing" limit of the atmosphere and is typically a few arcsec. A number of techniques have been employed to
enhance this resolution. They include maximum-entropy image reconstruction (Grasdalen *et al.* 1984, Strom *et al.* 1985), speckle interferometry (Beckwith *et al.* 1984,
Zinnecker, Chelli & Perrier 1987) and lunar occultation (Simon *et al.* 1985). The highest angular resolution is achieved by lunar occultation observations, but unfortunately,
occultations of YSOs are very infrequent.

The high degree of polarization of the near-infrared emission from YSOs would
suggest that a large fraction of this emission is scattered light arising from near the poles
of the young stars. Thus, one might expect that high resolution images of the near-
infrared emission from YSOs would reveal elongated structures oriented perpendicular
to their disc axis. In fact, in some cases (Castelaz *et al.* 1985) such polar reflection
lobes are seen. L1551 IRS-5 and HL Tau are two relatively nearby YSOs, both known
to have collimated stellar jets and large scale molecular outflows, that have been imaged
in the near-infrared (Grasdalen *et al.* 1984, Strom *et al.* 1985, Moneti *et al.* 1988). In
both sources the infrared emission is elongated with its major axis perpendicular to
the outflow axis and not parallel. The diameter of these structures are 300 AU for HL
Tau and 1 000 AU for L1551 IRS-5. Infrared speckle interferometry has confirmed the
presence of such structures in HL Tau and have provided evidence for several hundred
AU sized halos around a number of other YSOs (Beckwith *et al.* 1984, Zinnecker *et al.*
1987).

The 2 μm image of IRS-5 from Strom *et al.* (1985) is shown in Figure 2a. The
interpretation of the near-infrared emission in IRS-5 is that it arises from the inner
edge of a disc tilted slightly towards us; this geometry is illustrated in Figure 2b. Such

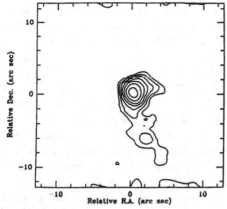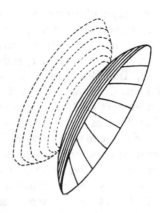

Figure 2. Contour map of the reconstructed 2 m image of L1551 IRS-5, the effective resolution is about 1.5" (left). A sketch of the geometry of a tilted, thick disc surrounding IRS-5 (right). The observed infrared emission arises from light scattered from the face of the disc tilted toward the observer, the scattered light from the other face is obscured by dust in the midplane of the disc. Both figures are from Strom *et al.* (1985).

---

an interpretation is supported by the observations of Campbell *et al.* (1988) who found that the position of the near-infrared emission was displaced to the south-west of the 2 cm radio continuum emission which marks the true position of IRS-5, as would be expected for the disc model illustrated in Figure 2b. Campbell *et al.* also found that at successively shorter wavelengths the emission was displaced further and further from IRS-5. Thus, at shorter wavelengths the scattered light accessible to view is at greater distances from IRS-5, due to the greater extinction of the disc.

The pattern of polarization found in optical, polarimetric images of HL Tau and other sources (Gledhill & Scarrott 1989) cannot be entirely explained in terms of a simple reflection nebula. An additional source of polarization is needed toward the central star. Gledhill & Scarrott have suggested that the light may be polarized as it passes though aligned grains in a circum-stellar disc. These polarization discs are discussed in more detail by Scarrott (this volume).

The highest spatial resolution observations of a YSO have been achieved through lunar occultation measurements of M8E-IR at 3.8 and 10 μm (Simon *et al.* 1985). From these observations, two components have been identified surrounding M8E IR. The smaller of the two components is the hotter and probably arises from the thermal emission of dust grains heated by the young star. This component is elongated with a minor axis of 11 AU and a major axis of approximately 100 AU and is interpreted as a disc seen nearly edge-on.

3.2 MOLECULAR LINE OBSERVATIONS

Shortly after the detection of bipolar molecular outflows, numerous efforts were made to map the dense gas in star forming regions to determine what role, if any, discs played in collimating the observed outflows. These spectral line observations could, in principle, have provided information on both the morphology and kinematics of discs. Though these initial observations were made with large centimetre and millimetre wavelength telescopes, they still suffered from relatively poor spatial resolution (5 000 to 10 000 AU), even in the closest star forming regions. Nevertheless, these observations revealed large, elongated and dense structures surrounding many YSOs and in most cases the structures were found to be oriented perpendicular to the outflow axes of the young stars (Cantó *et al.* 1981, Torrelles *et al.* 1983, Kaifu *et al.* 1984, Menten *et al.* 1984, Kawabe *et al.* 1984, Gusten, Chini & Neckel 1984, Little *et al.* 1985, Torrelles *et al.* 1985, Heyer *et al.* 1986, Torrelles *et al.* 1986a, Marcaide *et al.* 1988). These structures, sometimes called interstellar discs or toroids, typically have sizes of 0.1 to 1.0 pcs and masses of tens to hundreds of solar masses. A few of these interstellar discs show clear signs of rotation, but in general rotation has not been detected. However, the rotation speed for these large discs is expected to be comparable to, or smaller, than the line-widths of these regions, so it may not be surprising that rotation is not detected.

The fact that these interstellar discs are centred on YSOs and are aligned both perpendicular to the outflow axes and perpendicular to the magnetic fields in the clouds, would suggest that they are closely related to the star formation process of young stars. However, it is clear that these structures are too large and massive to be directly related to the accretion process in low mass stars or the formation of planetary systems. It is also unlikely that these interstellar discs play any role in the generation of energetic stellar winds, since observations have shown that in many YSOs the stellar winds are generated and collimated on much smaller size scales. It is possible, however, that these structures play a role in the continued collimation of the mass outflows at greater distances from the star. To detect circum-stellar discs will require observations on much smaller spatial scales.

The most promising instruments to map the dense gas around young stars at high angular resolution are the centimetre and millimetre wavelength interferometers and sub-millimetre telescopes. Yet even these techniques do not provide the angular resolution necessary to detect circum-stellar discs around YSOs more distant than 300 pc. Unfortunately, this excludes observations of circum-stellar discs around luminous YSOs, since they are more distant than 300 pc. Nevertheless, observations of luminous infrared sources have provided some interesting results.

High angular resolution observations of discs associated with luminous infrared sources have been obtained in NGC 2071 (Tauber *et al.* 1988, Kawabe *et al.* 1989), Cep A (Torrelles *et al.* 1986b), S106 (Bally, Snell & Predmore 1983, Bieging 1984, Mezger *et al.* 1987) and NGC 6334I (Jackson, Ho & Haschick 1989). In all four of these sources the observed discs are clearly rotating and have diameters of 20 000 to 80 000 AU and masses of 10 to 30 $M_\odot$. The rotation curves for the Cep A and NGC 6334I discs are well determined and can be best fit by including both a massive central object and a massive disc. In fact, in all of these sources the implied mass of the central star is comparable to the estimated mass of the disc. The exact role that these large discs play in the formation of the massive stars is not known.

The spatial resolution necessary to detect molecular emission from circum-stellar discs can only be obtained in nearby molecular clouds such as those in the Taurus and ρ Oph dark cloud complexes. There have been extensive observations of three sources in these regions: L1551 IRS-5 (Sargent *et al.* 1988), HL Tau (Sargent & Beckwith 1987) and IRAS 16293-2422 (Mundy, Wilking & Myers 1986, Sargent & Mundy 1988). The discs found in all three sources are similar in size, with diameters between 1400 and

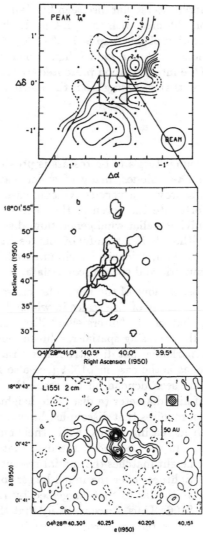

Figure 3. A composite figure showing the relative sizes and positions of the large interstellar disc (top, from Kaifu *et al.* 1984), the circumstellar disc (middle, from Sargent *et al.* 1988) and possibly the ionized inner edge of a torus (bottom, from Rodríguez *et al.* 1986) surrounding L1551 IRS-5.

2000 AU. In both HL Tau and L1551 IRS-5 the mass of the discs have been estimated to be about 0.1 $M_\odot$ and in IRAS 16293-2422 the mass of the disc has been estimated to be 1.3 $M_\odot$. In HL Tau the rotation curve of the disc is consistent with Keplerian rotation about a central mass of approximately 0.5 $M_\odot$, about five times more massive than the disc. In IRAS 16293-2422 the rotation curve suggests that the disc contains a larger fraction of the total mass of the system than in HL Tau.

An illustration of the disc structures that are seen in L1551 IRS-5 is presented in Figure 3. This figure shows the relationship between the circum-stellar disc seen by

Sargent *et al.* (1988) and the larger interstellar disc seen by Kaifu *et al.* (1984). A map of the radio continuum emission obtained with the VLA by Rodríguez *et al.* (1986) is also shown in Figure 3; the resolution of this map is 0.15″ or 2.4 AU. In the radio map, both the collimated, partially ionized stellar wind (aligned nearly east-west) and two compact sources can be seen. One interpretation presented by Rodríguez *et al.* is that the compact sources are emission that is arising from the ionized inner part of a torus with an inner radius of 25 AU.

4 Summary

The evidence for discs around young stars and protostars is provided by data obtained using a number of observational techniques, many of which have been described in this review. Together these data provide a very convincing case that most YSOs are surrounded by discs. It is unfortunate that even in the nearest star forming regions the size scales of interest for studying stellar wind generation, disc accretion and planet formation cannot be resolved. Although, in a handful of sources discs of 300 to 2 000 AU have been directly detected, in general, evidence for circum-stellar discs comes almost entirely from the indirect polarimetric and spectroscopic data.

What the high resolution observations of the molecular emission in the vicinity of young stars have revealed, is a hierarchy of size scales in which disc-like structures are present. At the largest scales, that of the "interstellar" discs, structures as large as 1 pc are detected. However, with increasing spatial resolution observations of smaller disc structures are seen. It is remarkable that disc structures on a variety of size scales are often seen in a single source, as is the case in L1551 (see also Rodríguez 1988) and the size of the structures that are seen depend on the resolution in use. In addition, it is interesting that the ratio of disc diameter to disc scale-height is similar at all size scales. To reconcile these results, requires that discs must have a morphology in which their scale-height increases with increasing radius, while their emission per unit area decreases with increasing radius. In such a model, the disc that is "seen" has a size dictated by where the disc scale-height matches the spatial resolution; for this size disc there will be large contrast in brightness over the much larger, but less emissive, out regions of the disc. This model is similar to that suggested by Harvey, Lester & Joy (1987) for the structures seen in S106. Thus, it seems likely that the circum-stellar and interstellar discs are physically connected and probably originate together as part of the star formation process.

References

Adams, F. C., Lada, C. & Shu, F. H., 1987. *Astrophys. J.*, **312**, 788.
Adams, F. C., Lada, C. & Shu, F. H., 1988. *Astrophys. J.*, **326**, 865.
Appenzeller, I., Jankovics, I. & Ostriecher, R., 1984. *Astron. Astrophys.*, **141**, 108.
Bally, J., Snell, R. L. & Predmore, R., 1983. *Astrophys. J.*, **272**, 154.
Bastian, U. & Mundt, R. 1985., *Astron. Astrophys.*, **144**, 57.
Beckwith, S., Zuckerman, B., Skrutskie, M. & Dyck, H., 1984. *Astrophys. J.*, **287**, 793.
Bertout, C., Basri, G. & Bouvier, J., 1988. *Astrophys. J.*, **330**, 350.
Bieging, J., Cohen, M. & Schwartz, P., 1984. *Astrophys. J.*, **282**, 699.
Bieging, J., 1984. *Astrophys. J.*, **286**, 591.
Campbell, B., Persson, S., Strom, S. E. & Grasdalen, G., 1988. *Astron. J.*, **95**, 1173.

Cantó, J., Rodríguez, L. F., Barral, J. & Carral, P., 1981. *Astrophys. J.*, **244**, 102.

Castelaz, M., Hackwell, J., Grasdalen, G., Gehrz, R. & Gullixson, C., 1985. *Astrophys. J.*, **290**, 261.

Croswell, K., Hartmann, L. & Avrett, E., 1987. *Astrophys. J.*, **312**, 227.

Davis, L & Greenstein, J., 1951. *Astrophys. J.*, **114**, 206.

Dyck, H. & Lonsdale, C., 1979. *Astron. J.*, **84**, 1339.

Edwards, S., Cabrit, S., Strom, S. E., Ingeborg, H., Strom, K. & Anderson, E., 1987. *Astrophys. J.*, **321**, 473.

Elsasser, H. & Staude, H., 1978. *Astron. Astrophys.*, **70**, L3.

Gledhill, T. & Scarrott, S., 1989. *Mon. Not. R. Astron. Soc.*, **236**, 139.

Grasdalen, G., Strom, S. E., Strom, K., Capps, R., Thompson, D. & Castelaz, M., 1984. *Astrophys. J. Lett.*, **283**, L57.

Gusten, R., Chini, R. & Neckel, T., 1984. *Astron. Astrophys.*, **138**, 205.

Hartmann, L. & Kenyon, S., 1985. *Astrophys. J.*, **299**, 462.

Hartmann, L. & Kenyon, S., 1987a. *Astrophys. J.*, **312**, 243.

Hartmann, L. & Kenyon, S., 1987b. *Astrophys. J.*, **322**, 393.

Harvey, P., Lester, D. & Joy, M., 1987. *Astrophys. J. Lett.*, **316**, L75.

Heckert, P. & Zeilik, M., 1981. *Astron. J.*, **86**, 1076.

Heyer, M., Snell, R. L., Goldsmith, P., Strom, S. E. & Strom, K., 1986. *Astrophys. J.*, **308**, 134.

Hodapp, K., 1984. *Astron. Astrophys.*, **141**, 255.

Jackson, J., Ho, P. T. P. & Haschick, A., 1988. *Astrophys. J. Lett.*, **333**, L73.

Kaifu, N., Suzuki, S., Hasegawa, T., Morimoto, M., Inatani, J., Nagane, K., Miyazawa, K., Chikada, Y., Kanzawa, T. & Akabane, K., 1984. *Astron. Astrophys.*, **134**, 7.

Kawabe, R., Morita, K., Ishiguro, M., Kasuga, T., Chikada, Y., Handa, K., Iwashita, H., Kanzawa, T., Okumura-Kawabe, S., Kobayashi, H., Takahashi, T., Murata, Y. & Hasegawa, T., 1989. preprint.

Kawabe, R., Ogawa, H., Fugui, Y., Takano, T., Takaba, H., Fujimoto, Y., Sugitani, K. & Fujimoto, M., 1984. *Astrophys. J. Lett.*, **282**, L73.

Kobayashi, Y., Kawara, K., Maihara, T., Okuda, H., Sato, S. & Noguchi, K., 1978. *Publ. Astron. Soc. Jpn*, **30**, 377.

Lada, C., 1985. *Annu. Rev. Astron. Astrophys.*, **23**, 267.

Little, L., Dent, W., Heaton, B., Davies, S. & White, G., 1985. *Mon. Not. R. Astron. Soc.*, **217**, 227.

Lynden-Bell, D. & Pringle, J., 1974. *Mon. Not. R. Astron. Soc.*, **168**, 603.

Marcaide, J., Torrelles, J. M., Gusten, R., Menten, K., Ho, P. T. P., Moran, J. & Rodríguez, L. F., 1988. *Astron. Astrophys.*, **197**, 235.

Menten, K. Walmsley, C., Krugel, E. & Ungerechts, H., 1984. *Astron. Astrophys.*, **137**, 108.

Mezger, P., Chini, R., Kreysa, E. & Wink, J., 1987. *Astron. Astrophys.*, **182**, 127.

Moneti, A., Forrest, W., Pipher, J. & Woodward, C., 1988. *Astrophys. J.*, **327**, 870.

Mundt, R. & Fried, J., 1983. *Astrophys. J. Lett.*, **274**, L83.

Mundy, L., Wilking, B. & Myers, S., 1986. *Astrophys. J. Lett.*, **311**, L75.

Nagata, T., Sato, S. & Kobayashi, Y., 1983. *Astron. Astrophys.*, **119**, L1.

Rodríguez, L. F., 1988. In *Galactic and Extragalactic Star Formation*, p. 97, eds. Pudritz, R. E. & Fich, M., Kluwer Academic Publishers, Dordrecht.

Rodríguez, L. F., Cantó, J., Torrelles, J. M. & Ho, P. T. P., 1986. *Astrophys. J. Lett.*, **301**, L25.

Rucinski, S., 1985. *Astron. J.*, **90**, 2321.

Rydgren, A. & Zak, D., 1986. *Publ. Astron. Soc. Pac.*, **99**, 141.

Sargent, A. & Mundy, L., 1988. In *Galactic and Extragalactic Star Formation*, p. 261, eds. Pudritz, R. E. & Fich, M., Kluwer Academic Publishers, Dordrecht.

Sargent, A. & Beckwith, S., 1987. *Astrophys. J.*, **323**, 294.

Sargent, A., Beckwith, S., Keene, J. & Masson, C., 1988. *Astrophys. J.*, **333**, 936.

Sato, S., Nagata, T., Nakajima, T., Nishida, M., Tanaka, M. & Yamashita, T., 1985. *Astrophys. J.*, **291**, 708.

Shu, F. H., Adams, F. C. & Lizano, S., 1987. *Annu. Rev. Astron. Astrophys.*, **25**, 23.

Simon, M., Peterson, D., Longmore, A., Storey, J. & Tokunaga, A., 1985. *Astrophys. J.*, **298**, 328.

Snell, R. L., Bally, J., Strom, S. E. & Strom, K., 1985. *Astrophys. J.,* **290**, 587.

Snell, R. L., Loren, R. & Plambeck, R., 1980. *Astrophys. J. Lett.,* **239**, L17.

Snell, R. L., 1987, In *Star Forming Regions*, IAU Symposium **115**, p. 213, eds. Peimbert, M. & Jugaku, J., Reidel, Dordrecht.

Strom, K., Strom, S. E., Wolff, S., Morgan, J. & Wenz, M., 1986. *Astrophys. J. Suppl. Ser.,* **62**, 39.

Strom, K., Strom, S. E., Edwards, S., Cabrit, S. & Skrutskie, M., 1989. preprint.

Strom, S. E., Strom, K., Grasdalen, G., Capps, R. & Thompson, D., 1985. *Astrophys. J.,* **90**, 2575.

Tauber, J., Goldsmith, P. & Snell, R. L., 1988. *Astrophys. J.,* **325**, 846.

Torrelles, J. M., Cantó, J., Rodríguez, L. F., Ho, P. T. P. & Moran, J., 1985. *Astrophys. J. Lett.,* **294**, L117.

Torrelles, J. M., Rodríguez, L. F., Cantó, J., Carral, P., Marcaide, J., Moran, J. & Ho, P. T. P., 1983. *Astrophys. J.,* **274**, 214.

Torrelles, J. M., Ho, P. T. P., Moran, J., Rodríguez, L. F. & Cantó, J., 1986a. *Astrophys. J.,* **307**, 787.

Torrelles, J. M., Ho, P. T. P., Rodríguez, L. F. & Cantó, J., 1986b. *Astrophys. J.,* **305**, 721.

Welty, A., Strom, S. E., Hartmann, L. W. & Kenyon, S. J., 1989. preprint.

Zinnecker, H., Chelli, A. & Perrier, C., 1987. In *Star Forming Regions*, IAU Symposium **115**, p. 71, eds. Peimbert, M. & Jugaku, J., Reidel, Dordrecht.

VLA observations of ammonia toward molecular outflow sources

José M. Torrelles[1], Paul T. P. Ho[2], Luis F. Rodríguez[2,4],
Jorge Cantó[3] and Lourdes Verdes-Montenegro[1]
[1]*Instituto de Astrofísica de Andalucía, Spain*
[2]*Harvard-Smithsonian Center for Astrophysics, USA*
[3]*Universidad Nacional Autonoma de Mexico*

1 Introduction

Since the discovery of bipolar molecular outflows, a significant observational effort has been made to study the role of the dense molecular cores ($n(H_2) \geq 10^4$ cm^{-3}) in the collimation processes. Dense molecular gas is almost always found in association with the central regions of a bipolar outflow. As a matter of fact, there is practically a one-to-one correspondence. This association of dense gas with the central parts of bipolar outflows supports the notion that the energy source of the outflows is a very young star (Torrelles *et al.* 1986a). There is also evidence that molecular toroids or discs with interstellar dimensions are present in several regions and that they play, at least on the scale of tenths of pc, an important role in the collimation and channelling of the high-velocity gas. See Rodríguez (1988) and Snell (this volume) for reviews.

In the last few years, our group has obtained Very Large Array (VLA) NH$_3$ observations toward regions of molecular outflows. These observations have revealed the morphology of the high-density molecular gas on scales of $\sim 3''$. This program allowed: (1) the study of dense gas as a possible focusing mechanism of bipolar outflows, (2) the study of local heating effects produced by star formation, and (3) the analysis of the kinematics of the regions. Here we present VLA NH$_3$(1,1) and NH$_3$(2,2) observations toward four regions with molecular outflows. These observations were obtained with the VLA of the National Radio Astronomy Observatory (NRAO)[5].

2 Individual Sources

2.1 CEPHEUS A

Our main VLA NH$_3$ results can be summarized as follows: (1) We find a very elongated structure with a deconvolved aspect ratio of 13 to 1 (Figure 1). (2) We observe that heating is dominated by the stellar activity centre, indicated by the position of the H$_2$O masers, with a temperature ≥ 70 K. (3) The observed velocity as a function of radial distance is flat to within $15''$ of the stellar activity centre. There is an important line broadening in the central $15''$ (Figure 2).

We interpret the observed motions as simple rotation of a very thin molecular toroid for the following reasons: (1) High degree of elongation is observed at several angular scales with different molecules (Figure 1). (2) The stellar activity centre is located at the kinematic centre where the relative rotation curve changes significatively (Figure 2). (3) The flat rotation curve, similar to that observed in the outer parts of spiral galaxies,

---

[4] On sabbatical leave from Instituto de Astronomía, Universidad Nacional Autonoma de Mexico.
[5] NRAO is operated by Associated Universities, Inc., under contract with the National Science Foundation.

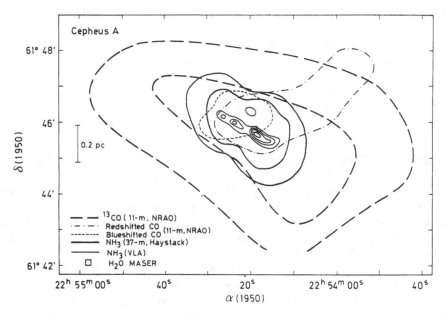

Figure 1. Superposition of different maps for Cepheus A: ^{13}CO (Sargent 1977), CO (Rodríguez *et al.* 1980), NH$_3$(1,1;m) (single-dish, Ho *et al.* 1982), NH$_3$(2,2;m) (VLA, Torrelles *et al.* 1986b).

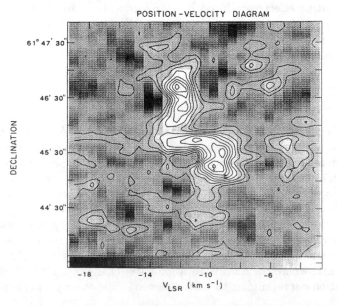

Figure 2. Spatial velocity diagram along the major axis of the elongated VLA-NH$_3$(2,2) structure shown in Figure 1. Contour levels are 1, 2, 3, 4, 5, 6, 7, 8, 9, 10× (20 mJy/beam). Synthesized beam= 14″. This diagram has been obtained by Hanning smoothing in velocity the data of Torrelles *et al.* (1986b).

and the magnitude of the rotational motions are in fact consistent with the observed mass within the disc (~ 90 M$_\odot$).

2.2 HH26-IR AND GGD12-15

Our NH$_3$(1,1) data on HH26-IR and GGD12-15 (Torrelles *et al.* 1989b) reveal a double-maximum structure with a size of $\simeq 55'' \times 30''$ (HH26-IR) and $\simeq 30'' \times 15''$ (GGD12-15) located at the geometrical centre of the bipolar CO outflows. Both VLA-NH$_3$ structures are oriented perpendicular to the axis of the bipolar outflows. These results suggest that the double-maxima structures can represent molecular toroids which help collimate the outflow. However, no significant velocity gradient is observed in the VLA-NH$_3$ structures, at least within our 1.2 km s^{-1} spectral resolution.

2.3 MONOCEROS R2

NH$_3$ single dish observations showed an elongated high-density structure with a NE-SW orientation and aligned perpendicular to the bipolar outflow (Torrelles *et al.* 1983). In Figure 3 we show, superposed, the integrated intensity VLA maps of the (1,1) (contour plot) and (2,2) (grey scale) ammonia main component lines. The overall spatial distribution of this high-density molecular gas shows an arc-like structure located $\sim 40''$ SW of the stellar activity centre indicated by several IRS's, H$_2$O masers, and an HII region.

The NH$_3$(2,2) arc-like structure is shifted NE with respect to the NH$_3$(1,1) arc-like structure (Figure 3). This implies a significant heating effect at the NE edge of this structure. By comparing the NH$_3$(2,2) and NH$_3$(1,1) emission we observe a temperature gradient, with the highest temperatures of ~ 200 K toward the inner NE edge, toward the stellar activity centre, and decreasing to ~ 20 K at the outer SW edge.

The arc-like structure can be interpreted as part of an inner molecular wall of a toroid's cavity produced by the pressure of the stellar wind driving the bipolar molecular outflow. This interaction can cause the heating observed at the inner NE edge of the arc-like structure.

3 Summary

In our opinion the high-density gas must simultaneously fulfil three different conditions to count as observational evidence for interstellar toroids or discs seen nearly edge-on: (1) a high degree of elongation, (2) a velocity gradient along the major axis of the elongated high-density molecular gas structure, which can be explained by rotating motions, (3) stellar activity centre located near the kinematic centre.

Cepheus A exhibits these three defining characteristics for interstellar toroids seen nearly edge-on. HH26-IR and GGD12-15 are good candidates for interstellar toroids, but higher spectral resolution observations are needed to study the kinematics in these two regions.

References

Ho, P. T. P., Moran, J. M. & Rodríguez, L. F., 1982. *Astrophys. J.*, **262**, 619.

Rodríguez, L. F., 1988. In *Galactic and Extragalactic Star Formation*, p. 97, eds. Pudritz, R. E. & Fich, M., Kluwer Academic Publishers, Dordrecht.

Rodríguez, L. F., Ho, P. T. P. & Moran, J. M., 1980. *Astrophys. J. Lett.*, **240**, L149.

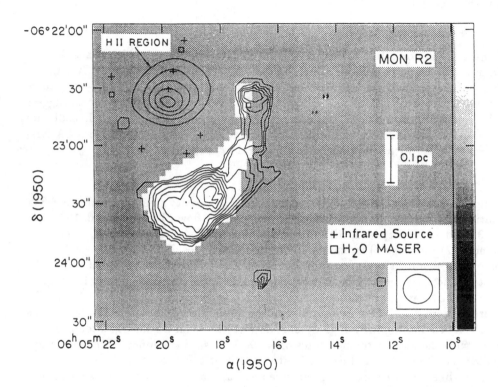

Figure 3. (Figure adopted from Torrelles *et al.* 1989a) Contour map and grey scale image of the integrated intensity of the (1,1) and (2,2) ammonia main component line respectively over the $V_{LSR}=$ 9.0-12.6 km s^{-1} range (synthesized beam= 15″). Contour levels are 1, 2, 3, 4, 5, 6, 7, 8, 9× (43 mJy/beam km s^{-1}). The grey scale is linear and ranges from -43 to 43 mJy/beam km s^{-1} from down to up of the grey bar (right of figure). Continuum map of the HII region is also superposed.

Sargent, A. I., 1977. *Astrophys. J.,* **218**, 736.

Torrelles, J. M., Ho, P. T. P., Moran, J. M., Rodríguez, L. F. & Cantó, J. 1986a. *Astrophys. J.,* **307**, 787.

Torrelles, J. M., Ho, P. T. P., Rodríguez, L. F. & Cantó, J., 1986b. *Astrophys. J.,* **305**, 721.

Torrelles, J. M., Ho, P. T. P., Rodríguez, L. F. & Cantó, J., 1989a. *Astrophys. J.,* submitted.

Torrelles, J. M., Ho, P. T. P., Rodríguez, L. F., Cantó, J. & Verdes-Montenegro, L., 1989b. *Astrophys. J.,* in press.

Torrelles, J. M., Rodríguez, L. F., Cantó, J., Carral, P., Marcaide, J., Moran, J. M. & Ho, P. T. P., 1983. *Astrophys. J.,* **274**, 214.

Derivation of the physical properties of molecular discs by an MEM method

John Richer and **Rachael Padman**

Mullard Radio Astronomy Observatory, Cambridge University

1 Introduction

Molecular discs are a necessary feature of models of collimated outflow from young stellar objects and protostars. One observational technique used to detect the signature of rotation is to map the *position-velocity (l-v)* diagram of the gas around the star in a high lying molecular line which traces dense material. For example, such maps have shown evidence for a rotating disc in the S106 system through HCO^+ and HCN (3-2) observations (Padman & Richer 1989), and for a rotating ammonia disc around a $30M_\odot$ star in NGC 6334 I (Jackson *et al.* 1989). In the latter case, there is a suggestion that the disc also contains appreciable mass ($\sim 30M_\odot$).

2 A simple disc model

Given such data, what can we learn about the disc structure? Previous approaches have been simply to fit Keplerian curves through the locus of peak emission (Sargent & Beckwith 1987, Jackson *et al.* 1989), and hence obtain a crude estimate of the stellar mass. With high quality data, we should be able to derive more information than this. We consider the case of a molecular disc in Keplerian rotation about a star. If the disc is optically thin, axisymmetric and edge-on, then the resulting *l-v* diagram $I(x,v)$ is completely determined by the mass and emissivity distributions, $M(r)$ and $\varepsilon(r)$:

$$I(x,v) = 2 \int_{r=x}^{\infty} \varepsilon(r)\phi(v - \frac{x}{r}v_k(r))\frac{r}{\sqrt{r^2 - x^2}}dr, \qquad (1)$$

where $\phi(v)$ is the normalised intrinsic line shape, and $v_k(r)$ is the Keplerian velocity, $(GM(r)/r)^{1/2}$. In reality, with a finite beam size and in the presence of noise, we actually observe the intensity $I_{obs}(x,v)$, where

$$I_{obs}(x,v) = B(x,v) * I(x,v) + \text{noise}. \qquad (2)$$

$B(x,v)$ is the telescope response.

For example, Figure 1(a) shows a model *l-v* diagram for a massless disc (with Gaussian emissivity distribution) orbiting a $20M_\odot$ star. Figure 1(b) shows the corresponding plot after convolution with the telescope beam.

3 Deriving the disc properties

We would like to invert equation (2) to derive the mass and emissivity distribution. We have adopted a maximum entropy approach which ensures *positivity* of $M(r)$ and $\varepsilon(r)$, and deals sensibly with the noise. A maximum likelihood (least-squares) approach could not do either of these things. Our first tests of the method have been to recover $\varepsilon(r)$ for a known $M(r)$, since this problem is *linear*. We create a simulated data set,

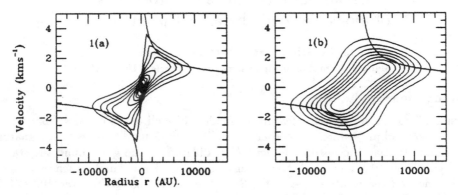

Figure 1. (a) Model *l-v* diagram for a massless disc around a 20M$_\odot$ star. (b) Same model, convolved with telescope beam. Also shown is the corresponding Keplerian rotation curve.

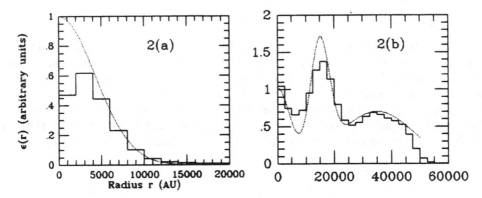

Figure 2. Two examples of recovery of $\varepsilon(r)$ from position velocity diagrams (true distribution is dotted line).

according to equations (1) and (2), and add Gaussian noise. Figure 2 shows typical MEM reconstructions from simulated data sets, with both the true distribution and MEM reconstruction plotted.

This method is powerful and *general*. We are extending it derive $\varepsilon(r)$ and M(r) *simultaneously* (a non-linear problem): this will allow us to analyse systems such as NGC 6334 I, where the disc mass is non-negligible. It should also be possible to include observations of several different molecular tracers into such a scheme.

We thank Steve Gull and John Skilling for providing the MEM package MEMSYS.

References

Jackson, J. M., Ho, P. T. P & Haschick, A. D., 1989. *Astrophys. J. Lett.*, **333**, 73.
Padman, R. & Richer, J. S., 1989. In *Sub-millimetre and Millimetre Wave Astronomy*, ed. Webster, A. S., Kluwer, Dordrecht, in preparation.
Sargent, A. I. & Beckwith, S., 1987. *Astrophys. J.*, **323**, 294.

Masers associated with discs around young stars

Geneviève Brebner
Lancashire Polytechnic

1 Introduction

The sample of CO outflows reported in Bally & Lada (1983) was searched for OH maser emission (Prestwich 1985). The stronger OH maser sources found in this search – augmented by two other sources – were then mapped using MERLIN. The maser distributions were compared with the molecular emission from the sources.

A clear-cut case of the association between OH maser emission and a molecular disc is found in G35.2-0.7N (Brebner et al. 1987). Its bipolar outflow is well collimated and an ammonia condensation is observed, clearly elongated in a direction perpendicular to the outflow direction. The OH masers are situated at, or near the exciting source of the region, and lie in an elongated distribution with an orientation which reflects that of the larger-scale ammonia disc.

2 Comparison of OH maser and CO outflow distributions

Comparisons could be made for eight sources – the masers in Orion-KL were mapped by Norris (1984). As observations of a molecular disc were unclear, or had not yet been attempted in many cases, a comparison was made between the orientations of the CO outflows and the OH maser distributions. The major axis of the OH maser distribution was deduced from a least squares fit and the angular differences between maser and outflow orientations are shown in (Figure 1). The largest source of error is in the estimates of the outflow direction, many of which had to be made by eye. (A selection effect is inherent: the sample could not have face-on discs, since the CO outflow direction could not be identified in such cases.)

If the OH masers were embedded in a disc surrounding the outflow source, we would expect the histogram to peak at 90°. There is a very slight tendency towards the perpendicular, but it is clear that at best it is weak.

However, the trend could be masked by uncertainties in the outflow direction. *e.g.* the CO outflow in W3 IRS5 has been mapped on two different scales: the more extended structure (Brackmann & Scoville 1983) is not aligned with the more compact structure (Claussen et al. 1984). The masers are oriented nearly orthogonally to the extended structure. The angle from the more recent recent results was used in the histogram.

3 Other results

The axis ratio of such a masing disc, estimated from the maser distributions alone, would be 1:0.6. However, this estimate makes no allowance for projection or selection effects.

The dynamical ages of the CO outflows are estimated to range from 0.4×10^4 to 4×10^4 years. Such a range suggests that OH maser emission is present from such sources throughout a large fraction of the CO outflow lifetime of 10^5 years (Lada 1985).

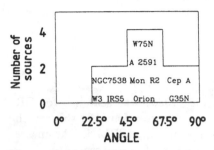

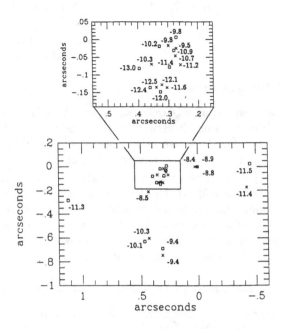

Figure 1. (above) A histogram of the angular difference between the CO outflow and the maser distributions.

Figure 2. (right) The 1667 MHz emission from AFGL 2591. TOP: An enlargement of the central condensation. (Cross and box – right and left hand circularly polarised emission. The velocity of each feature is shown in km/s.)

The velocity distribution of the masers can indicate the motions of the gas in which they are embedded. On the whole, masers seem to indicate turbulent motion which is not gravitationally bound. In the source AFGL 2591, however, the velocity distribution of the masers may be consistent with gravitationally bound motion, and the compact structure may even reflect rotation. The central condensation of maser features is shown in Figure 2. All the features in this condensation were strongly blended, but as different velocities were sampled, the position of peak emission was seen to change. This means that the overall distributional and velocity information of the condensation is correct, but the identification of any single feature is less certain. If the condensation did show gravitationally bound motion, the central mass would be 1 $M_\odot$.

In conclusion, OH masers show a tendency to be aligned perpendicular to the outflow direction. Such maser discs have apparent axial ratios of 1:0.6 and lifetimes of perhaps a few tens of thousands of years. The motions of the maser features are complicated, but, with one exception, they are unlikely to be due to gravitational effects.

References

Bally, J. & Lada, C. J., 1983. *Astrophys. J.*, **265**, 824.

Brackmann, E. & Scoville, N., 1980. *Astrophys. J.*, **242**, 112.

Brebner, G. C., Heaton, B., Cohen, R. J. & Davies, S. R., 1987. *Mon. Not. R. Astron. Soc.*, **229**, 679.

Claussen, M. J., *et al.*, 1984. *Astrophys. J. Lett.*, **285**, L79.

Lada, C. J., 1985. *Annu. Rev. Astron. Astrophys.*, **23**, 267.

Norris, R. P., 1984. *Mon. Not. R. Astron. Soc.*, **207**, 127.

Prestwich, A., 1985. *MSc thesis*, University of Manchester.

The nature of polarization discs around young stars

S. M. Scarrott
University of Durham

1 Introduction

Although circumstellar discs play an important role in many of the phenomena associated with star formation there is little direct evidence for them at optical wavelengths. In polarization studies of reflection nebulae surrounding young stars and protostars we have noticed a deviation in the expected polarization pattern that appear to indicate the presence of circumstellar discs. We call this feature the 'polarization disc'.

2 Examples and properties of polarization discs

Figure 1 shows a polarization map of the reflection nebulosity illuminated by the star HL Tau. At large distances from the star the polarization pattern has the expected centrosymmetric form but in inner regions the pattern deviates to form an anomalous band running across the illuminating star which itself is linearly polarized. We identify this inner pattern, the so called polarization disc, with a circumstellar disc of dusty material. Table 1 gives a comprehensive list of objects possessing such discs and indicates any additional peculiarities in the polarization data.

The properties of polarization discs are summarized below. Obviously not all of these properties are found in every object but they seem to represent various facets of the same phenomenon.

(1) The polarization disc consists of an anomalous band of polarization centred on the apical region of reflection nebulae.
(2) This band is normally present regardless of the visibility of the central source.
(3) When the central source is directly visible it is significantly linearly polarized in a similar manner to the disc.
(4) The position angle of the polarization of the central star varies with wavelength.
(5) Circular polarization in the central source and disc region has been found.
(6) The polarization disc is perpendicular to the axis of molecular outflows.

Table 1

Object	Illuminator	Features	Object	Illuminator	Features
HL Tau Neb	HL Tau	vs, jet	HH46/47	HH46IRS	bn, hh, irs
Pars 21	Pars 21	vs, cn, dc	NGC2261	R Mon	vs, cn, vr, cp, dc
HH34	HH34IRS	vs, jet, dc	GM29	PV Cep	vs, cn, dc, pr
Serpens	SVS2	vs, bn, dc	HH100	HH100IRS	irs? dc
NGC6729	R/T CrA	vs, cn, dc	HH102	L1551IRS	cn, jet, irs, mo, dc, pr

Key: vs – visible illuminating star, cn – cometary nebula, bn – bipolar nebula, irs – illuminator is an IR source, dc – within dark cloud, pr – polarization tends to follow bright rim of optical nebula, pc – circular polarization, hh – Herbig Haro objects, jet – stellar jet, vr – variable polarization disc, mo – molecular outflow

(7) The inner parts of the stellar jet and source (where visible) are polarized with orientations perpendicular to the jet axis.

(8) The extremities of the polarization disc show null points.

(9) The polarization disc may change orientation on time scales of months.

(10) There is a tendency for polarization orientations to become parallel to the bright rims of the lobes of optical nebulosity centred on the exciting object.

3 Origin of the polarization disc

Although it seems unequivocal that the polarization disc is a consequence of the dusty circumstellar disc, the polarizing mechanism giving rise to the feature is not immediately obvious. Single and multiple scattering of unpolarized radiation in a circumstellar environment will not generate the polarization disc without recourse to unrealistic geometries. It appears that competing polarization mechanisms are required – Scarrott *et al.* (1989) discuss polarized source and magnetized disc models and conclude that neither model could account for all observed features. They favour a hybrid model of a magnetized disc with the possibility of illumination from an intrinsically polarized source.

4 Conclusion

Although the interpretation of the physical processes giving rise to the phenomenon of polarization discs is complex it is quite clear that these optical polarization features are caused by circumstellar discs of dust and gas and they allow the extent and orientation of the disc to be ascertained. A detailed study of the polarization disc phenomenon will lead to a better understanding of the physical conditions in the region close to young stars.

Reference

Scarrott, S. M., Draper, P. W. & Warren-Smith, R. F., 1989. *Mon. Not. R. Astron. Soc., in press.*

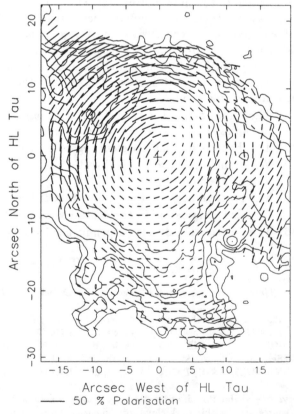

Figure 1. A linear polarization map of the nebulosity surrounding the young star HL Tau.

The correlation between the main parameters of the interstellar gas (including Salpeter's spectrum of masses) as a result of the development of turbulent Rossby waves

V. V. Dolotin and **A. M. Fridman**

Astronomical Council of the Academy of Sciences of the USSR

1 Basic equation

We derive an equation, which describes the dynamics of perturbations with small but finite amplitude in a uniformly rotating gaseous system with inhomogeneous density $\rho_0(r)$,

$$\left(\frac{\partial}{\partial t} + \frac{1}{2\Omega_0}\left[\nabla_\perp \tilde{\chi}, \nabla_\perp\right]_z\right)\left(\Delta_\perp \tilde{\psi} - \frac{w_0^2}{4\Omega_0^2}\Delta_\perp \tilde{\chi}\right) - \frac{w_0^2}{2\Omega_0}\frac{1}{r}\frac{\partial \tilde{\chi}}{\partial \phi} = 0, \qquad (1)$$

where $\tilde{\psi}$ is the perturbed part of the gravitational potential $\psi(r, \phi, t) = \psi_0(r) + \tilde{\psi}(r, \phi, t)$, Ω_0 is the angular velocity of rotation of the system, $w_0^2 \equiv 4\pi G\rho_0$ and $\chi \equiv \psi + g(\rho)$ with $g(\rho)$ obtained from $\frac{1}{\rho}\frac{\partial p(\rho)}{\partial x_i} = \frac{\partial g(\rho)}{\partial x_i}$ and $p = p(\rho)$, the barotropic equation. As $\Delta\psi = 4\pi G\rho$, then $\chi = \psi + g(\Delta\psi)$.

Equation (1), which is unknown in theoretical and mathematical physics, is obtained using the geostrophic approximation: $\Omega_0 \frac{d}{dt} \ll 1$.

2 Short- and long-wave limits

For long wave perturbations $\lambda \gg \lambda_J$ (λ_J is the Jeans wavelength) equation (1) may be reduced to the well known Charney-Obukhov equation in hydrodynamics (Charney 1948, Obukhov 1949) or the Hasegawa-Mima equation in a plasma (Hasegawa & Mima 1978) with a vector non-linearity. For short waves, $\lambda \ll \lambda_J$, equation (1) may be reduced to the non-linear equation for Rossby waves (Charney & Flierl 1981) with scalar and vector non-linearities, the ratio of which is the following:

$$\beta \equiv \frac{\text{scalar non-linearity}}{\text{vector non-linearity}} \sim \frac{a^2}{\rho^2}\cdot\frac{a}{R} \qquad \text{with} \qquad \rho \equiv \frac{c_0}{\Omega_0}.$$

If $\beta \gg 1$, then the stationary solution of the equation describes solitary vortices. If $\beta \ll 1$, then the non-linear equation for the short wave perturbations is classified as being of the same type as the non-linear equation for the long wave perturbation. The stationary solution of this equation describes solitary dipole vortices (modones). Binary objects, such as the binary nuclei of galaxies, binary galaxies and binary stars, may offer an astrophysical application for solutions of this last type.

3 Turbulence in the interstellar medium

The turbulence spectrum of this equation turns out to be as follows: $W_k^{(1)} \sim k_\perp^{-7/2}$, $W_k^{(2)} \sim k_\perp^{-9/2}$, where W_k is the energy density in k-space and $k_\perp$ is the cross (to the rotation axis) wave vector. For the energy density in the coordinate space, $W_\lambda \sim \lambda_k^\gamma$, where $\lambda_k \equiv \frac{2\pi}{k}$, $3.5 \leq \gamma \leq 4.5$ *i.e.* $\gamma \approx 4$. Numerical solutions (Hasegawa *et al.* 1979) have in fact found that $\gamma \approx 4$. For $W_\lambda \sim \lambda_k^4$ it is easy to obtain the observed correlations in the interstellar gas (Myers 1983):

$$v_\lambda \sim \lambda^{1/2}, \qquad \rho_\lambda \sim \lambda^{-1}.$$

and also Salpeter's spectrum (Massevitch & Tutukov 1988)

$$n_\lambda \sim m_\lambda^{-3/2}.$$

The linear approximation of the equations obtained describes the ordinary Rossby waves ($\lambda \ll \lambda_J$) and gravitational Rossby waves ($\lambda \gg \lambda_J$).

A detailed description of this work is to be published elsewhere (Dolotin & Fridman 1989).

References

Charney, I. G., 1948. *Geophys. Publ. Kosjones Vors. Videtship. Acad. Olso*, **17**, 3.

Charney, I. G. & Flierl, G. R., 1981. In *Evolution of Physical Oceanography*, eds. Warren, B. A. & Wunsch, C., The MIT Press, Cambridge, MA.

Dolotin, V. V. & Fridman, A. M., 1989. In *Nonlinear Waves in Physics and Astrophysics*, eds. Gaponov-Grekhov, A. V. & Rabinovitch, M. I., Springer-Verlag, New York.

Hasegawa, A., MacLennan, C. G. & Kodama, Y., 1979. *Phys. Fluids*, **22**, 2122.

Hasegawa, A. & Mima, K., 1978. *Phys. Fluids*, **21**, 87.

Massevitch, A. G. & Tutukov, A. V., 1988. *Stellar evolution: theory and observation*, Nauka, Moscow.

Myers, P. C., 1983. *Astrophys. J.*, **270**, 105.

Obukhov, A. M., 1949. *Izv.AN SSSR, Geography and Geophysics*, **13**, no. 4, 281.

Discs in cataclysmic variables and X-ray binaries

A. R. King

University of Leicester

Abstract I review recent progress in the study of accretion discs in cataclysmic variables (CVs) and X-ray binaries. Observations of CVs, especially eclipse mapping, give detailed agreement with steady-state disc theory. Coronae and winds are probably universal features of discs in such systems. Our present ignorance of the disc viscosity is the main barrier to progress in understanding time dependence and stability properties. Non-axisymmetric structure is particularly prominent in observations of low-mass X-ray binaries. This may be caused by the interaction of the mass transfer stream from the companion star with the disc.

1 Introduction.

Cataclysmic variables (CVs) are close binary systems, having periods of a few hours, in which a white dwarf accretes material from a main-sequence companion which fills the Roche lobe. If the white dwarf is replaced by a neutron star or black hole we have a low-mass X-ray binary (LMXB).

The formation of an accretion disc lying in the orbital plane is very likely under these circumstances since the accretion stream from the companion is highly supersonic and follows an essentially ballistic trajectory; its closest approach to the accreting object is a few $\times 10^9$ cm, larger than the radius of any likely accreting object. The resulting self-collisions of the stream imply dissipation. As this can remove energy much more effectively than angular momentum the matter arranges itself into a collection of orbits of lowest energy for fixed angular momentum, *i.e.* Kepler orbits, in the binary plane. Internal dissipation ("viscosity") causes the inner parts of the distribution to give up their angular momentum to the outer parts and spiral slowly inwards, thereby liberating gravitational energy to replace that dissipated as heat. This is usually radiated away efficiently, so that the matter does not build up significant internal pressure and remains closely confined to the orbital plane in ballistic orbits.

The theory of such accretion discs is treated extensively in the literature (*e.g.* Pringle 1981, Frank, King & Raine 1985), and the interested reader should consult these references for further details. This review summarizes the extent of observational support for this simple picture, and discusses those areas where refinement or revision may be needed.

2 Observational support

The basic driving mechanism for accretion discs is the viscosity. As this is ill-understood (see below), we are rather in the position of astronomers studying stellar structure and evolution before the discovery of thermonuclear processes. Just as then (*cf.* Eddington 1926) we are able to make a surprising amount of progress in checking theoretical ideas against observation; clearly we must concentrate on those aspects which do not involve the viscosity very directly. Before describing these it is important to understand the advantages of CVs and LMXBs for studies of discs. For CVs these include

* We now know the orbital periods of $\sim$ 100 systems, and some 20 of these are eclipsing (*cf.* Ritter 1987).

* This information gives us a good idea of the component masses and separation; binary evolution theory gives an idea of the expected accretion rates.
* Accretion dominates the luminosity at most wavelengths.
* The time-scales are particularly convenient for observation: orbital periods are a few hours (implying separations $\sim R_\odot$), Kepler periods are in the range 10s – 15 min and luminosity variations occur over a few days.

Far fewer LMXBs are known (~ 20 with known orbital periods), but they are important in providing a constraint on disc theory: this should predict essentially the same structure at disc radii large compared with that of the accreting star – unless irradiation of the disc by the central X-ray emission is important.

There are two observable properties of discs which are effectively independent of the viscosity: the kinematics of the disc material and (in a steady state) its effective temperature distribution. In both cases the most stringent tests come from observations of eclipsing systems.

2.1 ORBITAL MOTION IN THE DISC

The matter in a Keplerian disc has a velocity about the accreting star which varies as $R^{-1/2}$, where R is the distance from the star. Further, these orbits lie in the binary plane and have the same sense of rotation as the whole system. Spectroscopic observations of CVs show doubled emission lines, with the blue- and red-shifted components arising from gas with rotational velocities towards and away from the observer. In an eclipse by the companion, material at the edge of the disc moving towards the companion is eclipsed first. As the eclipse proceeds, the faster material closer to the accreting star is occulted, followed by that moving away from the companion with high then low velocities. The emission-line eclipse should show a "rotational disturbance", with the blue wing eclipsed first, in the order red to blue, followed by the red wing in the order in the order red to blue; eclipse egress should of course occur in the reverse order. These effects are observed in many CVs, and were among the original indications of orbiting gas (Greenstein & Kraft 1959, Kraft 1962).

A significant number of systems deviate from this simple picture, however, which may be an indication that some intrinsic mechanism, such as Stark broadening, dominates the Keplerian broadening, or possibly that the velocities are influenced by a magnetic field on the white dwarf (Williams 1989).

2.2 EFFECTIVE TEMPERATURE DISTRIBUTION

In a steady state, energy is dissipated at a rate $3GM\dot{M}/(8\pi R^3)$ per unit surface area at disc radii $R \gg$ the radius of the central star. By definition this equals σT_{eff}^4, where T_{eff} is the effective temperature of the disc at radius R. Thus T_{eff} should vary as $R^{-3/4}$, implying that the disc's centre should be bright and blue and its edges fainter and redder. If the binary system has high enough inclination the centre of the disc will be eclipsed. Thus observing at two or more wavelengths should reveal light curves of rather different character: at long wavelengths the disc is fairly uniform in brightness and gives a broad shallow eclipse, while the concentration of the blue light implies a deep narrow eclipse. Just this is seen in several systems.

One can in principle use the eclipses as a way of mapping the surface brightness distribution of the disc and checking the quantitative prediction for $T_{\rm eff}$. As a by-product this method gives information about the binary dimensions and the accretion rate. It constitutes an almost viscosity-independent check on accretion disc theory: while $T_{\rm eff}$ is indeed independent of viscosity in a steady-state disc, the observed quantity is the colour temperature. In principle, the step from this to $T_{\rm eff}$ involves the atmosphere of the disc and hence the viscosity, although in practice this is not a severe restriction. The best way of extracting the information from the disc is the maximum-entropy eclipse mapping method developed by Horne and his collaborators (*e.g.* Horne & Marsh 1986). Generally good agreement with the expected $R^{-3/4}$ behaviour is found, with derived accretion rates in reasonable accord with evolutionary expectations.

3 Viscosity

Dissipation is the basic driving force in a disc, yet we are still ignorant of its nature. It is usually thought of as some kind of turbulent eddy viscosity, although there is no proof of this. Since a kinematic viscosity can be expressed as the product of a velocity and a length-scale, the best we can manage is to put limits on plausible values of these quantities. It is unlikely that the characteristic velocity can be supersonic over much of the disc, as shocks are likely to develop and make it subsonic. (Note that this reasoning could be wrong in restricted regions of the disc.) The length-scale probably cannot exceed the disc semi-thickness $H(\sim R/\text{Mach number of Kepler flow})$ although again this could be wrong if the process is highly anisotropic. Thus we write

$$\nu = \alpha c_s H,$$

with α dimensionless. Apart from the hope that $\alpha < 1$ this gains us nothing: in particular there is absolutely no reason to suppose that α is constant in time or space. The disc's surface density, Σ, evolves according to a non-linear diffusion equation involving the vertical average of ν:

$$\frac{\partial \Sigma}{\partial t} = \frac{3}{R}\frac{\partial}{\partial R}\left\{R^{1/2}\frac{\partial}{\partial R}\left[\nu\Sigma R^{1/2}\right]\right\}.$$

Thus Σ diffuses on a locally determined time-scale

$$t_{\rm visc} \sim l^2/\nu,$$

where

$$l \sim \Sigma\bigg/\left|\frac{\partial \Sigma}{\partial R}\right|.$$

It is clear that our ignorance of ν means that we have no good theory of the time dependence or stability of accretion discs. Various functional forms have been suggested for ν in order to produce global disc behaviour, and thus luminosity variations, mimicking those of dwarf novae (a subset of CVs). While these forms reflect considerable ingenuity, they have no theoretical basis. The one fairly firm piece of evidence we have is that the observed time-scales (days) for disc evolution in dwarf nova outbursts must imply $\alpha \sim 1$ at least some of the time, since l cannot exceed the disc radius ($\sim$ few $\times 10^{10}$ cm).

4 Surface layers

There is no reason to suppose that dissipation is absent from the optically thin layers near the disc faces. The resulting effects must also depend on α, and so are quantitatively uncertain. However, some qualitative consequences can be seen.

First, extensive dissipation would upset the relation between colour and effective temperature referred to above in the discussion of eclipse mapping. The fact that observations do give an $R^{-3/4}$ behaviour suggests that this effect cannot be very strong, at least in some cases. Indeed, the consequent flattening or even inversion of the vertical temperature gradient may lead to better agreement with observation than disc spectra constructed by placing stellar atmospheres on each annulus of the disc, thus implicitly assuming no dissipation occurs in the atmosphere. These typically give absorption features which are too deep (Wade 1988).

A second important effect is evident from a simple analytic treatment of the energy balance in optically thin layers (see the contribution by Czerny & King in this volume). An atmosphere in which dissipation occurs cannot in general remain in vertical hydrostatic balance, since at low densities radiative cooling cannot cope with the dissipation. Coronae and wind outflow from much of the disc surface result. Irradiation by the central X-ray source in LMXBs amplifies this effect, to the point where one half, or more, of the mass overflow from the companion star may be blown away rather than accreted.

Shaviv & Wehrse (1986) found an example of this effect in a particular case by attempting to construct self-consistent disc atmospheres including dissipation. Their full calculations (to be published) demonstrate the possibility of a third important effect. In monitoring luminosity variations of CVs, especially dwarf novae, it is usual to assume that observing at one wavelength gives at least the direction of the changes in accretion rate, *i.e.* an increase in the V-band flux implies an increase in $\dot{M}$. Depending on what is assumed about α, this may be incorrect: in some models, the flux at particular wavelengths actually decreases as $\dot{M}$ increases; this of course results from a change in the overall spectral shape.

5 Non-axisymmetric structure

In addition to eclipse modulations, optical light curves of CVs frequently display other periodic features. Most notable is a large hump (amplitude $\sim 50\%$) lasting for up to half the binary period. In eclipsing systems this reaches its maximum just prior to ingress. It has long been realised that this light comes from the impact of the mass transfer stream on the edge of the accretion disc at what has come to be called the "bright spot", the modulation resulting from foreshortening as the system rotates.

From the remarks made earlier, the disc structure in LMXBs might be expected to be similar. X-ray observations however, reveal some surprising features (White & Mason 1985). Since X-rays dominate the emission of these systems, they must be produced deep in the potential well, very close to the neutron star or black hole. One might therefore expect the X-ray light curve to be essentially that of a point source, perhaps periodically totally occulted. Some systems instead show partial eclipses, and indeed continuous (periodic) modulation, indicating that the source is extended. The likely cause is that the observed X-rays are not coming directly from the central object, but are scattered into our line of sight by a corona above the disc. This is supported by the fact that systems showing this effect are on average fainter than other LMXBs. The required scattering optical depth of the corona (a few percent) is in line with that expected from the considerations of the previous section (Czerny & King, this volume, and in preparation). The continuous periodic modulation shows that the corona has a stable non-axisymmetric structure.

A further type of periodic behaviour occurs in LMXBs. In some systems the main modulation consists of dips in X-ray intensity in a narrow phase range. In high resolution the dips are seen to consist of many very short deep events, caused by a mixture of photoelectric absorption and scattering. Initially this behaviour was ascribed to occultations of the central X-ray source by structure at the outer edge of the accretion disc (see White & Mason 1985 for a review). This structure would have to move on unphysical orbits however; both ballistic and pressure-supported matter out of the disc plane will intersect it half-way along their orbits about the central mass. Thus only rather smooth structure is possible at the disc edge, implying occultations lasting for large fractions of a binary period, unlike those observed.

An alternative model, suggested by Frank, King & Lasota (1986), makes use of the fact that the accretion stream is probably somewhat thicker in vertical extent than the edge of the disc: some of it then skims over the disc faces towards smaller radii. As in the original picture leading to disc formation, much of this material is likely to circularize at the radius $R_{\text{circ}} \sim 10^{10} P^{2/3}$ cm ($P = $ binary period in units of 5 hr). Newly arriving material will collide with this ring of matter and be shock-heated to $\sim 10^7$ K, thus rising out of the disc plane. In the presence of the central X-ray flux, this matter is 2-phase unstable: most of the matter collapses into cool absorbing cloudlets, which orbit out of the disc plane. As these pass across the line of sight to the central X-ray source they produce the observed occultations and absorptions. If some of the gas stream continues to follow its original ballistic trajectory rather than circularizing, structure at other binary phases may result. There is some evidence (see White & Mason 1985) that the phases of dominant dipping activity in LMXBs are indeed separated by the expected amount.

6 Conclusions

There is strong observational evidence supporting the viscosity-independent predictions of simple thin disc theory particularly when applied to CVs. Other properties which do not depend strongly on the form of viscosity, such as the existence of coronae and winds, also agree with observation. It appears that radical revisions of this picture to accommodate X-ray observations of LMXBs are not needed if due account is taken of the possibility that some of the accretion stream may not join the disc at its edge. Our present ignorance of disc viscosity however limits real progress in understanding other features, such as time dependence and stability.

References

Eddington, A. S., 1926. *The Internal Constitution of the Stars*, Cambridge University Press, Cambridge.

Frank, J., King, A. R. & Lasota, J. P., 1986. *Astron. Astrophys.*, **178**, 137.

Frank, J., King, A. R. & Raine, D. J., 1985. *Accretion Power in Astrophysics*, Cambridge University Press, Cambridge.

Greenstein, J. L. & Kraft, R. P., 1959. *Astrophys. J.*, **130**, 99.

Horne, K. & Marsh, T., 1986. In *The Physics of Accretion on to Compact Objects*, p. 1, eds. Mason, K. O., Watson, M. G. & White, N. E., Springer-Verlag, New York.

Kraft, R. P., 1962. *Astrophys. J.*, **135**, 408.

Pringle, J. E., 1981. *Annu. Rev. Astron. Astrophys.*, **19**, 137.

Ritter, H., 1987. *Astr. Astrophys., Suppl. Ser.*, **70**, 335.

Shaviv, G. & Wehrse, R., 1986. *Astron. Astrophys.*, **159**, L5.

Wade, R. A., 1988. *Astrophys. J.*, **335**, 394.

White, N. E. & Mason, K. O., 1985. *Space Sci. Rev.*, **40**, 167.

Williams, R. E., 1989. *Astron. J.*, in press.

A disc instability model for soft X-ray transients containing black holes

Shin Mineshige and **J. Craig Wheeler**
University of Texas at Austin, USA

1 Introduction

Low-mass X-ray binaries (LMXB) are semi-detached binary systems consisting of a mass-losing late-type star and a compact object (neutron star or black hole) which is surrounded by an accretion disc fed by mass loss from the late-type companion. Soft X-ray transients are unique in this group by showing outbursts with recurrence time of 0.5 – 50 years, rise time scale 2 – 10 days, and decline time scale of order of a month (for recent reviews, see *e.g.* White *et al.* 1984; van Paradijs & Verbunt 1984; Priedhorsky & Holt 1987). Two models are proposed for outbursts of soft X-ray transients: the disc instability model (Cannizzo *et al.* 1985), and the mass-transfer burst model (Hameury *et al.* 1986).

2 Thermal Instability of Accretion Discs

As the first step, we integrate the vertical structure of the disc in LMXB following the method described in Mineshige & Osaki (1983). We scale the viscosity parameter $\alpha = \alpha_0(h/r)^n$, where α_0 and n are numerical constants and h represents the semi-thickness of the disc. We also assume that the effects of X-ray illumination of the outer disc by the central disc are negligible. We find that for relevant accretion rates, the disc suffers a thermal instability due to the ionization and recombination of the hydrogen and the helium, leading to intermittent accretion onto the central compact object, similar to models for the outbursts in dwarf novae (Osaki 1974; Meyer & Meyer-Hofmeister 1981)

3 Global Propagation of the Thermal Instability

We calculate the global propagation of the thermal instability in the disc. Particular interest is focused on reproducing the light curve of the possible black hole binary A0620-00 (McClintock & Remillard 1986). In this system there is observational evidence that the X-ray luminosity is below the detection limit of $\sim 10^{32}$ erg s^{-1} in quiescence (Long *et al.* 1981).

The results are summarized as follows:

(1) For $n \leq 1.0$, the inner region never suffers the thermal instability, yielding a relatively small light variation, $\Delta V < 1 - 6$ mag, and shorter recurrence time, $\tau_{rec} \sim 50 - 200$ days, for $M = 10$ M$_\odot$.

(2) For $n \geq 1.5$, the thermal instability can propagate over the whole disc and the disc shows bigger light variations, $\Delta V > 8$ mag, and longer recurrence time, $\tau_{rec} \sim 4 - 16$ yrs, for $M = 10$ M$_\odot$. These models give a good fit to the observed light curve of A0620-00.

(3) If we reduce the central mass to 1 $M_\odot$ as would be appropriate to a central neutron star, the recurrence time also decreases by a factor of more than 10 compared with the models with 10 $M_\odot$.

4 Conclusions

(1) The disc-instability model can account for the basic features of outbursts in soft X-ray transients.
(2) The central mass of $\sim$ 10 $M_\odot$ is favoured over $\sim$ 1 $M_\odot$ to reproduce the observed recurrence time of A0620-00 ($\sim$ 60 yr), *i.e.* the compact object in A0620-00 is more likely to be a black hole.
(3) Hardening in the spectrum of A0620-00 is observed at the beginning of outbursts (White *et al.* 1984). This suggests that the bi-modal behaviour observed in the spectra of the black-hole candidates such as Cyg X-1, GX 339-4, and A0620-00 might be related to the thermal instability of the discs.

Further details are given in Huang & Wheeler (1989) and Mineshige & Wheeler (1989).

References

Cannizzo, J. K., Wheeler, J. C. & Ghosh, P., 1985. In *Proc. Cambridge Workshop on Cataclysmic Variables and Low-Mass X-Ray Binaries*, p. 307, eds. Lamb, D. Q. & Patterson, J, Reidel, Dordrecht.
Hameury, J. M., King, A. R. & Lasota, J. P., 1986. *Astron. Astrophys.*, **162**, 71.
Huang, M. & Wheeler, J. C., 1989. *Astrophys. J.*, **343**, in press.
Long, K. N., Helfand, D. J. & Grabelsky, D. A., 1981. *Astrophys. J.*, **248**, 925.
McClintock, J. E. & Remillard, R. A., 1986. *Astrophys. J.*, **308**, 110.
Meyer, F. & Meyer-Hofmeister, E., 1981. *Astron. Astrophys.*, **104**, L10.
Mineshige, S. & Osaki, Y., 1983. *Publ. Astron. Soc. Jpn*, **35**, 377.
Mineshige, S. & Wheeler, J. C., 1989. *Astrophys. J.*, **343**, in press.
Osaki, Y., 1974. *Publ. Astron. Soc. Jpn*, **26**, 429.
Priedhorsky, W. C. & Holt, S. S., 1987. *Space Sci. Rev.*, **34**, 291.
van Paradijs, J. & Verbunt, F., 1984. In *High Energy Transients in Astrophysics*, p. 49, ed. Woosley, S. E., AIP Conf. Proc. 115, New York.
White, N. E., Kaluzienski, J. L. & Swank, J. H., 1984. In *High Energy Transients in Astrophysics*, p. 31, ed. Woosley, S. E., AIP Conf. Proc. 115, New York.

X-ray variability from the accretion disc of NGC 5548

J. S. Kaastra

Laboratorium voor Ruimteonderzoek, Leiden, The Netherlands

1 Observations

X-ray observations of the spectrum and variability of the Seyfert galaxy NGC 5548 were obtained with 2 instruments aboard the European X-ray satellite *EXOSAT*. The *low energy* (LE) experiment was an imaging device with a spatial resolution of 18″ (FWHM) on axis. It operated in the energy range of 0.05 – 2 KeV and has no intrinsic energy resolution, but multi-colour photometry was possible using different filters. We used data obtained with the 300 nm Lexan (3Lx), 400 nm Lexan (4Lx), Aluminium-Parylene (Al/Pa) and Boron (B) filter. The 3Lx and B filters in particular, have distinctly different spectral responses to AGN spectra. The *medium energy* (ME) experiment consisted of an array of 8 passively collimated proportional counters. It had no intrinsic position resolution, but spectra with moderate energy resolution in the energy range 1 – 50 KeV could be obtained.

The data set consists of 3 long observations in 1984 and 1986. Both components show correlated variability on a typical time scale of half a day. The variability amplitude is low: a few tenths. There is evidence for a delay of the hard X-rays with respect to the soft X-rays of 1 – 2 hour.

2 Spectral fit

The spectra were fitted by a power law plus a soft excess, which we modelled by a modified blackbody spectrum. In fact, other two-parameter models for the soft excess (like a simple blackbody or thermal radiation) also yield acceptable fits; the modified blackbody is chosen in order to be consistent with the disc model to be discussed below. The power law has a typical photon index of 1.7 and a 2–10 KeV luminosity of 6×10^{36} W. The modified blackbody component has a temperature in the range of 0.06–0.11 KeV and a typical luminosity of 4×10^{37} W. Due to the limited spectral resolution of the LE instrument, the modified blackbody temperature cannot be determined with more accuracy.

3 Accretion disc model

The spectral data can be explained in terms of an accretion disc surrounding a black hole. The disc model considered is a standard Shakura & Sunyaev (1973) disc with an accretion rate of 0.1 $M_{\odot}$/year, $\alpha = 0.1$ and a central mass of 10^7 $M_{\odot}$. The accretion disc emits a modified blackbody spectrum (due to electron scattering) locally, with a strong maximum in the emissivity near a distance of 5 Schwarzschild radii from the black hole (due to the temperature maximum of the disc). This radiation can be identified with the soft X-ray excess. The accretion rate and central mass given above follow from the observed soft X-ray temperature and flux.

The hard X-ray power law component can be understood if the central part of the disc is covered by a hot corona which scatters the soft photons emerging from the disc below. The radius of the corona (1.33 inner radii of the disc or 4 Schwarzschild radii) follows from the derived accretion disc parameters and the hard X-ray flux.

The inner, radiation pressure dominated, part of the disc is unstable (*cf.* Shakura & Sunyaev 1976). In principle, instabilities on both the thermal and the viscous time scale are expected to occur, but the thermal instability is the fastest growing mode. The basic thermal time scale of the disc at the position of maximum X-ray emissivity, computed from our spectral parameters, is consistent with the observed variability time scale of half a day. In addition, the delay of the hard X-ray component as compared to the soft X-ray component is consistent with the sound travelling time from 5 Schwarzschild radii (where most of the soft X-rays are produced) to the corona.

4 Discussion

There are more than ten other known examples of (variable) soft X-ray excesses in Seyfert galaxies, *e.g.* Mkn 841 (Arnaud *et al.* 1985), Fairall 9 (Morini *et al.* 1986). In particular, the spectral and temporal characteristics of Mkn 841 are very similar to those of NGC 5548. We conclude that X-ray observations contribute much to our understanding of the accretion disc dynamics in active galactic nuclei. The relative simple model of a standard Shakura-Sunyaev disc surrounded by a hot corona in the inner parts is sufficient to model both the spectral and the variability data of NGC 5548.

More details on the topics discussed in this contribution will appear in Kaastra & Barr (1989).

References

Arnaud, K. A., Branduari-Raymont, G., Culhane, J. L., Fabian, A. C., Hazard, C., McGlynn, T. A., Shafer, R. A., Tennant, A. F. & Ward, M. J., 1985. *Mon. Not. R. Astron. Soc.,* **217**, 105.
Kaastra, J. S. & Barr, P., 1989. *Astron. Astrophys.,* Submitted.
Morini, M., Scarsi, L., Molteni, D., Salvati, M., Perola, G. C., Piro, L., Simari, G., Boksenberg, A., Penston, M. V., Snijders, M. A. J., Bromage, G. E., Clavel, J., Elvius, A. & Ulrich, M. H., 1986. *Astrophys. J.,* **307**, 486.
Shakura, N. I. & Sunyaev, R. A., 1973. *Astron. Astrophys.,* **24**, 337.
Shakura, N. I. & Sunyaev, R. A., 1976. *Mon. Not. R. Astron. Soc.,* **175**, 613.

Viscously heated coronae and winds around accretion discs

M. Czerny[1] and **A.R. King**
University of Leicester

1 Introduction

The existence of rarefied matter in some galactic X-ray sources (so called accretion disc corona sources, or ADC sources) is well established by observations (for a review see Mason 1986). Several mechanisms have been already proposed to explain this phenomenon. Here we show that viscous heating, if present in atmospheres of accretion discs, can also produce coronae and/or winds.

2 Origin of the wind

The energy equation for a hydrostatic atmosphere of an accretion disc in the presence of α-viscosity is:

$$\frac{3}{2}\sqrt{\frac{GM}{r^3}}\alpha\frac{k\rho T}{\mu m_{\mathrm{H}}} \;+\; \frac{\kappa\rho F}{m_{\mathrm{e}}c^2}\big(\langle E\rangle - 4kT\big) \;-\; \rho^2\Lambda(T) \;=\; 0 \qquad (1)$$

Here the first term represents viscous heating, the second is due to Comptonization, and the third represents cooling by atomic processes (bremsstrahlung, line-cooling *etc.*). The radiative flux F is either the flux produced locally in the disc (the *non-illuminated* case) or the sum of the flux produced locally and that received from the central source (the *illuminated* case). As the last term in equation (1) decreases faster than the first two with decreasing density, only viscous heating and Comptonization determine the temperature at sufficiently low optical depths. For increasing disc radii viscous heating decreases more slowly than the Comptonizing flux in either case, which results in a rise of the surface temperature, and a hot corona is formed. Beyond some radius, Comptonization cannot balance viscosity at all, and cooling by expansion is required, *i.e.* a wind region develops. For accretion rates lower than $\sim 3 \times 10^{-3}\alpha\dot{M}_{\mathrm{crit}}$, where $\dot{M}_{\mathrm{crit}}$ is the Eddington accretion rate, the wind region covers the whole disc surface.

3 Conclusions

Our main conclusions concerning the properties of viscously heated coronae and winds are:

(1) The optical depth of the hydrostatic corona is small, typically of order 10^{-4} to 10^{-3}.
(2) In the non-illuminated case, mass loss from the disc is relatively more important for low accretion sources, as the Comptonizing flux is then small. This may be connected with the observed increasing importance of the power-law component with respect to the big bump in AGNs when total luminosity decreases, and also with the spectral behaviour of Cyg X-1 in high and low states.

---

[1] Present address: Warsaw University Observatory, Poland

(3) For very low accretion rates, the wind originates from the entire surface of the disc.

(4) In the illuminated case, the mass loss is much stronger than predicted by Begelman *et al.* (1983), who included Comptonization only. Viscously heated winds may therefore be responsible for ADC sources.

(5) The mass loss rate is a strong function of the accretion rate, the energy of illuminating photons, assumed disc shape, and the viscosity parameter. On the other hand the optical depth of the wind depends strongly on the viscosity parameter, but relatively weakly on other parameters. Our model may therefore explain the constancy of the relative X-ray eclipse depths in the X-ray transient *EXO 0748-676* (Parmar *et al.* 1986).

More detailed results and discussion will be presented elsewhere (Czerny & King 1989a, b).

References

Begelman, M. C., McKee, C. F. & Shields, G. A., 1983. *Astrophys. J.*, **271**, 70.
Czerny, M. & King, A. R., 1989a. *Mon. Not. R. Astron. Soc.*, **236**, 843.
Czerny, M. & King, A. R., 1989b. *Mon. Not. R. Astron. Soc.*, submitted.
Mason, K. O., 1986. In *The Physics of Accretion onto Compact Objects*, p. 29, eds. Mason, K.O., Watson, M. G. & White, N. E., Springer-Verlag, New York.
Parmar, A. N., White, N. E., Giommi, P. & Gottwald, M., 1986. *Astrophys. J.*, **308**, 199.

Optical emission line profiles of symbiotic stars

K. Robinson and **M. F. Bode**
Lancashire Polytechnic
J. Meaburn and **M. J. Whitehead**
University of Manchester

1 Introduction

High resolution (≤ 10 km s^{-1}) spectra of a selection of symbiotic stars have been obtained using the Manchester Echelle Spectrograph and IPCS on the 2.5 m Isaac Newton Telescope on La Palma, as a preliminary to a high resolution survey of all known symbiotic stars now being conducted from the INT and ESO. In several cases, the [OIII] 5007 Å region shows complex structure, probably originating in extended outflows. However, the Hα line in many objects shows a well-known double-peaked profile (see *e.g.* Anderson *et al.* 1980). This is very reminiscent of those associated with dwarf novae, where observations through eclipse indicate that the emission originates in the accretion disc surrounding the white dwarf component of the semi-detached binary (see *e.g.* King, this volume).

If this were also the case with symbiotic systems, then theoretical modelling of the line profiles would enable us to constrain the all-important binary parameters. Our preliminary aim, however, is to determine whether such line profiles can be reconciled with emission from accretion discs at all. Anderson *et al.* (1980) concluded that the case still remained ambiguous.

2 The model

In order to construct theoretical line profiles from accretion discs, we have adapted the optically thick disc model of Horne & Marsh (1986) producing double-peaked profiles which have a deep "V"-shaped central reversal. These match the observed profiles more closely than earlier (optically thin) models (*e.g.* Smak 1969). Disc dimensions for any given object are defined by the binary star separation, masses, and the type of accretor assumed (main-sequence star or white dwarf) using standard theory (see *e.g.* King 1989). The model as formulated is, of course, unable to reproduce the obvious asymmetry of the profiles, which would be explicable in terms of emission from a hot spot on the disc, giving rise to the more intense peak.

3 Comparison with observations, and conclusions

The stars we have chosen to model from our observational results (CH Cyg and RS Oph) had clear double-peaked profiles, and have been suggested to have accretion discs (see *e.g.* Duschl 1983 and Bruch 1986 respectively).

Figure 1(a) shows comparison with our observations of CH Cyg on 1986 June 28. A main-sequence accretor, $M_1 = 1$ $M_\odot$, with $M_1/M_2 = 0.2$ and a binary period, $P = 28.2$ days is assumed (values from the accretion disc model of Duschl 1983). The disc inclination must be high to give the observed depth of the central reversal; here we have taken $i = 81°$. It can clearly be seen that the model peak-to-peak separation is far too

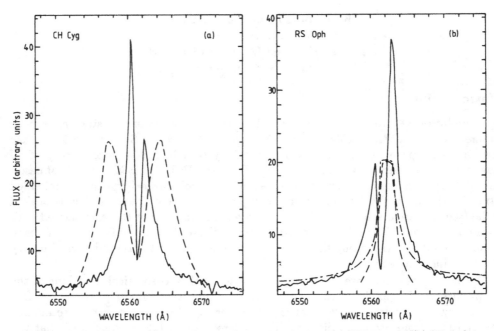

Figure 1. Comparison of the 1986 June 28 Echelle spectra of (a) CH Cyg and (b) RS Oph. Dashed lines represent main-sequence accretor model, and dot-dash white dwarf accretor.

high. To reduce this separation would necessitate making the outer radius of the disc unphysically large.

Figure 1(b) shows a similar comparison for RS Oph, whose binary parameters are better defined than for CH Cyg, (i) with a main-sequence accretor, and (ii) with a white dwarf. In both cases $M_1 = 1 \, M_\odot$, $M_1/M_2 = 1$, $i = 23.3°$, $P = 230$ days (from radial velocity measurements by Garcia 1986). The resulting model fit is very poor. At such a low inclination, the central reversal is almost lost. It has also proved impossible to increase i within the constraints of Garcia's observations, and reasonable masses, to a point at which the reversal depth is more acceptable.

We therefore conclude that the double-peaked profiles are not compatible with emission from an accretion disc in either CH Cyg or RS Oph, and it is likely that similar profiles in many other symbiotics do not arise in this way either. Exploration of line emission from a partially ionised wind is now underway.

References

Anderson, C. M., Oliverson, N. A. & Nordsieck, K. H., 1980. *Astrophys. J.*, **242**, 188.
Bruch, A., 1986. *Astron. Astrophys.*, **167**, 91.
Duschl, W. J., 1983. *Astron. Astrophys.*, **119**, 248.
Garcia, M. R., 1986. *Astron. J.*, **91**, 400.
Horne, K. & Marsh, T. R., 1986. *Mon. Not. R. Astron. Soc.*, **218**, 761.
King, A. R., 1989. In *Classical Novae*, p. 17, eds. Bode, M. F. & Evans, A., Wiley, Chichester.
Smak, J., 1969. *Acta Astron.*, **19**, 155.

The effect of formation of Fe II in winds confined to discs for luminous stars

G. Muratorio[1] and M. Friedjung[2]

[1] *Observatoire de Marseille, France*
[2] *Institut d'Astrophysique de Paris, France*

1 Self-absorption curves

In previous work (Friedjung & Muratorio 1987, Muratorio & Friedjung 1988), we developed methods using *self-absorption curves* (SACs) to study stars having Fe II emission lines in their spectra. Such a curve is obtained by plotting $\log(F\lambda^3/gf)$ against $\log(gf\lambda)$, where F is the total flux, λ the wavelength, g the lower level statistical weight, f the oscillator strength. $gf\lambda$ is proportional to the optical thickness. If no selective excitation mechanisms exist for particular levels, and the levels inside a term have populations proportional to their statistical weights, such a plot for emission lines of the same multiplet will have points lying on the same self-absorption curve. The shape of the curve is characteristic of the nature of the medium where the line is formed. Shifting the curves for different multiplets (which should have the same shape) relative to each other so as to superpose them, will give at the same time the relative populations of their upper and also their lower terms. Until now, we have calculated SACs for various simplified cases, and a comparison was made with observations of luminous stars whose spectra contained many Fe II emission lines. It was found that observations of certain Magellanic cloud stars could not be fitted by spherically symmetric wind models. Another line emitting medium seemed to be present (a slab or a thin disc with constant opening angle), which is also suggested by the continuum energy distributions.

2 Non-spherically symmetric case

In order to improve both the models and agreement with observation, we now study theoretical SACs for line formation in disc-like equatorial winds with large opening angles. This is done by changing the limits of integration in the formulae of Friedjung & Muratorio (1987). All present cases have a constant wind velocity. Figure 1 shows SACs for a low velocity wind in a disc of varying opening angle α viewed pole on, compared with the spherical wind SAC and the former very thin disc SAC. Figure 2 shows how the SAC varies with viewing angle. Figure 3 shows high velocity wind cases.

3 Comparison with observations

In Figure 4, we compare a theoretical $\alpha = 120°$, pole on low velocity wind SAC with UV measurements for the VV Cep type star KQ Pup. The SACs for the different multiplets were horizontally and vertically shifted so as to best coincide, enabling us to obtain the relative upper term populations (Figure 5) corresponding to a Boltzmann law lower term population for a temperature of 5000 K. This star is a binary containing a cool giant and a hot main sequence component which could have a dense equatorial wind or an accretion disc (Muratorio *et al.* in preparation).

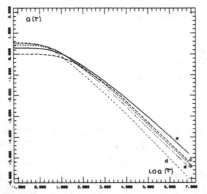

Figure 1. SACs for low velocity winds. Curves are for: (a) $\alpha = 120°$; (b) 90°; (c) 30°; (d) spherical wind SAC and (e) former very thin disc SAC.

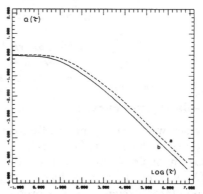

Figure 2. SACs for low velocity winds with $\alpha = 30°$ viewed (a) pole and (b) from an inclination of 60° to the plane of the sky.

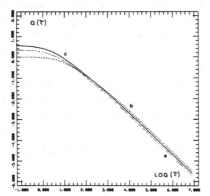

Figure 3. SACs for a high velocity wind. (a) $\alpha = 10°$ seen pole on; (b) $\alpha = 30°$ inclined 60° to plane of sky; (c) a spherical high velocity SAC.

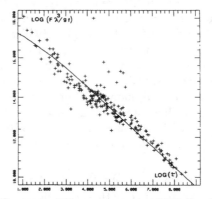

Figure 4. UV measurements of KQ Pup compared with the theoretical $\alpha = 120°$, pole on, low velocity wind SAC.

References

Friedjung, M. & Muratorio, G., 1987. *Astron. Astrophys.*, **188**, 100.

Muratorio, G. & Friedjung, M., 1988. *Astron. Astrophys.*, **190**, 103.

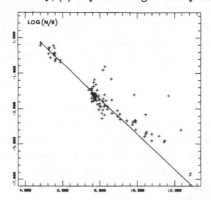

Figure 5. The relative upper term populations corresponding to a Boltzmann relative lower term population law for a temperature of 5000 K.

Observational evidence for accretion discs
in active galactic nuclei

Matthew Malkan

University of California, Los Angeles, USA

Abstract Most Seyfert 1 nuclei and quasars show strong excess continuum flux in the blue and ultraviolet, relative to an extrapolation of their spectra at longer wavelengths. The arguments for identifying this "Big Blue Bump" as thermal emission from an optically thick accretion flow are outlined. Further (less secure) arguments are presented that the flow is flattened, possibly in a disc. The close agreement between simple accretion disc models and the observations is summarized. Several modifications needed to make disc models more realistic are discussed. Finally, the extent to which these models are constrained by observations, such as detections of "Soft X-Ray Excesses", and prospects for obtaining future observational evidence of AGN accretion discs is considered.

1 UV excess

Almost from the first multi-frequency observations of Seyfert 1 nuclei and quasars, it was realized that their optical and ultraviolet spectra were far flatter than their infrared spectra, which have typical slopes of -1.2 ($f_\nu \sim \nu^{-1.2}$). The different variability properties of the infrared and optical/ultraviolet continuum further suggest that they are produced by physically separated components (Cutri *et al.* 1985). The blue component (also known as the "UV excess" or "Big Blue Bump") has a flux density rising with frequency in the optical, and a broad maximum somewhere in the ultraviolet. It falls (probably rather steeply) in the far- or extreme-UV. A falling high-frequency tail may be observed in the soft X-rays. This characteristic shape strongly suggests thermal emission from optically thick gas (Shields 1978, Malkan & Sargent 1982).

What is the source of this thermal energy? One fact leads me to believe that these UV photons are primary, rather than secondary (*i.e.* that this energy first appears as radiation in the ultraviolet; it did not originate as higher-energy radiation which is simply reprocessed by optically thick gas clouds). This is because in luminous quasars the thermal UV component contains the majority of the total quasar energy output. In particular, the UV component usually has more power than the X-rays. For a conservative example, suppose we integrate the UV flux down to 1000 Å only, and compare this with the X-ray flux integrated all the way from 1 KeV to 0.5 MeV. For a typical luminous radio-quiet quasar's α_{ox} (the logarithmic slope from 2500 Å to 2 KeV) of -1.5 to -1.6, the UV component is 3 to 6 times more luminous than the X-rays. Thus even if 100% of the X-ray power were absorbed by cool optically thick gas clouds (which completely covered the source), it would still not provide sufficient power to explain the strength of the UV component. The discrepancy is an order of magnitude larger, when we consider that the typical covering fraction of optically thick gas around the X-ray continuum is probably not 100%, but actually less than 10%.

In addition to this energetics argument, there is spectroscopic evidence that the UV component is not thermalized re-radiation of the non-thermal X-rays. As calculated by Ferland & Rees (1988), if the UV bump were entirely reprocessed X-rays, we would expect a strong Lyman jump and a deep gash in the soft X-ray spectrum, since at

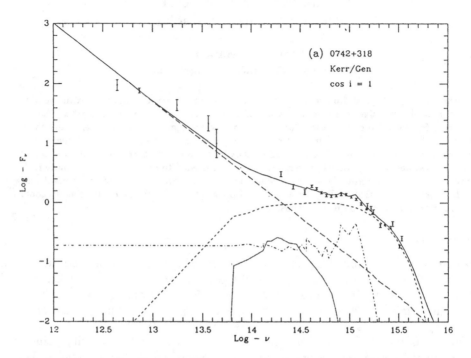

Figure 1. Fit to the energy distribution of the bright quasar 0742+318. The data are shown as vertical error bars; accretion disc model – dashed line; recombination and Fe II – dot-dash line; and starlight – dark line.

frequencies below a few KeV the optically thick gas would be highly absorptive. Most of the hard X-rays, by contrast, would be reflected by Compton scattering (see also Lightman & White 1988). Since such features are not present in the spectra of most quasars (Mushotzky 1984), this suggests that much less than 100% of the thermal UV power could actually be reprocessed X-rays. Rather than invoke two separate mechanisms which would each produce part of the thermal UV bump, it is more parsimonious to attribute it to a single, primary energy output of the central engine. On the Massive Black Hole hypothesis, it is probably produced by dissipation in accreting gas. Although it is more difficult to determine the geometry of the accreting gas, several general considerations suggest it is disc-like, or at least flattened.

2 Simple accretion disc models

We know that some deviations from spherical accretion are required to dissipate enough kinetic energy to make the energy generation process efficient. The many observational indications that the central engines of quasars are inherently bipolar also suggest that the accretion flow has significant angular momentum by the time it reaches the small radii where most of the energy is produced. The broadness of the thermal spectrum (Malkan 1983) requires that we are viewing gas at a large range of temperatures simultaneously and that the gas is probably spread over a wide range of radii. The lack of

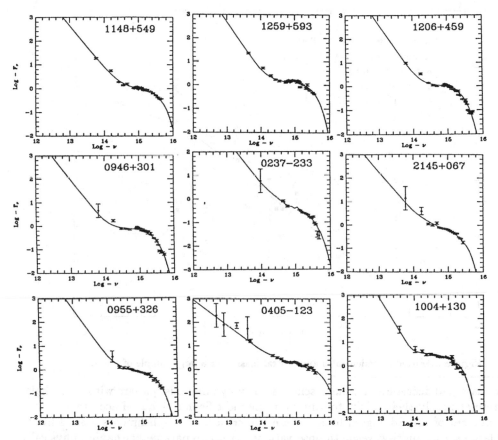

Figure 2. Examples of accretion disc model fits to the optical/UV energy distributions of nine typical quasars.

soft X-ray absorption indicates that the optically thick gas at large radii hardly ever obstructs our view to the inner hotter regions.

All of these considerations motivate the hypothesis that the component is thermal emission from an optically thick, geometrically thin accretion disc. Aside from the model's physical plausibility, it is relatively straightforward to test by calculating spectra with a minimum of free input parameters, for comparison with observations. Figure 1 illustrates the fit to the energy distribution of the bright quasar 0742+318. The accretion disc dominates the spectrum above $\log \nu = 14.5$; at frequencies below $\log \nu = 14.0$, the flux is primarily from a power law of slope -1.2. The other components shown are energetically unimportant. The thin disc model gives excellent 2-parameter (black hole mass and accretion rate) fits to the optical/UV energy distributions of all (~ 100) quasars and Seyfert 1 nuclei examined thus far (Sun & Malkan 1989). Nine typical examples from that paper are reproduced in Figure 2. The resulting parameters for the large AGN sample are illustrated in Figure 3. These fitted values of black

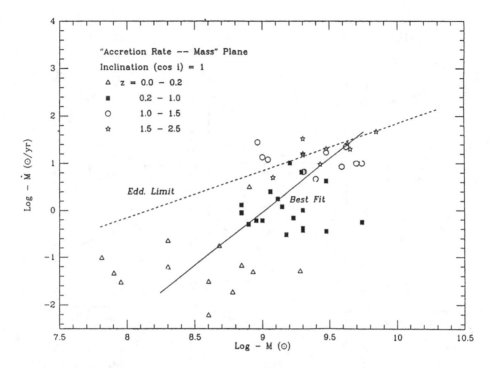

Figure 3. Fitted accretion rates and central masses for a large sample of AGN.

hole mass and accretion rate make sense: some very luminous quasars with black holes accrete at nearly their Eddington limits, while most Seyfert 1 nuclei accrete at only a few percent of the Eddington limit. These two critical accomplishments – good fits to the large set of multi-wavelength observations, with two parameters having physically reasonable values – have not yet been achieved by any non-thermal explanations of the UV continuum.

3 Tests of the accretion disc model

Einstein advised us that "Everything should be as simple as possible, but not simpler." The thin disc model is as simple as any viable explanation of the optical-UV continuum could be – but there are some observational tests to find out if it is too simple.

3.1 ACCRETION DISC ATMOSPHERES

Several theoretical investigations have considered the vertical structure and opacity of the accretion disc atmosphere, to make more realistic predictions of the continuum radiation it should emit. Two expectations from this work seem, at first glance, in conflict with observations:

(1) Since electron scattering dominates the opacity in the inner disc, its thermal emission should be significantly polarized when viewed at high indication. The predicted

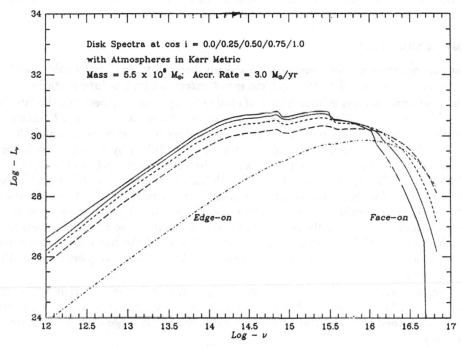

Figure 4. Disc spectra with atmospheres.

maximum linear polarization of 5% (Webb & Malkan 1986) is higher than the highest observed polarization, which is 1-2%. A simple way out of this dilemma is that we never view the entire disc surface at glancing angles ($i > 70°$). For example, Coleman & Shields (1989) calculated that a disc with surface irregularities (hills and valleys) would have a maximum linear polarization of 1–2%.

(2) Hydrogen bound-free opacity in the disc atmosphere should produce a Lyman limit absorption edge in its spectrum. As illustrated by the predicted spectra in Figure 4, when the disc is viewed close to face-on, its Lyman edge should not be smeared out by Doppler shifts. Observations seem to contradict this expectation, since no Lyman edge intrinsic to a quasar has been conclusively detected. Again, this expectation could be too naïve. The appearance of the Lyman limit depends critically on the (highly uncertain) vertical structure of the disc (Sun & Malkan 1987). Since substantial viscous energy is probably released at small-to-moderate optical depths, the vertical temperature gradient could be much smaller than that in a stellar atmosphere. This shallow vertical gradient in the source function could simultaneously explain both the low observed polarization and the weakness of the Lyman limit jump. There is a similar disagreement between models and the observations of accretion discs in cataclysmic variable binaries (Wade 1984). Detections of strong UV continuum polarization or Lyman limit absorption in quasars would have been dramatic support for the presence of thin discs. However the converse is

not true. Until we are more confident in our theoretical predictions of the vertical structure in thin discs, we will not be able to infer much from the absence of Lyman limit jumps or large UV polarization in quasar spectra.

3.2 HIGH ENERGY FLUX

Several properties of quasar spectra have already been observed, and could possibly probe the accreting gas. Here the problem is not detectability, but interpretation.

Observational progress is being made on closing the spectral gap between the far-UV and soft X-rays. Reimers *et al.* (1989) recently detected a quasar at $z = 2.75$ down to the short-wavelength limit covered by IUE, corresponding to a rest wavelength of 330 Å. Unfortunately, the observed far-UV fluxes must be multiplied by a large, but highly uncertain factor (~ 6) to correct for absorption by intergalactic HI clouds along the line-of-sight. Detailed spectroscopy by HST with high resolution and high signal/noise ratio will be needed to determine column densities for the absorbing clouds, for accurate de-blanketing of the continuum. At the other end of this critical gap, a growing number of AGN observations, especially by EXOSAT and Einstein, have revealed excess flux at energies below 1 KeV, compared to the extrapolation of the flat hard X-ray slope (Pounds 1985). This "soft X-ray excess" appears to be the high frequency tail of the Big Blue Bump.

To explain accretion disc flux at such high energies probably requires a re-examination of the assumption that the radiation from every part of the disc surface is roughly described by a blackbody. In discs with sufficiently large viscosity, the inner regions may

(a) emit a "modified" blackbody due to the dominance of electron scattering opacity (Czerny & Elvis 1987);
(b) be hot enough for Comptonization to modify the emergent spectrum; or
(c) become optically thin, and therefore extremely hot. In this case, the spectrum is difficult to predict, but would probably be dominated by free-free emission in the X-rays.

Two additional effects that also make the observed disc spectrum harder are:

(d) At high accretion rates, the inner disc may thicken into a funnel. Multiple reflections in the throat of the funnel lead to very high effective temperatures, which can be detected when the disc is viewed near face-on Madau (1988).
(e) When the disc is viewed at high inclination, radiation from its hot inner regions is strongly Doppler boosted (Sun & Malkan 1987).

Several disc model calculations have investigated each of these effects individually, although there are still hardly any detailed fits to actual observations of the soft-X-ray excess. The preliminary indications are that the observations of the high-energy tail of the Big Blue Bump could be well fitted with either: (b), a combination of (a) and (e), or (d) in super-Eddington cases. Distinguishing among these possibilities to identify a unique "best" explanation for the soft-X-ray excess will be more difficult, and may require further input data beyond its spectral shape (*e.g.* variability properties).

3.3 VARIABILITY

The overall flux, and often the spectral shape of the Big Blue Bump often varies on time-scales of weeks to years. The most rapid and largest changes tend to be seen in the less luminous Seyfert nuclei. There is evidence in many AGN (*e.g.* Maraschi *et al.* 1987) that the amplitude of variability increases, and the time-scale decreases with frequency. Thus the optical/ultraviolet continuum often hardens when it brightens (but see also Clavel *et al.* 1989). This seems qualitatively consistent with an accretion disc origin; a detailed quantitative comparison between theory and the observed spectral variability is underway. In at least one case, the variations were consistent with a simple accretion disc model (Sitko 1986). New predictions of the fastest variability time-scales in α-discs, as a function of wavelength, are presented by Siemiginowska & Czerny (1989). From a comparison with a small sample of observations, they appear to favour viscosity parameter values of 0.001 to 0.01.

3.4 FLOURESCENT LINES

Recent spectroscopy by GINGA provides evidence of interaction between the non-thermal X-rays and some thermal gas (Ohashi 1988). The detection of fluorescent Fe Kα emission lines in the X-ray spectra of several Seyfert 1 nuclei indicates that some X-rays are absorbed and reprocessed by optically thick gas in the nucleus. The equivalent widths of the K lines vary widely, but generally require a large column density of absorbing gas ($N_H > 10^{23}$ cm^{-2}). In many cases these implied column densities are far larger than the upper limits for gas along our line-of-sight. This special geometry, where thick absorbing gas is widely present, but not in our line-of-sight, suggests that the absorption and re-emission could occur in an accretion disc. If so, the flux and central wavelength of the K line may yield unique information about the disc structure and inclination.

3.5 INFRARED EMISSION

The infrared spectra of many quasars and Seyfert 1 nuclei show a broad flux excess centred on $3 - 5$ μm. It is illustrated by the $0.1 - 100$ μm energy distribution in Figure 5, from Edelson & Malkan (1986). The flux measurements of the quasar 3C 273 (here in νf_ν units) are fitted by a model which includes a parabola to describe the "near-infrared bump". Two recent papers attribute it to thermal emission in the outer parts of a disc which intercepts and re-radiates 5 or 10% of the continuum originating at smaller radii. Sanders *et al.* (1989) postulate that the near-infrared emission is quasi-blackbody thermal emission from hot dust grains. To be sufficiently hot ($T = 500 - 1000$ K) to emit at such short wavelengths, the dust must be near the central energy source, and yet in relatively cool, dense surroundings so that it does not all evaporate too quickly. To avoid significant reddening of the central UV continuum, Sanders *et al.* suggest that the dust resides in a disc which usually does not block our view of the inner regions. An alternate hypothesis by Collin-Souffrin (1987) attributes the near-infrared bump to thermal emission from dense atomic and molecular gas in the outer regions of an accretion disc.

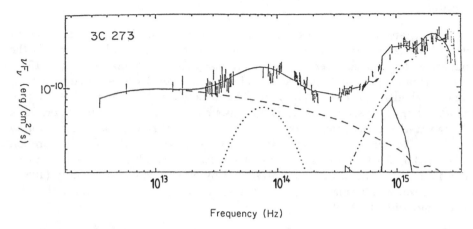

Figure 5. The energy distribution of 3C 273 (error bars). The solid line is a model fit, which includes a parabola (dotted line) to describe the "near infrared bump".

On either hypothesis, the thermal emission would probably include some spectral features. For example, the hot dust grains should produce some of the "unidentified" infrared lines, including the 3.3 μm feature. The dense gas postulated by Collin-Souffrin could produce emission lines from simple molecules, the Ca II infrared triplet and the many optical multiplets from excited levels of Fe II. Observational searches for correlations such as these are underway. In both cases, the observed energy is too large to be generated locally: it must be re-emission of 5 – 10% of the continuum generated closer to the centre. This requires either that (a) a significant portion of the central continuum emission arises well out of the accretion plane (*e.g.* in jets, tori, or coronae); or (b) the surface of the outer disc is flared, or warped, so that it intercepts 5 – 10% of the sky as seen from the inner disc. Option (a), at least, would imply that quasars with relatively strong near-infrared bumps, like 3C 273, have discs which are more likely to be viewed close to face-on.

Today it is not clear which of these lines of theoretical and observational investigation (or some completely different one, such as emission line profiles) is most likely to yield conclusive evidence for or against the presence of accretion discs in active galactic nuclei. I can only hope that Nature would not have chosen this power source without leaving some "smoking gun" for us to find, if we know where to look.

Many of the results discussed above resulted from collaboration with Dr. Wei-Hsin Sun, who provided some of the unpublished figures. This work was supported by NSF grant AST-86-14510.

References

Clavel, J. *et al.*, 1989. *Mon. Not. R. Astron. Soc.*, in press.
Coleman, R. & Shields, G. A., 1989. *Astrophys. J.*, in press.
Collin-Souffrin, S., 1987. *Astron. Astrophys.*, **179**, 60.
Cutri, R. M., Wisniewski, W. Z., Rieke, G. H. & Lebofsky, M. J., 1985. *Astrophys. J.*, **296**, 423.
Czerny, B. & Elvis, M., 1987. *Astrophys. J.*, **321**, 305.
Edelson, R. A. & Malkan, M. A., 1986. *Astrophys. J.*, **308**, 59.

Ferland, G. J. & Rees, M. J., 1988. *Astrophys. J.,* **332**, 141.

Lightman, A. P. & White, T. R., 1988. *Astrophys. J.,* **335**, 57.

Madau, P., 1988. *Astrophys. J.,* **327**, 116.

Malkan, M. A., 1983. *Astrophys. J.,* **268**, 582.

Malkan, M. A. & Sargent, W. L. W., 1982. *Astrophys. J.,* **254**, 22.

Maraschi, L. *et al.,* 1987. Multi-frequency Observations of the Sy 1 Galaxy 3C 120, in *Evidence of Activity in Galaxies*, IAU Symp. **121**, p215, Byurukan.

Mushotzky, R. F., 1984. *Adv. Space Res.*, **3**, No. 10-13, 157.

Ohashi, T., 1988. In *Physics of Neutron Stars and Black Holes*, p301, ed. Tanaka, Y., Universal Academy Press, Tokyo.

Pounds, K. A., 1985. In *Galactic and Extragalactic Compact X-Ray Sources*, p261, eds. Tanaka, Y. & Lewin, W. H., ISAS, Tokyo.

Reimers, D. *et al.,* 1989. *Astron. Astrophys.,* in press.

Sanders, D. B., *et al.,* 1989. *Astrophys. J.,* in press.

Shields, G. A., 1978. *Nature,* **272**, 706.

Siemiginowska, A. & Czerny, B., 1989. In *Theory of Accretion discs*, eds. Duschl, W. & Meyer, F., NATO Conference Series, in press.

Sitko, M. L., 1986. In *Continuum Emission in Active Galactic Nuclei*, p29, N.O.A.O., Tucson, AZ.

Sun, W.-H. & Malkan, M. A., 1987. In *Supermassive Black Holes*, ed. Kafatos, M., Cambridge University Press, Cambridge.

Sun, W.-H. & Malkan, M. A., 1989. *Astrophys. J.,* in press.

Wade, R. A., 1984. *Mon. Not. R. Astron. Soc.,* **208**, 381.

Webb, J. & Malkan, M. A., 1986. In *The Physics of Accretion onto Compact Objects*, p. 15, Lecture notes in Physics, eds. Mason, K. O., Watson, M. G. & White, N. E., Springer-Verlag, New York.

The fuelling of active galactic nuclei by non-axisymmetric instabilities

I. Shlosman[1], J. Frank[2] and M. C. Begelman[3]
[1] *Caltech, Pasadena, U.S.A.*
[2] *Max-Planck-Institut für Astrophysik, F.R.G.*
[3] *JILA, University of Colorado, U.S.A.*

1 Introduction

A two stage mechanism for fuelling AGN is proposed which makes use of stellar dynamical and gas dynamical instabilities (Shlosman *et al.* 1989). First, a stellar bar sweeps the interstellar material inwards as a consequence of the gas losing angular momentum to the bar. In a second stage, the gaseous disc accumulated in the nuclear region of the host galaxy, goes bar unstable again and the material flows further in. The main criterion for the occurence of the second instability is that the mass of the gaseous disc must exceed some critical fraction of the total mass of the host galaxy. This critical mass fraction is of the order of a tenth or so according to our estimates. The inflowing gas may eventually join a viscosity-driven accretion disc if a black hole was already present or lead to its formation. If the host galaxy is relatively gas rich, but the disc formed during the first stage does not exceed the critical mass, or if the inflow of gas is halted around resonances, a nuclear starburst may follow. This mechanism may explain the association of bars and rings with nuclear activity and the dichotomy between AGN and starburst nuclei.

2 Gas in galaxies

The atomic and molecular gas content of spiral galaxies is observed to be in the range $10^8 - 10^{10}$ $M_\odot$ and to peak at Hubble types Sb-Sbc (Haynes & Giovanelli 1984, Verter 1987). Moreover, in a galaxy of moderate mass ($M_G \sim 10^{12}$ $M_\odot$) some ~ 10 $M_\odot$ are released per year as a result of normal stellar evolution (Begelman *et al.* 1984). This is sufficient to power an AGN by accretion on to a massive black hole especially if the activity has a short duty cycle (Begelman *et al.* 1984). The problem is that most of this material is released in the main body of the host galaxy at distances of several kpc from the nucleus and has too much angular momentum. Even if the host is an elliptical galaxy with a small fraction ϵ of its support against gravity provided by rotation, the material released at some radius r_0 is likely to circularize at $r_{circ} \sim \epsilon^{1/2} r_0$. For reasonable assumptions the timescale for viscous inflow exceeds the Hubble time if $r_{circ} \gtrsim 100$ pc (Shlosman & Begelman 1987). In spiral hosts a large scale stellar bar, which may be a relict or tidally induced, may sweep the interstellar medium inwards typically to distances ~ 0.1 of the bar radius if no inner Lindblad resonances are encountered on the way (van Albada & Roberts 1981, Schwarz 1985, Matsuda *et al.* 1987). Further evolution by viscosity would still be far too slow.

3 Model

However, gas accumulated in the nuclear region is rotationally supported and may become vulnerable to non-axisymmetric instabilities if its own gravity dominates the dynamics. We have modelled the situation by a gaseous disc of characteristic length scale a embedded in a fixed stellar potential of radial scale b. A well known semi-empirical global stability criterion (Ostriker & Peebles 1973) can then be used to obtain first estimates of the conditions required for the gaseous disc to go unstable again. It is easy to show that the gaseous disc is likely to go unstable if its mass exceeds a critical fraction of the total galactic mass which is approximately $M_{gas} \gtrsim 0.6 M_G (a/b)^{1/2}$. Therefore, if a spiral host containing $\sim 20\%$ of its mass in gas developed a bar which swept all the gas into a disc with $a \sim 0.1b$, then this disc would be prone to global instabilities. Similarly, the gas accumulated at the centre of an elliptical host with $\epsilon \sim 0.01$ would become globally unstable when the mass collected in the central disc exceeded $\sim 20\%$ of the total.

If a spiral host contains less than the critical fraction of gas, or if the large scale stellar bar is inefficient in sweeping the material in because it is weak or has inner Lindblad resonances, the second stage of instability in the gas disc will not occur. It is then likely that further evolution leading to the formation of a black hole and/or its fuelling would be suppressed. The gas accumulated in the nuclear regions may then yield a starburst galaxy in agreement with recent evidence for most starbursts being barred (Jackson *et al.* 1987). The higher frequency of AGN and starburst galaxies in intermediate morphological classes (Haynes & Giovanelli 1984, Verter 1987) may be the result of their relatively high gas content.

In this picture the presence of a large scale bar in a spiral host is a favourable pre-condition for both starburst and AGN activity. Hence a statistical association between these is natural (Jackson *et al.* 1987, Simkin *et al.* 1980, MacKenty 1989). However the second stage of small-scale dynamical instability is needed for further evolution to an AGN. Thus evidence of dynamical instabilities on ~ 100 pc scales is expected in the nuclear regions of active spirals and ellipticals.

References

Begelman, M. C., Blandford, R. D. & Rees, M. J., 1984. *Rev. Mod. Phys.,* **56**, 255.

Haynes, M. P. & Giovanelli, R., 1984. *Astron. J.,* **89**, 758.

Jackson, J. M., Barrett, A. H., Armstrong, J. T. & Ho, P. T. P., 1987. *Astron. J.,* **93**, 531.

MacKenty, J., 1989. *Astrophys. J.,* (in press).

Matsuda, T., Inoue, M., Sawada, K., Shima, E. & Wakamatsu, K., 1987. *Mon. Not. R. Astron. Soc.,* **229**, 295.

Ostriker, J. P. & Peebles, P. J. E., 1973. *Astrophys. J.,* **186**, 467.

Schwarz, M. P., 1985. *Mon. Not. R. Astron. Soc.,* **212**, 677.

Shlosman, I. & Begelman, M. C., 1987. *Nature,* **329**, 810.

Shlosman, I., Frank, J. & Begelman, M. C., 1989. *Nature,* **338**, 45.

Simkin, S., Su, H. & Schwarz, M. P., 1980. *Astrophys. J.,* **237**, 404.

van Albada, G. D. & Roberts, W. W., 1981. *Astrophys. J.,* **246**, 740.

Verter, F., 1987. *Astrophys. J. Suppl. Ser.,* **65**, 555.

The circum-nuclear disc in the Galactic centre

Wolfgang J. Duschl

Universität Heidelberg, FR Germany

1 Introduction

A striking feature of the distribution of mass in the central region of the galaxy is the existence of a cavity with a radius of about 2 pc surrounded by a gas disc. The transition from the disc to the cavity region is comparatively sharp (for a recent review, see Genzel & Townes 1987). One asks how, in this environment, the gas disc maintains its sharp edge and the central cavity is kept (almost) free from gas?

Here, I propose that the inner edge of the circum-nuclear disc results from a change in the radial accretion velocity of the material due to a change in the gradient of the underlying galactic mass distribution. This mechanism also produces a self-sustaining cavity. Furthermore the sharpness of the edge gives information about the mass distribution in the very central regions of the galaxy.

2 The disc

The circum-nuclear disc is described by a geometrically thin, axisymmetric accretion disc with constant kinematic viscosity. The underlying mass distribution is assumed to be spherically symmetric. The accretion time scale, τ_{accr}, for arbitrary spherically symmetric mass distributions is

$$\tau_{\mathrm{accr}}(s) = \frac{1}{\nu(s)} \left| \frac{\left[s^2 \Omega(s) \right]'}{\Omega'(s)} \right|$$

(Duschl 1988). Here, $\nu(s)$ is the kinematic viscosity at a (cylindrical) radial distance s from the centre, $\Omega(s)$ is the angular velocity deduced from the mass distribution and a prime denotes a derivative with respect to s. The disc is not self-gravitating.

3 The sharpness of the inner edge

The mass, $M(r)$, within (spherical) radius r, is approximated by analytical expressions. In the following, I discuss two cases, M_A and M_B:

$$M_A(r) = \begin{cases} M_1 & \text{if } r \leq r_0, \\ M_2 \cdot x^a & \text{if } r > r_0, \end{cases} \qquad M_B(r) = M_1 + M_2 \cdot x^a.$$

Here x is the radial distance from the centre in units of 1 pc. To fit the observed $M(r)$, I choose: $M_1 = 10^{6.47}$ M$_\odot$, $M_2 = 10^{6.12}$ M$_\odot$, $r_0 = 1.79$ pc, and $a = 1.38$.

The mass distribution M_A has a very sharp transition from the constant contribution of the central object to the power law distribution while M_B shows a much smoother transition. Both models give acceptable fits to the observational data, although M_A is perhaps a better representation than M_B.

In the case of M_A,

$$\frac{\tau_{\mathrm{accr}}(s) \cdot \nu(s)}{s^2} = \begin{cases} \frac{1}{3} & \text{if } s \leq r_0, \\ \left(\frac{a+1}{3-a} \right) & \text{if } s > r_0. \end{cases}$$

This means that the corresponding radial velocity increases strongly inwards, and consequently the surface density drops drastically. In this approximation an essentially infinitely sharp edge of the disc is produced.

In the case of M_B ($\mu = M_2/M_1$):

$$\frac{\tau_{\mathrm{accr}}(s) \cdot \nu(s)}{s^2} = \left| \frac{1 + (1+a)\mu x^a}{3 + (3-a)\mu x^a} \right|.$$

Thus, a smooth change in the gradient of the mass distribution still gives rise to a quite strong change in radial velocity over a small radial distance, *i.e.* a sharp inner edge.

So one has the contribution from the change of the radius which brings an inward decrease of the accretion time scale ($\propto s^{-2}$), *i.e.* an increase of the radial velocity towards the centre. Additionally one has the increase due to the change of the radial gradient of the power law, *i.e.* due to the transition to inner regions where a highly condensed central object dominates $M(r)$.

This second effect sharpens the edge: the sharpness of the edge depends on the length scale over which $M(r)$ changes. One can go a step further and use the sharpness of the edge to measure the change in the gradient of the mass distribution.

4 Conclusions

It is shown that a change in the gradient of the mass distribution can cause the sharp edge of the circum-nuclear accretion disc. The smaller the region over which the gradient of the potential changes, the sharper the edge will be. So it is possible to determine the change in the gradient of the potential, *i.e.* the gradient of the mass distribution, by measuring the radial extent of the disc edge. The fact that one observes a sharp edge means in this model that the transition from the domain of the central object to the power law mass distribution is correspondingly sharp. This means that the mass within about the central 2 pc of the galaxy is small compared to the mass of the central object.

So we have an independent confirmation that the mass distribution in the galactic centre is such that the central $\sim 10^6$ M$_\odot$ are confined in a region much smaller than about 1 pc.

It is important to notice that a mechanism like that proposed in this paper produces a self-sustaining cavity in the central region of the galaxy. The material falling from the disc's inner edge towards the centre has – because of it's higher radial velocity – only a comparatively short time before it will be swallowed by a central mass concentration. Because of the clumpiness of the material (even in the disc) it is not clear whether inside the cavity any type of accretion disc picture is sensible or whether radial motion of the material of the then disrupted clumps predominates. The observations of the streamers in the central cavity point strongly in the latter direction.

References

Duschl, W. J., 1988. *Astron. Astrophys.*, **194**, 33.
Genzel, R. & Townes, C. H., 1987. *Annu. Rev. Astron. Astrophys.*, **25**, 377.

Non-axisymmetric instabilities in thin self-gravitating differentially rotating gaseous discs

J. C. B. Papaloizou[1] and **G. J. Savonije[2]**
[1] *Queen Mary College, London*
[2] *Universiteit van Amsterdam, The Netherlands*

Abstract We discuss the linear theory of non-axisymmetric normal modes in self-gravitating gaseous discs. These instabilities occur when the disc is stable to axisymmetric modes. They can have co-rotation situated either inside or outside the disc. The profile of the ratio of vorticity to surface density is found to be important in determining the properties of the normal modes. These modes may be important for redistributing the angular momentum in the disc.

1 Introduction

Discs and rings in which the internal self-gravity plays an important role are important in astronomy. Examples are the rings around Saturn and Uranus, (Goldreich & Tremaine 1982), and the Milky Way and other spiral galaxies (Toomre 1977, 1981). They may also exist around active galactic nuclei and T. Tauri stars. In both of these cases, instabilities may be important for driving mass accretion and angular momentum transport (Paczynski 1977, Lin & Pringle 1987). An understanding of non-axisymmetric instabilities is clearly important because they may play a significant role in determining the structure and evolution of all of these objects. In this paper we discuss the linear theory of stability as applied to self-gravitating gaseous discs. We find various kinds of instabilities, some of which are generalizations of those found in the non-self-gravitating case (Papaloizou & Pringle, 1984, 1985, 1987). These are essentially due to the unstable interaction of waves on either side of co-rotation. However, when self-gravity is included, there are other modes which have co-rotation outside the system. For these, the behaviour of the ratio of surface density to vorticity plays an important role.

2 Basic Equations

The basic equations governing the dynamics of a two-dimensional self-gravitating disc in cylindrical coordinates (r, ϕ) are the equations of motion

$$\frac{\partial v_r}{\partial t} + v_r \frac{\partial v_r}{\partial r} + \frac{v_\phi}{r}\frac{\partial v_r}{\partial \phi} - \frac{v_\phi^2}{r} = -\frac{1}{\Sigma}\frac{\partial P}{\partial r} - \frac{\partial \psi}{\partial r},$$ (1)

and

$$\frac{\partial v_\phi}{\partial t} + v_r \frac{\partial v_\phi}{\partial r} + \frac{v_\phi}{r}\frac{\partial v_\phi}{\partial \phi} + \frac{v_\phi v_r}{r} = -\frac{1}{\Sigma r}\frac{\partial P}{\partial \phi} - \frac{1}{r}\frac{\partial \psi}{\partial \phi},$$ (2)

where the velocity $\mathbf{v} = (v_r, v_\phi)$, Σ is the surface density, P is the vertically integrated pressure, and ψ the gravitational potential. In addition we have the continuity equation

$$\frac{\partial \Sigma}{\partial t} + \frac{1}{r}\frac{\partial}{\partial r}[r\Sigma v_r] + \frac{1}{r}\frac{\partial}{\partial \phi}[r\Sigma v_\phi] = 0.$$ (3)

For the work discussed in this paper we suppose there is a simple equation of state such that P is a function of Σ alone. This enables us to define a local sound speed, c, such that

$$c^2 = dP/d\Sigma. \tag{4}$$

We suppose the gravitational potential to be composed of two parts such that

$$\psi = \psi_e + \psi_G, \tag{5}$$

where ψ_e is an external fixed potential due to for example an external point mass, and ψ_G is the potential arising from the disc, which is given by the Poisson integral such that

$$\psi_G = -G \int_0^R \int_0^{2\pi} \frac{\Sigma(r', \phi')r'd\phi'dr'}{\sqrt{r^2 + r'^2 - 2rr'\cos(\phi - \phi')}}, \tag{6}$$

where R is the radius of the outer boundary of the disc.

3 Non-axisymmetric normal modes in a self-gravitating gaseous disc

3.1 LINEARIZATION

We here discuss some of the different kinds of non-axisymmetric normal modes and instabilities that may occur in an initially axisymmetric, differentially rotating, disc for which

$$\mathbf{v} = (0, r\Omega),$$

where Ω is the angular velocity. In the unperturbed axisymmetric disc, P, Σ, and ψ are functions of r alone. We linearize the basic equations by writing

$$v_r = v_r'(r)\exp(im\phi + i\sigma t),$$
$$v_\phi = r\Omega + v_\phi'(r)\exp(im\phi + i\sigma t),$$
$$\Sigma = \Sigma(r) + \Sigma'(r)\exp(im\phi + i\sigma t),$$
$$P = P(r) + c^2\Sigma'(r)\exp(im\phi + i\sigma t)$$

and

$$\psi = \psi(r) + \psi'(r)\exp(im\phi + i\sigma t),$$

where m is the azimuthal mode number and σ the complex eigenfrequency.

Here, perturbations to the axisymmetric state are assumed small so that the basic equations may be linearized in them. For an unstable mode, the imaginary part of σ is negative, the linearized equations of motion (1) and (2) then give

$$i\bar\sigma v_r' - 2\Omega v_\phi' = -\partial W/\partial r \tag{6}$$

and

$$i\bar\sigma v_\phi' + v_r'\left(\frac{1}{r}\frac{dh}{dr}\right) = -imW/r, \tag{7}$$

where $h = r^2\Omega$, $W = c^2\Sigma'/\Sigma + \psi_G'$ and $\bar\sigma = \sigma + m\Omega$.

The linearized continuity equation (3) is then

$$i\bar{\sigma}\Sigma' = -\frac{1}{r}\frac{d}{dr}[r\Sigma v_r'] - im\Sigma\frac{v_\phi'}{r}. \tag{8}$$

The perturbed velocities may be eliminated from equations (6) – (8) in order to obtain a relation for Σ' in terms of W alone which may be expressed in the operator form

$$\Sigma' = L(W), \tag{9}$$

where

$$L(W) = -\frac{1}{r}\frac{d}{dr}\left\{\frac{r\Sigma}{D}\left[\frac{dW}{dr} + \frac{2m\Omega\bar{\sigma}W}{\kappa^2 r}\right]\right\} + \frac{2m\Omega\bar{\sigma}\Sigma}{\kappa^2 r D}\left[\frac{dW}{dr} + \frac{2m\Omega\bar{\sigma}W}{\kappa^2 r}\right]$$
$$+ \frac{mW}{r\bar{\sigma}}\frac{d}{dr}\left(\frac{\Sigma r}{h'}\right) - \frac{4m^2\Omega^2\Sigma W}{r^2\kappa^4}. \tag{10}$$

Here κ^2 is the square of the epicyclic frequency defined by $\kappa^2 = (2\Omega/r)dh/dr$, $h' = dh/dr$ and $D = \bar{\sigma}^2 - \kappa^2$.

The perturbed gravitational potential may be found by linearizing the Poisson integral in the form

$$\psi' = \psi_G' = -G\int_0^R K_m(r, r')\Sigma'(r')r'dr', \tag{11}$$

where

$$K_m(r, r') = \int_0^{2\pi} \frac{\cos(m\phi)d\phi}{\sqrt{r^2 + r'^2 - 2rr'\cos\phi}}.$$

From equation (11) we deduce that

$$W = \frac{c^2\Sigma'}{\Sigma} - G\int_0^R K_m(r, r')\Sigma'(r')r'dr'. \tag{12}$$

Equations (10) and (12) form a pair of simultaneous equations for Σ' and W. Assuming we can invert the operator L, we find $W = L^{-1}(\Sigma')$ and we obtain a single equation for Σ' in the form

$$L^{-1}(\Sigma') = \frac{c^2\Sigma'}{\Sigma} - G\int_0^R K_m(r, r')r'\Sigma'(r')dr'. \tag{13}$$

The operator L may be singular, however, and we must investigate the conditions where this occurs more closely.

3.2 THE INVERTIBILITY OF L

We first note that if σ is real, L may be singular at the co-rotation resonance for which $\bar{\sigma} = \sigma + m\Omega = 0$. However, this requires that $d(r\Sigma/h')/dr$ is non-zero at the co-rotation point. We remark that we adopt the convention that m and Ω are positive in this paper, so that σ must be negative for a co-rotation resonance to be possible. The quantity $d(r\Sigma/h')/dr$ is the gradient of the ratio of the surface density to vorticity or inverse of the vortensity. If this gradient is zero at co-rotation, then co-rotation is ineffective as a singularity. We shall assume this to be the case for the time being.

Other potential singularities occur where $D = \bar{\sigma}^2 - \kappa^2 = 0$, which defines the Lindblad resonances. The inner Lindblad resonance occurs where $\bar{\sigma} = \kappa$, and the outer Lindblad resonance where $\bar{\sigma} = -\kappa$. However, it is not difficult to show that the presence of these resonances does not introduce singularities into L, viewed as a second order differential operator, if the co-rotation resonance is ineffective as described above. That is for a ring, or disc with vanishing surface density at any boundary, the equation $\Sigma' = L(W)$ has a non-singular solution for W provided the equation $L(W) = 0$ does not have a non-trivial regular solution and, of course, Σ' is sufficiently well behaved. We remark that if Σ vanishes at any boundary, going to zero, for example as a power of the distance to that boundary, σ is real and the co-rotation singularity is ineffective, L is a self-adjoint operator with weight r. That is, for any two regular functions W_1 and W_2, we have

$$\int_0^R r W_1^* L(W_2) dr = \int_0^R r W_2 L^*(W_1^*) dr.$$

We may seek conditions under which $L(W) = 0$ has a non-trivial regular solution. This condition implies $\Sigma' = 0$, and hence from equation (8)

$$v'_\phi = -\frac{1}{im\Sigma} \frac{d(r\Sigma v'_r)}{dr}.$$

From equations (6) and (7) we deduce, by eliminating W, that

$$i\bar{\sigma} v'_r - 2\Omega v'_\phi = \frac{d}{dr}\left(\frac{\bar{\sigma}}{m} v'_\phi - i\frac{dh}{dr}\frac{v'_r}{m}\right).$$

Hence we find that $U = \Sigma r v'_r$ satisfies

$$\Sigma r \frac{d}{dr}\left(\frac{r}{\Sigma}\frac{dU}{dr}\right) = U\left[m^2 + \frac{m}{\bar{\sigma}}\Sigma r \frac{d}{dr}\left(\frac{1}{r\Sigma}\frac{dh}{dr}\right)\right]. \tag{14}$$

From equation (14) it is easy to see that if co-rotation is not on a disc boundary, there is no acceptable solution for U with an associated regular W at the boundaries, if either the vortensity gradient is zero, or sufficiently small. In any case, provided Σ is differentiable this gradient always has a small effect for large enough m. As our discussion applies for general m, we may quite generally assume that no acceptable solution for U exists and that L is invertible in what follows.

Because L is a second order differential operator, it may be inverted by the standard Green's function technique. We construct the Green's function by considering the eigenvalues, λ_n, and associated eigenfunctions, u_n, $n = 1, 2, 3 \ldots$, defined by the equation

$$L(u_n) = \lambda_n u_n.$$

Because of the self-adjointness of L, these eigenfunctions are orthogonal and may be defined such that

$$\int_0^R r u_m^* u_n dr = \delta_{nm}.$$

If there is a Lindblad resonance in the disc where $\bar{\sigma}^2 = \kappa^2$, it is easy to see that L has both positive and negative eigenvalues. We denote the positive sequence only by

λ_n, $n = 1, 2, 3 \ldots$, and the negative sequence by μ_n, $n = 1, 2, 3 \ldots$; λ_n tends to infinity and μ_n tends to minus infinity as n tends to infinity. The eigenfunctions associated with the λ_n are non-oscillatory in the region between the Lindblad resonances and oscillatory exterior to an outer Lindblad resonance and interior to an inner Lindblad resonance. The eigenfunctions associated with the μ_n are oscillatory only in the region between Lindblad resonances.

In terms of these eigenfunctions, the Green's function may be written as

$$H(r', r) = \sum_{n=1}^{\infty} \frac{u_n^*(r') u_n(r)}{\lambda_n} + \sum_{n=1}^{\infty} \frac{w_n^*(r') w_n(r)}{\mu_n}, \qquad (15)$$

where u_n and w_n denote eigenfunctions associated with λ_n and μ_n respectively. Since $L(W) = \Sigma'$, we have

$$W = \int_0^R r' \Sigma'(r') H(r', r) dr'. \qquad (16)$$

Combining equations (13) and (16), we obtain a single integral equation in the form

$$\frac{c^2 \Sigma'}{\Sigma} = \int_0^R r' \Sigma'(r') S(r', r) dr', \qquad (17)$$

where the kernel $S(r, r') = G K_m(r, r') + H(r, r')$ is symmetric in r and r' for real σ.

3.3 DISCUSSION OF THE EIGENVALUE PROBLEM

The eigenvalue problem as originally formulated contains σ as an eigenvalue. However, this is contained in equation (17) in a complicated manner. In addition, if real σ are considered, σ must be fixed so that the co-rotation singularity is ineffective. In this situation we find it more convenient to consider σ as fixed initially and to work on the eigenvalue problem

$$\frac{\bar{\lambda} c^2 \Sigma'}{\Sigma} = \int_0^R r' \Sigma'(r') S(r', r) dr', \qquad (18)$$

where $\bar{\lambda}$ is now the eigenvalue. $\bar{\lambda}$ can be thought of as determining the scale of the sound speed, or pressure, for which a normal mode exists for a prescribed σ. We remark that an equilibrium can be preserved while this scale is varied by introducing a suitable fixed background potential. It turns out that by considering the variation of the eigenvalues $\bar{\lambda}$ with other parameters in the problem, we can infer the existence of instabilities. Equation (18) can be rewritten so that the $\bar{\lambda}$ are the discrete eigenvalues of a self-adjoint Fredholm operator for which well known variational principles exist, (see Courant & Hilbert 1953). To do this, set

$$y - \Sigma' \left(\frac{r c^2}{\Sigma} \right)^{1/2},$$

so that equation (18) becomes

$$\bar{\lambda} y(r) = \int_0^R I(r, r') y(r') dr',$$

where

$$I(r,r') = S(r,r')rr'\left[\frac{\Sigma(r)}{rc(r)^2}\right]^{1/2}\left[\frac{\Sigma(r')}{r'c(r')^2}\right]^{1/2}$$

is a self-adjoint Fredholm kernel.

In general the spectrum will consist of a decreasing positive sequence $\bar{\lambda}_i$, $i = 1, 2, 3 \ldots$, and an increasing negative sequence $\bar{\mu}_i$, $i = 1, 2, 3 \ldots$, both sequences having accumulation points at zero. From the variational principle $\bar{\lambda}_1$ is the maximum possible value of

$$Q = \frac{\int_0^R \int_0^R rr'\Sigma'(r')\left(\Sigma'(r)\right)^* S(r,r')dr dr'}{\int_0^R rc^2|\Sigma'|^2/\Sigma dr} \tag{19}$$

with respect to variations of Σ'. Further, the nth eigenvalue, $\bar{\lambda}_n$ is the maximum of Q with respect to variations of Σ' subject to the constraint of orthogonality with weight (rc^2/Σ) to all lower eigenfunctions associated with the positive eigenvalue sequence. Using these principles, we discuss the eigenvalue spectrum of a differentially rotating gaseous disc.

4 Unstable normal modes

4.1 UNSTABLE MODES ASSOCIATED WITH EXTREMA IN THE RATIO OF VORTICITY TO SURFACE DENSITY

In order to apply the above formalism, σ must be real and the co-rotation singularity ineffective. One way of achieving this is to have co-rotation located at an extremum of the ratio of vorticity to surface density. In this situation one can show that there will in general be an infinite sequence of positive eigenvalues, $\bar{\lambda}_i$, $i = 1, 2, 3 \ldots$, with an accumulation point at zero. To see this, we remark that after using equations (17) and (15), equation (19) can be written as

$$Q = \left\{ G\int_0^R \int_0^R rr'\Sigma'(r')\left(\Sigma'(r)\right)^* K_m(r',r)dr dr' \right.$$
$$\left. + \sum_{n=1}^\infty \left[\left|\int_0^R \Sigma' r u_n^* dr\right|^2 \frac{1}{\lambda_n} + \left|\int_0^R \Sigma' r w_n^* dr\right|^2 \frac{1}{\mu_n} \right] \right\} \Big/ \int_0^R \frac{rc^2|\Sigma'|^2}{\Sigma}dr,$$

as long as a Lindblad resonance exists in the disc. Using the fact that the first (self-gravity) term is positive definite, and considering the variational principles, but restricting ourselves to trial functions for Σ' that are linear combinations of the u_n, which are orthogonal to the w_n, one readily verifies the existence of the infinite positive sequence of eigenvalues $\bar{\lambda}_i$, $i = 1, 2, 3 \ldots$ and associated eigenfunctions, because all the λ_n are positive.

This discrete sequence of normal modes can be understood as follows. The local dispersion relation governing neutral disturbances in a gaseous disc derived in the WKB approximation (see Toomre 1964) is

$$(\sigma + m\Omega)^2 = \kappa^2 - 2\pi G\Sigma|k| + c^2 k^2, \tag{20}$$

where k is the radial wavenumber. The eigenvalues derived from equation (18) and the associated variational principle (19) correspond to finding the values of $\bar\lambda$ for a fixed σ for which the equation

$$(\sigma + m\Omega)^2 = \kappa^2 - 2\pi G\Sigma|k| + \bar\lambda c^2 k^2, \qquad (21)$$

describes a normal mode. We may expect these satisfy a condition of the form

$$\int_{r_1}^{r_2} k(r,\bar\lambda)dr = n\pi + \delta, \qquad (22)$$

where $k(r,\bar\lambda)$ is obtained by solving (21), n is an integer, δ is a constant, and r_1 and r_2 are boundaries of a wave propagation zone in the disc. As long as at least one Lindblad resonance exists in the disc, the region between this resonance and the disc boundary, which does not contain co-rotation, will be a propagation region, even when self-gravity is unimportant. We then expect there to be critical scales of the sound speed, or values of $\bar\lambda$, for which normal mode conditions (22) are satisfied in this region. These then belong to the discrete spectrum we have derived rigorously above, assuming that co-rotation is located at an extremum of the ratio of vorticity to surface density (so that there is no singularity there).

The above argument establishes that a spectrum of neutral modes exists for different scales of the sound speed. We now show that instability will exist for parameters in the neighbourhood of those appropriate to a neutral mode. We suppose a neutral mode exists, the sound speed to be rescaled so $\bar\lambda = 1$, and then let $c^2 \to c^2 + \delta c^2$, and $\sigma \to \sigma + \delta\sigma_R - i\gamma$, where $\delta\sigma_R$ is the real part of the corresponding change to σ and γ is the small growth rate, assumed positive. To evaluate γ, we return to the eigenvalue problem defined by equations (9) and (12) which apply for complex σ. From these one may easily show that

$$
\begin{aligned}
X &= \int_0^R r W^* L(W) dr \\
&= \int_0^R \frac{r\Sigma}{D}\left[\frac{dW}{dr} + \frac{2m\Omega\bar\sigma W}{\kappa^2 r}\right]\left[\frac{dW^*}{dr} + \frac{2m\Omega\bar\sigma W^*}{\kappa^2 r}\right] dr \\
&\quad + \int_0^R |W|^2 \left[\frac{m}{\bar\sigma}\frac{d}{dr}\left(\frac{r\Sigma}{h'}\right) - \frac{4m^2\Omega^2\Sigma}{r\kappa^4}\right] dr,
\end{aligned}
\qquad (23)
$$

and

$$X = \int_0^R r W^* \Sigma' dr = \int_0^R \frac{c^2 r}{\Sigma}|\Sigma'|^2 dr - G\int_0^R\int_0^R rr' K_m(r,r')\Sigma'(r')\left(\Sigma'(r)\right)^* dr dr'. \quad (24)$$

Here the integrals are taken over the disc. Letting $c^2 \to c^2 + \delta c^2$, $\sigma \to \sigma + \delta\sigma_R - i\gamma$, $\Sigma' \to \Sigma' + \delta\Sigma'$, $W \to W + \delta W$, and taking the first order variations of equations (23) and (24), we find

$$\frac{\partial X}{\partial\sigma}(\delta\sigma_R - i\gamma) = \int_0^R \frac{r|\Sigma'|^2\delta c^2}{\Sigma} dr. \qquad (25)$$

To evaluate $\partial X/\partial\sigma$ for real σ, we use the Landau prescription for $\bar\sigma^{-1}$, which arises in the potentially singular co-rotation term, in the limit $\gamma \to 0$.

Thus $\bar{\sigma}^{-1} = P(\bar{\sigma}^{-1}) + \pi i \delta(\bar{\sigma})$, where P denotes the principal value, and δ is Dirac's delta function. Then

$$\frac{\partial X}{\partial \sigma} \to \frac{\partial X}{\partial \sigma} - i\pi \left[\frac{|W|^2}{|m\Omega'||\Omega'|} \frac{d^2}{dr^2} \left(\frac{r\Sigma}{h'} \right) \right]_{\mathrm{CR}} = \frac{\partial X}{\partial \sigma} - iF,$$

where $\partial X / \partial \sigma$ is now evaluated after having used the principal value prescription to work out X. Accordingly it is real, and the subscript CR denotes evaluation at the original co-rotation point of the neutral mode. Then equation (25) gives

$$\gamma = -F \int_0^R \frac{r|\Sigma'|^2 \delta c^2}{\Sigma} \left[F^2 + \left(\frac{\partial X}{\partial \sigma} \right)^2 \right]^{-1}. \tag{26}$$

From equation (26) it is apparent that we can choose the sign of δc^2 so that γ is positive as assumed. This establishes that weakly unstable modes can exist with co-rotation points close to extrema in the gradient of the ratio of vorticity to surface density and that when co-rotation is at such an extremum we can have marginal stability. This situation appears to occur in the numerical calculations described by us elsewhere in these proceedings.

4.2 RESONANT INSTABILITIES

A second type of instability, which we refer to as a resonant instability, corresponds to the interaction between waves located on either side of an evanescent zone centred on co-rotation. The interaction is one between positive and negative energy waves and is therefore destabilizing. Instabilities of this kind have been found in the case of non self-gravitating discs, (Papaloizou & Pringle 1984, 1985, 1987). However, the evanescent zone narrows as self-gravity becomes important. From equation (20), one finds the boundaries of the evanescent zone are given by

$$(\sigma + m\Omega)^2 = \kappa^2 \left(1 - \frac{1}{Q^2} \right),$$

where the Toomre stability parameter $Q = \kappa c/(\pi G\Sigma)$. As $Q \to 1$ we see that the evanescent zone shrinks to zero width. Thus we expect the resonant instabilities to strengthen as self-gravity becomes important.

Resonant instabilities are most readily found when the gradient of the ratio of the vorticity to surface density is negligible or zero. We therefore consider a model for which $\frac{dh}{dr}/(r\Sigma)$ is constant apart perhaps for regions of very small radial extent near the disc boundaries. We return to the eigenvalue problem defined by equations (18) and (19). Because $\frac{dh}{dr}/(r\Sigma)$ is constant, we can allow σ to be real and to vary continuously such that co-rotation moves through the bulk of the disc without singularities. We may then monitor the eigenvalue, $\bar{\lambda}$, as a function of σ.

Consider the behaviour of the largest eigenvalue when co-rotation is close to the inner boundary of the disc. We suppose m is large enough so that the outer Lindblad resonance is well inside the disc. The largest $\bar{\lambda}$ then corresponds to choosing the scaling of the sound speed so that the lowest order mode fits between the Lindblad resonance and the outer edge. If σ now increases (*i.e.* becomes less negative) so that co-rotation moves further into the disc, the outer propagation region decreases in extent. From the WKB relation (22), in which the appropriate short wave branch of the dispersion relation is used, we expect $\bar{\lambda}$, or the sound speed scaling, to decrease accordingly. Suppose now co-rotation is chosen to be close to the outer disc boundary. The propagation region is now in the inner region, interior to the inner Lindblad resonance. As before the largest eigenvalue, $\bar{\lambda}$, corresponds to a sound speed scaling such that the lowest order mode just fits in. Further, if σ decreases, then by similar arguments to those used above, the propagation region decreases in extent, and $\bar{\lambda}$ should decrease. In this case $\bar{\lambda}$ is an increasing function of σ, while in the previous case with co-rotation near the inner boundary $\bar{\lambda}$ was a decreasing function of σ.

For the eigenvalue problem considered for general complex σ, $\bar{\lambda}$ is an analytic function of σ. So for some intermediate value of σ we expect a minimum at which $d\bar{\lambda}/d\sigma = 0$. Suppose this occurs for $\sigma = \sigma_0$, and $\bar{\lambda} = \bar{\lambda}_0$. We may make a Taylor expansion about these values, thus

$$\bar{\lambda} = \bar{\lambda}_0 + \left[\frac{1}{2}\frac{\partial^2\bar{\lambda}}{\partial\sigma^2}\right]_{\sigma_0} (\sigma - \sigma_0)^2 \ldots .$$

From this we expect complex σ, corresponding to instability for $\bar{\lambda} < \bar{\lambda}_0$. At $\sigma = \sigma_0$, we have the merging of two modes. One is associated with the outer propagation region and the other with the inner propagation region. We note that in the above discussion we considered the mode with the largest $\bar{\lambda}$. However, it also applies to the mode with the second largest value of $\bar{\lambda}$ and then the third largest and so on. We therefore expect many kinds of resonant modes.

The above analysis is most easily demonstrated for a slender annulus model rather than a complete disc. There may be a central point mass in this case so that the rotation is essentially Keplerian.

We write $r = R_0 + x$, where R_0 is the radius of the centre of the annulus. Then for a slender annulus, with radial extent much less than R_0, we approximate

$$\bar{\sigma} = \sigma + m\Omega(R_0) + m\left[\frac{d\Omega}{dr}\right]_{R_0} x$$

and take Σ to be a symmetric function of x. In addition, we approximate dh/dr to be constant and equal to its value at the ring centre. Then in equation (10), we replace $d/dr \to d/dx$ and all quantities apart from $\bar{\sigma}$ and Σ are replaced by their values at the ring centre. In the slender ring approximation the self-gravity kernel may be approximated by

$$K_m(r,r') = \frac{2}{R_0}K_0\left(\frac{m|x - x'|}{R_0}\right),$$

K_0 denoting the usual Bessel function. With these replacements, and setting $r = r' = R_0$ where they appear explicitly, we may define the eigenvalue problem for $\bar{\lambda}$ as given by equations (18) and (19).

It is easily seen that when co-rotation is at $r = R_0$, the eigenfunctions must be either even or odd in x, because Σ is an even function of x. In either case we have

$$\frac{d\bar{\lambda}}{d\sigma} = 0$$

when co-rotation is at $r = R_0$ by symmetry. The Taylor expansion argument then shows this is a state of marginal stability as far as variation of $\bar{\lambda}$ is concerned.

4.3 MODES WITH CO-ROTATION POINT EXTERIOR TO THE DISC

We next discuss a class of unstable modes for which the co-rotation resonance is outside the disc. Although these are more readily discussed in the well known softened gravity model (see Erickson 1974, and below), we first describe a situation for which they can occur in a gaseous disc.

We return to the eigenvalue problem defined by equations (18) and (19) for the sound speed scaling parameter $\bar{\lambda}$. If $\bar{\sigma}$ is real and co-rotation is just beyond the outer boundary, there is no singularity, and we may find a spectrum of neutral normal modes. The different values of $\bar{\lambda}$ correspond to scalings of the sound speed for which normal modes exist. We suppose that m is such that if co-rotation is at the outer boundary, there is an extensive region in the disc interior to the inner Lindblad resonance. Suppose in addition that the sound speed increases on entering the disc from outside and the Toomre Q value is only slightly greater than unity just inside the boundary. Then we can arrange a situation such that the disc region near co-rotation is a WKB propagation zone. If the sound speed increases there can be a transition to evanescence before the inner Lindblad resonance is reached. A propagation zone then occurs interior to the resonance. This is just the Q gradient effect discussed by Lau, Lin & Mark (1976). It is easy to show from the eigenvalue problem, equation (18), that for any mode, if the Q and Ω profiles are fixed, $\bar{\lambda}$ ultimately increases as the sound speed scaling decreases so that a spectrum of modes with $\bar{\lambda} = 1$ and with Q close to unity at the outer boundary may always be found. From the discussion above, we expect there to be two propagation zones, one near the outer boundary and one interior to the inner Lindblad resonance.

We now show, just as for the resonant modes, that the interaction between waves in these two regions is expected to be unstable. To do this, we take the mode with $\bar{\lambda} = 1$ and examine how σ changes consequent on a small constant change δc^2 to c^2. We use equations (23) and (24) which give

$$\frac{\partial X}{\partial \sigma} \delta\sigma = \int_0^R \frac{\delta c^2 r |\Sigma'|^2}{\Sigma} dr.$$

We reasonably suppose for a thin disc that $|dW/dr| \gg |mW/r|$. Then

$$\frac{\partial X}{\partial \sigma} = \int_0^R \frac{-2r\Sigma\bar{\sigma}}{D^2} \left|\frac{dW}{dr}\right|^2 dr - \int_0^R \frac{m}{\bar{\sigma}^2} |W|^2 \frac{d}{dr}\left(\frac{r\Sigma}{h'}\right) dr. \tag{27}$$

If the disturbance is located primarily interior to the Lindblad resonance, then the first term in equation (27) is dominant, provided $(\Sigma r/h')$ varies smoothly, and we have $\partial X/\partial\sigma < 0$. On the other hand if the disturbance is located in the outer propagation zone, being exponentially small in the inner zone, the second term in equation (27) may dominate. In fact this may become unbounded as co-rotation approaches the boundary from the outside, if Σ vanishes linearly with distance to the boundary. Then $\partial X/\partial\sigma > 0$. Thus for a mode located in the inner region and constant δc^2, we have $d\delta c^2/d\sigma < 0$, while for one located in the outer region $d\delta c^2/d\sigma > 0$. This indicates that these modes will interact unstably. A resonance between them should occur at a minimum in the sound speed scaling and instability results if this scaling is reduced further. For this to be possible the gradient of the ratio of surface density to vorticity must play an important role in the outer disturbance which then acts like a positive energy disturbance which can interact unstably with the negative energy disturbance which in a gaseous disc exists predominantly interior to the inner Lindblad resonance.

Modes of this type have recently been studied in the context of thin annuli in a state of almost Keplerian rotation by Papaloizou & Lin (1989) using the softened gravity model. In this model there is zero pressure but the gravitational potential is softened so that the potential between two mass points is proportional to $(r^2 + b^2)^{-1/2}$ rather than the usual r^{-1}. This model has wave propagation properties similar to those in the gaseous disc in the region between the Lindblad resonances; but it does not allow short waves to propagate outside this region as they would in a gaseous disc. Instead these waves are absorbed at Lindblad resonances.

Unstable modes resulting from the interaction of a disturbance associated with the gradient of the ratio of surface density to vorticity and one associated with the inner Lindblad resonance were demonstrated to occur analytically and were found numerically. Instabilities of the resonant type were found when the inner Lindblad resonance was interior to the inner boundary of the annulus. However, they persisted when the resonance was inside the annulus and the inner disturbance damped. We note that damping of the inner disturbance is destabilising for a mode which is predominant in the exterior region because the disturbances act as if they carried opposite signs of energy. We therefore expect instabilities to exist in the gaseous disc if the inner boundary condition is replaced by a radiation condition. In other words an inner boundary is not essential to the instability. Further, the work on the softened gravity disc also showed that it was not necessary for the surface density to fall to zero at the outer boundary. Provided it falls to a uniform value at high Q the exterior region is evanescent and unable to affect the interior situation. In this situation it is possible to find unstable modes which are essentially independent of the boundaries.

5 Discussion

We have discussed three types of non-axisymmetric normal modes that can occur in thin self-gravitating discs. The first type were associated with extrema in the ratio of surface density to vorticity. The second type were associated with resonances between modes located in propagation zones on either side of co-rotation. Both these types of modes occur in non-self-gravitating discs and therefore for high Q. However, the instabilities are expected to be stronger when self-gravity becomes important because the width of the intermediate evanescent region decreases.

The third kind of mode we have discussed has co-rotation exterior to the disc. This situation can only occur when self-gravity is included. The instability is due to an unstable interaction between a disturbance associated with the gradient of the ratio of surface density to vorticity and a disturbance interior to the inner Lindblad resonance. It has also been found in a softened gravity model disc. In that case the instability survives even if there is damping at the inner Lindblad resonance. Further, it is not necessary that the surface density fall to zero at the outer boundary. This may level off at a lower value. This instability may also exist when the disc is stable to axisymmetric modes. Further progress in this area depends on non-linear calculations of the effects of these instabilities. Some calculations of this type are reported by us elsewhere in these proceedings. It seems likely instabilities will result in angular momentum transfer outwards and mass flow inwards.

References

Courant, R. & Hilbert, D., 1953. *Methods of Mathematical Physics*, New York, Interscience.
Erickson, S. A., 1974. *PhD thesis*, MIT.
Goldreich, P. & Tremaine, S., 1982. *Annu. Rev. Astron. Astrophys.*, **20**, 249.
Lau, Y. Y., Lin, C. C. & Mark, J. W. K., 1976. *Proc. Natl. Acad. Sci. U.S.A.*, **73**, 1379.
Lin, D. N. C. & Pringle, J. E., 1987. *Mon. Not. R. Astron. Soc.*, **225**, 607.
Paczynski, B., 1977. *Acta Astron.*, **28**, 91.
Papaloizou, J. C. B. & Lin, D. N. C., 1989. *Astrophys. J.*, in press.
Papaloizou, J. C. B. & Pringle, J. E., 1984. *Mon. Not. R. Astron. Soc.*, **208**, 721.
Papaloizou, J. C. B. & Pringle, J. E., 1985. *Mon. Not. R. Astron. Soc.*, **213**, 799.
Papaloizou, J. C. B. & Pringle, J. E., 1987. *Mon. Not. R. Astron. Soc.*, **225**, 267.
Toomre, A., 1964. *Astrophys. J.*, **139**, 1217.
Toomre, A. 1977. *Annu. Rev. Astron. Astrophys.*, **15**, 437.
Toomre, A., 1981. In *Structure and Evolution of Normal Galaxies*, p. 111, eds Fall, S. M. & Lynden-Bell, D., Cambridge University Press, Cambridge.

Non-linear evolution of non-axisymmetric perturbations in thin self-gravitating gaseous discs

G. J. Savonije[1] and J. C. B. Papaloizou[2]

[1] *Universiteit van Amsterdam, The Netherlands*
[2] *Queen Mary College, London*

1 Introduction

We briefly discuss the results of numerical calculations in which we study the non-linear evolution of (initially small) non-axisymmetric perturbations in a thin self-gravitating gaseous disc. We neglect the thickness of the disc and replace the mass density and thermal pressure by a surface density Σ and a pseudo-pressure P in the plane of the disc. Furthermore, we adopt a polytropic relation $P = K\Sigma^a$ between these two variables, with $a = 2$ throughout our calculations. The matter in the disc rotates about a central point mass $M_c = 0.75M$, while the disc itself contains a mass $M_d = 0.25M$, where M is the total system mass, so that self-gravity of the disc material is non-negligible.

2 Disc structure

We adopt the following exponential density distribution for the axisymmetric equilibrium disc:

$$\Sigma_0(r) = \Sigma_c \exp(1 - \frac{r}{r_0}),$$

where we have introduced cylindrical coordinates (r, ϕ) centred on the central mass. For computational convenience we do not calculate the interior of the disc and limit ourselves to the region $r_0 \leq r \leq R$.

The angular velocity of matter in the disc $\Omega_0(r)$ is determined from the dynamical equilibrium condition:

$$-C_s^2 \frac{d\Sigma}{dr} + r\Sigma_0(r)\Omega_0^2(r) - \Sigma_0(r)\frac{d\Phi}{dr} = 0,$$

where Φ is the gravitational potential. The polytropic constant K and sound speed $C_s = \sqrt{2K\Sigma}$ are chosen such that the minimum value of Toomre's Q parameter in the disc is 1.2. The equilibrium disc is thus stable, but not far from gravitational instability.

Subsequently we introduce a small non-axisymmetric perturbation in the surface density:

$$\Sigma(r, \phi) = \Sigma_0(r)\left[1 + p\sin\frac{\pi(r - r_0)}{(1 - r_0)}\cos(m\phi)\right],$$

where $p = 10^{-2}$ and the azimuthal index m is adopted to be 2 here. Although the azimuthally averaged structure of the disc is stable according to the local Toomre criterion, we find that the imposed small non-axisymmetry renders the disc unstable.

3 Numerical method

We transform the standard set of non-linear hydrodynamic partial differential equations for a 2-D gaseous disc into a set of non-linear algebraic difference equations by introducing a (r, ϕ) grid of (100×64) cells. The equations are written in a conservative form, whereby the total mass and angular momentum are conserved. We solve this set of equations by a Lax-Wendroff finite-difference scheme with second order accuracy in both time and space, using a Cyber 205 computer.

We encountered a numerical problem in treating the small wavelength disturbances. As a result of non-linear effects disturbances of smaller and smaller wavelengths are generated on the grid. In a real gas the disturbances would eventually be dissipated on a microscopic scale. However, in a numerical simulation this process cannot be described and disturbances with a wavelength of order the grid spacing will build up. The second order finite difference scheme shows large dispersion errors for these extreme short wavelength disturbances, and the physical solution is severely mutilated by the accumulation of spurious waves. It took quite some experimentation to develop a numerical smoothing procedure that damps the spurious short wavelength disturbances and does not damage the physical solution. Part of the observed transport of mass and angular momentum through the disc (§4) could be caused by the artificial smoothing. However, we took care to keep this to a minimum and checked that in the case when the non-axisymmetric perturbations remain small (*i.e.* when $Q \gg 1$), mass and angular momentum were locally conserved.

The gravitational potential is calculated by a direct method using fast Fourier transforms in the azimuthal direction, where we have periodic boundary conditions. In the radial direction, we apply boundary conditions that are numerically convenient, *i.e.* we assume the radial velocity, V_r, to be zero at both the inner and outer boundary.

4 Results

We introduce units for which the gravitational constant G, the total system mass M and the outer radius of the disc R are all equal to unity. The Keplerian rotation period (and also, approximately the real rotation period) at the outer boundary is thus equal to 2π. We performed some 1500 (double) time-steps, which corresponds to a dimensionless period of about 30. During this time, matter near the inner disc boundary has revolved for about 40 times. For each time-step the value of the surface density Σ at a fixed azimuth ($\phi = 0$) is stored for each radial zone. This allows us to determine (by means of Fourier transforms) how the power of the disturbances is distributed as a function of frequency and radial coordinate in the disc.

Figure 1 shows a plot of equal power contours derived from the results sampled during the time $t = 12 - 22$. Two dominant modes with frequency $\sigma = 10.1$ and $\sigma = 3.8$ appear to have grown rapidly beyond the noise. The high frequency mode appears to have a co-rotation point (where the pattern speed of the mode equals the angular velocity of the disc matter) close to the inner boundary, whereas the lower frequency mode has its co-rotation point near the middle of the disc. The co-rotation point near the inner boundary is located near a maximum in the vorticity over density distribution, while the co-rotation point near the middle of the disc lies in a region where this distribution is flat. Later a third low-frequency mode ($\sigma = 2.5$) becomes prominent which has its co-rotation point near the outer edge of the disc. Note that the co-rotation point of the $\sigma = 10.1$ mode almost coincides with the inner Lindblad resonance of the $\sigma = 3.8$ mode. This phenomenon appears to be quite common in our

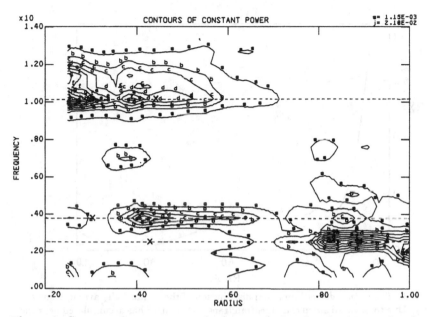

Figure 1. Contours of constant power in the perturbed surface density as a function of dimensionless frequency and radial distance from the disc centre. The contours (labelled a – j) differ by a constant increment. Three dominant modes are indicated by dashed lines on which the positions of the co-rotation point and the Lindblad resonances are given by a black dot and crosses respectively.

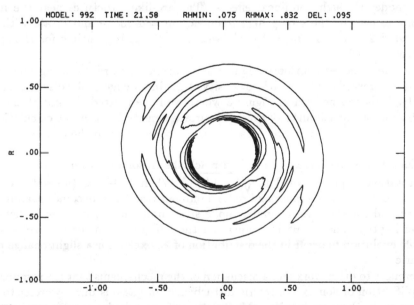

Figure 2. Contours of constant surface density at time 21.6. Contour values are between 0.075 and 0.832.

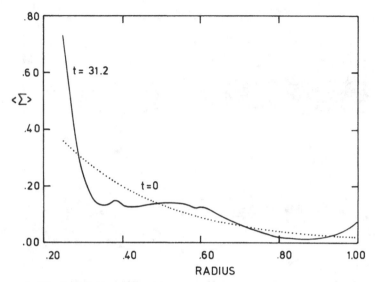

Figure 3. The initial (dotted) and final distribution of the azimuthally averaged surface density. Due to outward angular momentum transport, matter has accumulated near the inner disc boundary at $r = 0.25$.

calculations.

Figure 2 shows iso-density contours of the disc. It can be seen that the initial $m = 2$ symmetry is preserved during the run, though some power is shifted to $m = 4$, $m = 6$ and higher modes through non-linear effects. The arc-like structures near the inner boundary are caused by the strong high frequency $m = 2$ mode and rotate with the pattern speed $\sigma/2 = 5$. The $m = 2$ mode with $\sigma = 3.8$ is responsible for the spiral structure near $r = 0.6$.

The non-linear modes transport angular momentum outwards, causing matter in the disc to spiral inwards and the surface density at the inner edge of the disc to increase substantially. This can be seen in Figure 3 where we have plotted the initial and the final distribution of the azimuthally averaged surface density $\Sigma(r)$ in our calculation.

The growth-rate of the perturbations appears very sensitive to the Q values in the disc. This is as expected, since the width of the evanescent zone around co-rotation through which the waves have to tunnel to reach the 'resonance cavity' at the other side of co-rotation is proportional to $\sqrt{1 - 1/Q^2}$. When $Q \rightarrow \infty$ (no self-gravity), the whole region between the inner and outer Lindblad resonance becomes evanescent. Although we find for this case a similar mode spectrum, the growth rate is severely reduced. During the same simulation time as for the self-gravitating case there appears no noticeable evolution to occur in the distribution of Σ, except for a slight change near the boundaries.

For references to other work and a discussion of the mechanisms that are responsible for the amplification of non-axisymmetric disturbances in gaseous discs we refer to our review elsewhere in these proceedings. More details of the calculations will appear in a separate publication.

Eccentric gravitational instabilities
in nearly Keplerian discs

Steven P. Ruden[1], Fred C. Adams[2] and Frank H. Shu[1]
[1] *University of California, Berkeley, USA.*
[2] *Harvard-Smithsonian Center for Astrophysics, Cambridge, USA*

1 Introduction

There is growing evidence for circumstellar discs associated with young stellar objects (YSOs). Motivated by observational evidence suggesting that these discs produce significant luminosity, $L_D \sim L_*$, and have moderate masses, $M_D \sim M_*$, (Adams, Lada & Shu 1988), we explore the possibility that the accretion mechanism ultimately owes its origin to the growth of spiral gravitational instabilities. As a start, we study the growth and structure of linear, *global*, gravitational disturbances in star/disc systems.

2 The physics of $m = 1$ modes

For simplicity, we take the unperturbed discs to be infinitesimally thin and in centrifugal equilibrium; we characterize the surface density and temperature profiles in the disc as power-laws in radial distance from the star. Since the potential well of the star dominates that of the disc everywhere except near the disc's outer edge, the rotation curve is nearly Keplerian throughout most of the disc's radial extent.

Our study concentrates on modes with azimuthal wave number $m = 1$, since these modes can be global in extent and may also be the most difficult modes to suppress in unstable protostellar discs. Modes with $m = 1$ correspond to elliptic streamlines (*i.e.* eccentric particle orbits), which play a unique role in Keplerian potentials, a fundamental point explicitly recognized by Kato (1983). In an exactly Keplerian potential, circular streamlines of zero pressure are neutrally stable to kinematic perturbations that make them ellipses. The ellipses do not precess in space, so the angular orientations of their semi-major axes remain spatially fixed. If the potential is not exactly Keplerian (or if the pressure is not exactly zero), the distorted streamlines at different radii will precess at different angular rates, which poses a difficulty for pattern coherence. A small restoring force such as self-gravity is needed to give the different pression rates a single uniform value (*i.e.* to make the disturbance a *normal mode*).

Modes with $m = 1$ are especially interesting for YSO discs because the center of mass of the perturbed system does not lie at the position of the star. A coordinate system centered on the star does *not* correspond to an inertial frame of reference and gives rise to a non-inertial contribution to the total potential of the system – the *indirect potential*. We find that the proper inclusion of this effect is crucial to the possibility of obtaining unstable eigenmodes. Physically, the indirect potential allows angular momentum to be transferred between the disc and the star and allows $m = 1$ instabilities to grow when all other modes are effectively stabilized by a relatively large value of the Toomre stability parameter Q (Toomre 1964).

3 Results and future considerations

We have calculated modes in discs with sizes R_D from 10 to 10^4 times R_*, the radius of the star, and masses M_D from 1/3 to 1 times M_*, the mass of the star (*e.g.* Figure 1). Our results indicate that YSO discs will be unstable to the growth of eccentric distortions that have growth rates comparable to the orbital frequency at the outer edge of the disc, *i.e.* the distortions grow on nearly a dynamical timescale. We find that robust growth occurs as long as the Toomre Q parameter at co-rotation is less than about 3. Depending on the existence of a Q barrier near the co-rotation radius, the presence of this disturbance may lead to mass accretion or to the formation of a binary companion from the disc (or both). It has not escaped our attention that similar mechanisms might be at work in other astrophysical contexts, *e.g.* the bodies and nuclei of normal and active galaxies. Lop-sidedness and a wobbling of the central regions may be characteristic dynamical features of many systems. However, quasi-linear and non-linear analyses are needed to follow up these possibilities.

This work was funded in part by NSF grant AST83-14682 and in part by a NASA Center for Star Formation Studies.

References

Adams, F. C., Lada, C. J. & Shu, F. H., 1988. *Astrophys. J.*, **326**, 865.
Kato, S., 1983. *Publ. Astron. Soc. Jpn*, **35**, 249.
Toomre, A., 1964. *Astrophys. J.*, **139**, 1217.

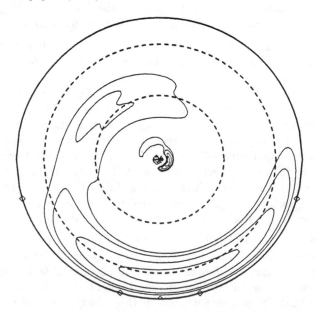

Equidensity Contours

Figure 1. The fundamental $m = 1$ normal mode. The two dashed circles mark the location of the co-rotation and outer Lindblad resonances.

Gravity mode instabilities in accretion tori

W. Glatzel

Max-Planck-Institut für Astrophysik, FRG

1 Introduction

Since their discovery by Papaloizou & Pringle (1984) non-axisymmetric instabilities in accretion tori have been discussed by many authors. It has been found that the instabilities are driven by shear and operate – depending on flow and perturbation parameters – through sound waves, surface waves or Kelvin-Helmholtz type modes. The spectrum of internal gravity waves which is associated with finite entropy gradients has not yet been studied and will be described in a forthcoming paper (Glatzel 1989). A brief summary of the main results is given here.

2 Basic assumptions

In order to allow for an analytical treatment we adopt cylindrical geometry and consider the limit of thin shells which rotate differentially in their own or an external gravitational field. The entropy distribution is required to guarantee a parabolic density stratification. Maximum density occurs when the effective gravity vanishes – its zeros determine the boundaries of the configuration. We assume incompressibility and neglect the self-gravity of the perturbations. Using an additional technical approximation, which has qualitatively no consequence for the modal structure, the perturbation equation is reduced to Whittaker's equation and the dispersion relation can be written in terms of confluent hypergeometric functions.

3 The modal structure

In a medium at rest a two-fold infinite set of gravity modes is found moving parallel to the boundaries in opposite directions. Modes occur in pairs corresponding to a symmetric and an anti-symmetric eigenfunction, where the symmetric mode owes its existence to the non-monotonic density stratification.

If a shear flow is superimposed, the pattern speed of the neutrally stable modes is distorted and the modal structure may be interpreted by a decrease of the pattern speed of the initially, *i.e.* for zero flow velocity, symmetric modes and an increase of the pattern speed of the initially antisymmetric modes. This distortion gives rise to a variety of mode crossings. Mode crossings without a critical layer in the flow unfold into avoided crossings of neutral modes. In contrast to the "antisymmetric" modes, the "symmetric" modes reach a critical layer in the flow and do not remain neutral. If their critical layer is close to the boundaries, they are damped. With increasing shear, the critical layer is shifted towards regions of smaller local Richardson number, the damping rate decreases and turns into a (non-resonant) growth rate, if the local Richardson number at the critical layer is smaller than $\sim 1/4$. For symmetry reasons "symmetric" modes having the same wavenumber and moving in opposite directions in a medium at rest have to cross. These crossings unfold into resonance instability bands.

Since the pattern speed of the gravity modes is intrinsically low, the instabilities occur – in a counter-intuitive way – predominantly at low shear rates and the number of unstable modes increases with decreasing shear. This might be important for the excitation of gravity modes in differentially rotating stars. In rotating shear layers an upper limit for the shear rate exists, above which no instability by internal gravity waves is found. The maximum growth-rate, σ, of the instabilities is roughly proportional to the shear-rate, $\bar{V}_0$:

$$\sigma \approx 10^{-1}\bar{V}_0.$$

4 Instability mechanisms

Two kinds of instabilities appear in the spectrum of gravity modes: Non-resonant instability, which is due to an interaction of a mode with the mean flow caused by over-reflection (Lindzen & Barker 1985), and resonance instability, due to an interaction and energy exchange between two modes (Glatzel 1987). For both of them, the symmetry of the buoyancy frequency profile plays an important role. Over-reflection instability requires wave regions on either side of the critical layer, which can only be provided by a "symmetric", *i.e.* non-monotonic profile of the Brunt-Väisälä frequency. "Symmetric" modes are necessary to provide shear-driven mode crossings and resonance instabilities. They are found only for a "symmetric" profile of the buoyancy frequency. No instability occurs for a monotonic profile of the buoyancy frequency.

References

Glatzel, W., 1987. *Mon. Not. R. Astron. Soc.*, **228**, 77.
Glatzel, W., 1989. in preparation.
Lindzen, R. S. & Barker, J. W., 1985. *J. Fluid Mech.*, **151**, 189.
Papaloizou, J. C. B. & Pringle, J. E., 1984. *Mon. Not. R. Astron. Soc.*, **208**, 721.

The stability of viscous supersonic shear flows Critical Reynolds numbers and their implications for accretion discs

W. Glatzel
Max-Planck-Institut für Astrophysik, FRG

1 The model

We consider the linear stability of a plane shear layer, including the effects of compressibility and viscosity, as the simplest model for viscous supersonic shear flows occurring in accretion processes. Details of this investigation can be found in Glatzel (1989). Measuring lengths and velocities in units of half of the thickness of the shear layer and the flow velocity at its edge respectively, the flow may then be described by dimensionless numbers, the influence of compressibility is described by the Mach number, M, and viscosity by two Reynolds numbers, Re_ν and Re_μ, corresponding to shear and volume viscosity respectively.

2 Instabilities and critical Reynolds numbers

We distinguish two types of modes, viscous modes and sonic modes, according to their physical origin: shear viscosity and compressibility. Shear-driven pairing of viscous modes, and distortion of the pattern speed of sonic modes, leads to mode crossings among the sonic, and between viscous and sonic modes, which unfold into bands of instability. The viscous-sonic resonances provide a new example of viscous instability; the role of viscosity is merely to provide an additional discrete spectrum, while shear is needed to produce mode crossings. The instability is ultimately caused by resonant exchange of energy between the crossing modes.

Critical Reynolds numbers for some resonances are plotted in Figure 1 as a function of the Mach number, M, for zero volume viscosity ($Re_\mu = 3Re_\nu$). For low and high Mach numbers, the critical Reynolds numbers tend to infinity, since the instabilities occur only in the supersonic regime and the strength of the instabilities decreases with increasing Mach number.

3 Application to accretion discs

Provided the instabilities considered lead to a turbulent state and generate a pure turbulent shear viscosity, we may regard the minimum critical Reynolds number as a measure of the turbulent Reynolds number. With a Reynolds number for the turbulent viscosity we can estimate the value of the viscosity parameter, α, used in accretion disc theory:

$$\alpha = \frac{M^2}{Re_\nu}.$$

For the most unstable flow at $M = 4.86$, we find $\alpha = 0.28$; even higher values are obtained for higher Mach numbers. Thus we conclude that turbulent viscosity generated by sonic instabilities could possibly account for even the highest values of $\alpha(= 0.1 - 1)$ implied by some observations.

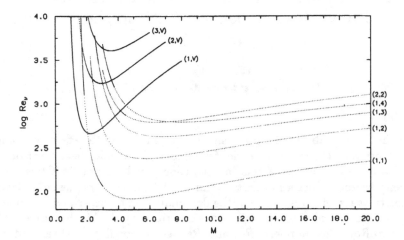

Figure 1. Critical Reynolds numbers for the sonic and the viscous-sonic resonance instabilities indicated as a function of the Mach number. Note that the minimum critical Reynolds number is determined by the (1,V) viscous-sonic resonance for low Mach numbers.

We emphasize that the values of α derived in this way must be regarded as upper limits. If the instabilities also generate a turbulent volume viscosity, or if other damping mechanisms are present, the minimum critical Reynolds number for shear viscosity will increase and the value of α will be smaller. In other words, the minimum critical Reynolds number given in Figure 1 is an upper limit for the turbulent shear viscosity. Furthermore, the model considered here is too simple to allow for more than a qualitative picture and order-of-magnitude estimates. It is meant rather as a first step to show that sonic instabilities can provide the α values needed to explain the observations and, in the absence of reliable simulations of the turbulent flow, how α values may be derived from the instabilities in a simple way.

Reference

Glatzel, W., 1989. *J. Fluid Mech.*, **202**, 515.

Asymptotic analysis of overstable convective modes of uniformly rotating stars

Umin Lee
University of Tokyo, Japan

1 Introduction and summary

We have investigated non-radial oscillations of uniformly rotating stars in a search for the cause of pulsations observed in early-type variables. One of the most interesting results we have obtained is that the convective motions in the core can excite non-radial oscillations in the envelope through the effects of rotation (Lee & Saio 1986). This mechanism for excitation of oscillations in the envelope also works for differentially rotating stars (Lee 1988). Our results indicate that non-radial instabilities are a promising candidate for the cause of pulsations in early-type stars.

In the course of our investigations, it became apparent that resonance phenomena between modes are ubiquitous in the non-radial oscillations of rotating stars. For example, resonance can not only cause a convective mode in the core to excite g modes in the envelope, but is also responsible for avoided crossings between envelope g modes, where modes exchange characteristics with each other (Lee & Saio 1987). In this paper we investigate these resonance phenomena systematically using an asymptotic method for low frequency non-radial oscillations.

2 Asymptotic analysis

We obtain the dispersion relations, using an asymptotic method, which include the coupling effects between non-radial modes with different indices $\lambda_{l,m}$. They are formally written as

$$D_1(\omega)D_2(\omega) = \epsilon, \tag{1}$$

where

$$D(\omega) = \tan(\pm \int k_r dx - \chi_1 + \chi_0), \tag{2}$$

and ϵ is a positive coupling coefficient. For low frequency oscillations the local wave number k_r is given by

$$k_r^2 = \frac{\lambda_{l,m}}{r^2}\frac{N^2}{\omega^2}, \tag{3}$$

where N is the Brunt-Väisälä frequency. The dispersion relation (1) is solved to give real or complex eigenfrequencies of the non-radial oscillations of rotating stars. If the eigenfrequencies are real, the non-radial modes are stable, while they are unstable if their eigenfrequencies are complex.

The eigenfrequencies of non-radial modes of a rotating star are depicted as functions of Ω in Figure 1. In this figure, we can see how the resonances between the modes arise. For further details, see Lee & Saio (1989).

3 Stability

We can discuss the stability of non-radial modes in resonance using the dispersion relation (1) (see Cairns 1979, Glatzel 1987). If

$$\frac{\partial D_1}{\partial \omega} \cdot \frac{\partial D_2}{\partial \omega} > 0, \qquad (4)$$

the non-radial modes in resonance are stable and avoid crossings. If this quantity is negative, on the other hand, the non-radial modes could be unstable. Lee & Saio (1989) showed that the condition (4) is not fulfilled for resonances between envelope g modes and convective modes in the core, and also between convective modes with different indices $\lambda_{l,m}$.

Following the procedure employed by Cairns (1979), we can show that the quantity

$$\omega \frac{\partial D(\omega)}{\partial \omega} \times (\text{amplitude})^2 \qquad (5)$$

can be regarded as the mode energy. Instability of the non-radial mode is therefore brought about when the energies of a pair of resonant modes have opposite signs.

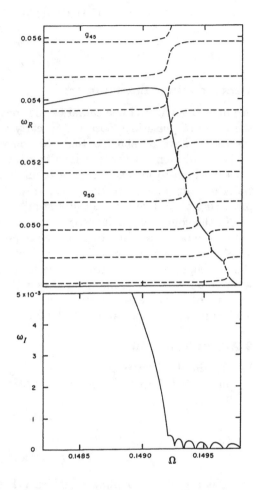

Figure 1. Eigenfrequencies of convective modes in the core and g modes in the envelope. $\epsilon = 10^{-3}$ for this case.

References

Cairns, R. A., 1979. *J. Fluid Mech.*, **92**, 1.
Glatzel, W., 1987. *Mon. Not. R. Astron. Soc.*, **228**, 77.
Lee, U., 1988. *Mon. Not. R. Astron. Soc.*, **232**, 711.
Lee, U. & Saio, H., 1986. *Mon. Not. R. Astron. Soc.*, **221**, 365.
Lee, U. & Saio, H., 1987. *Mon. Not. R. Astron. Soc.*, **224**, 513.
Lee, U. & Saio, H., 1989. *Mon. Not. R. Astron. Soc.*, in press.

Polytropic models in very rapid rotation

Vassilis S. Geroyannis
University of Patras, Greece

1 The numerical method

We consider polytropic models having very rapid differential rotation. In the computations, we implement a Complex-Plane Strategy, abbreviated CXPS, developed recently by the author (Geroyannis 1988; for clarity and convenience, we use hereafter the definitions and symbols of that paper). CXPS owes its high reliability to the fact that it proceeds to the functional determination of a model by integrating the structural differential equations in the complex plane and up to a point ξ_E far beyond the well-known 'first root' ξ_1 (in this implementation, we select $\xi_E \approx 2\xi_1$). Thus, we know in advance, the distortion due to rotation over an extended space surrounding the (initially spherical) configuration. From the numerical analysis viewpoint, the latter means that to compute a particular state of rotation, we always interpolate in the respective structural function tables, even in the case of very rapid rotation which causes very high distortion. The fact that we never need extrapolate beyond the end of the function tables keeps the error in the computations very small.

Table 1

Parameter	n/F_r			
	3.00/0.75	3.00/0.75	3.25/0.50	3.25/0.50
$\lvert T/W \rvert$	0.268	0.374	0.358	0.463
M_*/M_t	0.133	0.787	0.158	0.871
V_*/V_t	0.011	0.193	0.010	0.228
ξ_{e*}	2.129	4.659	2.023	7.111
ξ_{e*}/ξ_{et}	0.289	0.621	0.270	0.615
$(\xi_{p*})_{max}$	0.670	2.681	0.919	4.530
$(\xi_{p*})_{min}$	0.214	0.175	0.200	0.154
$\langle \xi_{p*} \rangle$	0.324	1.171	0.375	1.711
$\langle \xi_{p*} \rangle/\xi_{pt}$	0.053	0.196	0.050	0.230
$(\rho_*)_{max}$	1.199	6.539	1.379	12.85
$(\rho_*)_{min}$	0.970	0.970	0.968	0.968
$\langle \rho_* \rangle$	1.079	2.621	1.153	3.716
$(\rho_o)_{max}$	0.970	0.970	0.968	0.968
$\langle \rho_o \rangle$	0.075	0.169	0.062	0.165
$\langle \rho_o \rangle/\langle \rho_* \rangle$	0.070	0.065	0.053	0.044

2 Results

Table 1 gives characteristic results for two models; for first model: $[n = 3.00, F_r = 0.75]$, and for the second: $[n = 3.25, F_r = 0.50]$. The 'reduction factor' F_r measures the sharpness of the differential-rotation pattern; so, the value $F_r = 1.00$ corresponds to the sharpest pattern of differential rotation, and the value $F_r = 0.00$ corresponds to a uniform rotation. In Table 1, the subscripts $*$, o, and t denote the value of a quantity for the 'nucleus', for the 'halo', and for the global configuration, respectively. All the densities are measured in units of central density ρ_c. The first model is evaluated in two states of rotation; one with $|T/W|$ close to 0.27 where dynamic loss of stability against non-axisymmetric perturbations occurs, and the other with $|T/W|$ close to 0.36 where secular overstability with respect to axisymmetric perturbations arises. For the second model, we also calculate a rotation state with $|T/W|$ close to 0.36 and finally, a state with $|T/W|$ close to 0.46 where dynamic loss of stability against axisymmetric perturbations occurs.

In conclusion, we find that the nuclei computed attain shapes from thick-disc-like to concave-hamburger ones, or even to toroidal ones. Moreover, these nuclei become very dense and massive in comparison with the densities and masses of the respective halos.

Reference

Geroyannis, V. S., 1988. *Astrophys. J.*, **327**, 273.

Distribution and kinematics of gas in galaxy discs

R. Sancisi

Kapteyn Astronomical Institute, University of Groningen, The Netherlands

Abstract Three topics are briefly discussed concerning the gas distribution and kinematics in spiral galaxies. The first concerns the relative location of neutral hydrogen, HII regions, dust, molecules and non-thermal radio continuum emission in spiral arms. The second is the asymmetrical structure and the presence of large non-circular motions in spiral galaxies, as shown by the observations of M 101. Finally, attention is drawn to the presence of spiral arm structure and to some puzzling HI features in the outermost parts of gaseous discs. Observational evidence seems to indicate that infall of gas has important effects on the kinematics of discs and on their evolution.

1 Structure of spiral arms

Detailed, multifrequency observations of recent years of two nearby spiral galaxies, M 51 and M 83, have led to a new picture of the relative distributions of the various ingredients of the interstellar medium. In the classical schematic picture of spiral arm structure the HI is concentrated on the inner side of spiral arms on the dust lanes, which mark the location of spiral shocks. The observations of HII and HI regions by Allen *et al.* (1986) and Tilanus & Allen (1989) show that both HI and HII are displaced from the dust lanes toward the outer parts of the arm, although there is no small scale agreement between the distribution of HI and HII. The radio continuum ridge in M 51 coincides with the dust lanes (Tilanus *et al.* 1988). Its profile across the arm is much broader than expected. Observations of the CO emission with the Owens Valley Radio Observatory telescope by Vogel *et al.* (1988) show that the molecules are in the region of the dust. In summary, it seems that HI and HII are displaced downstream with respect to the region of the shock where the dust and the emission from CO and the non-thermal continuum originate. The explanation provided in the framework of density wave theory, is that the gas in molecular form streams across the arms, it forms stars and is dissociated (HI) or ionized (HII). Some new, very interesting information on the kinematics of this molecular gas has been recently obtained by Vogel *et al.* (1988), who claim to have found in their CO observations of M 51 velocity streaming across the arms of up to 60-80 km s^{-1}.

2 M 101 – asymmetry and non-circular motions

M 101 was the first external galaxy in which HI spiral structure was discovered (see Allen & Goss 1979). The HI closely follows the optical structure and also shows a striking large-scale asymmetry. The galaxy seems, however, to become less lopsided in its outer parts where a very extended, low density HI envelope appears (Huchtmeier & Witzel 1979). The structure of this HI is shown (see Figure 1) by recent 21-cm observations at Westerbork (van der Hulst & Sancisi 1988). These observations have also revealed the presence of gas moving at high velocities perpendicular to the disc of M 101. Two large HI complexes have been found with masses of order $10^7 - 10^8$ M$_\odot$ and velocities of up to 160 km s^{-1} with respect to the "local" rotating gas of the disc. Since they are redshifted with respect to the disc, they may be falling toward the disc of

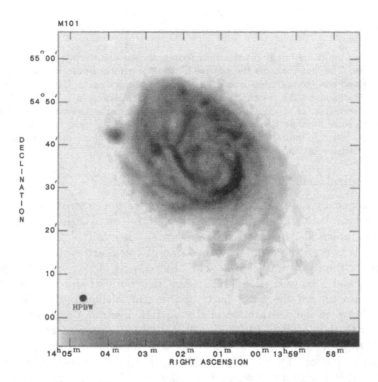

Figure 1. HI map of M 101 at 80" resolution (van der Hulst & Sancisi 1988). At the assumed distance of 7.2 Mpc, $1' = 2$ kpc.

M 101 on the near side or they may be moving away from the disc on the far side. The presence of holes and troughs in the HI layer in the direction of the HI clouds suggests that the second possibility is the more likely. Various explanations have been proposed. A plausible one is that of infall of large gas clouds punching the disc of M 101 and subsequently forming the observed high velocity features. The infalling material could be intergalactic gas clouds, which have, however, never been detected in intergalactic space, or, more likely, one of those gas-rich dwarf companions which are often seen in the neighbourhood of bright galaxies. A very interesting example in this connection is that of NGC 3359 (see below).

There are other remarkable large-scale deviations from circular motion in the disc of M 101. Some of them may be related to the high-velocity phenomenon. They show up, as first noted by Rogstad (1971), in the velocity field (Figure 2, see also Bosma *et al.* 1981) as pronounced wiggles, which roughly follow the spiral arm pattern. In particular there is a remarkably strong velocity gradient in the area about 8 arcmin to the south-west of the centre, where two bright HI arms seem to cross. The velocity increase across these two arms is almost 50 km s^{-1}. There is a remarkable symmetry between this velocity jump south-west of the centre and the high velocity features on

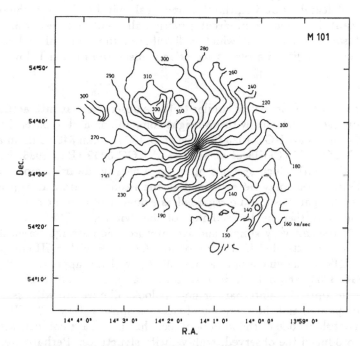

Figure 2. Map of the intensity-weighted mean velocity field of M 101 at 80″ resolution (from observations of van der Hulst & Sancisi 1988).

the north-east. Furthermore, they seem to be connected by a velocity wiggle running like an arc between these two locations and dominating the kinematics on the southeast side of M 101. This suggests a relationship between them. Whatever the event and the mechanism causing the high velocity phenomenon might have been, it has also strongly perturbed the large-scale kinematics of the gaseous disc of this galaxy.

3 Spiral structure in outer parts

HI gas layers occasionally extend far beyond the optical disc in spiral galaxies, but the density distributions in these outer parts have usually appeared noisy, featureless and dull. No detailed, high quality maps of the density and kinematic structure exist, except for a few objects. In some of these, as for example M 101 (van der Hulst & Sancisi 1988), NGC 628 (Shostak & van der Kruit 1984) and NGC 6946 (Tacconi & Young 1986), observations with sufficiently high sensitivity and angular resolution have finally revealed the presence of spiral structure to the outermost parts. In general the spiral pattern continues from the inner, optical parts to the faint HI outer parts. There is no apparent change or discontinuity at the edge of the optical disc or where the warp begins (see NGC 628). The spiral arms are difficult to trace: they are not very regular and are made up of a number of weak features.

In a number of cases (*e.g.* M 101, see Figure 1) the density distribution is strongly lopsided. Surprisingly, however, the HI layer of M 101 in the very outer parts and at the lowest intensity levels appears to become rather symmetrical again.

The asymmetrical distribution, the thin and weak spiral features in the outer parts and the significant deviations from circular motion all point at a recent arrival of this material or at some recent event which redistributed the material and affected the kinematics. The possibility of a recent accretion is discussed further below.

4 Gas infall

Clumpy and irregular HI structures similar to those observed near interacting systems, are found around some galaxies with no bright companions and apparently no strong tidal effects. Examples are NGC 1023 (Sancisi *et al.* 1984), Mkn 348 (Simkin *et al.* 1987, Sancisi 1988), NGC 3067 (Carilli *et al.* 1989) and NGC 3359 (Ball 1986, Broeils 1989 private communication). Such systems often have nearby, dwarf companions. Their puzzling HI distributions may be explained as the results of tidal disruption of gas-rich, dwarf irregular galaxies and be regarded as evidence of current merger processes. The companions may then be the remnant of the "victim". The case of the spiral galaxy NGC 3359 seems very simple and its interpretation straightforward. Broeils' observations show an extended HI layer around NGC 3359 with a HI companion and HI bridge. The HI companion contains about 10^8 M$_\odot$, which is approximately 2% of the HI mass of NGC 3359 (and only 0.1% of its total mass). It is not at all clear whether it has a very faint optical counterpart or not. It looks like we are witnessing here the capture of a companion and the accretion of gas by a spiral galaxy. The HI companion of NGC 3359 is perhaps one of the kind of object that may have collided with the disc of M 101 and produced the observed, high-velocity structures. Perhaps we are seeing in both cases, at somewhat different stages, infall of intergalactic gas and building up of galactic discs.

I am grateful to F. Briggs, J.M. van der Hulst, A. Broeils and J. Kamphuis for helpful and stimulating discussions.

References

Allen, R. J. & Goss, W. M., 1979. *Astr. Astrophys., Suppl. Ser.*, **36**, 135.
Allen, R. J., Atherton, P. D. & Tilanus, R. P. J., 1986. *Nature*, **319**, 296.
Ball, R., 1986. *Astrophys. J.*, **307**, 453.
Bosma, A., Goss, W. M. & Allen, R. J., 1981. *Astron. Astrophys.*, **93**, 106.
Carilli, C. L., van Gorkom, J. H. & Stocke, J. T., 1989. *Nature*, **338**, 134.
Huchtmeier, W. K. & Witzel, A., 1979. *Astron. Astrophys.*, **74**, 138.
van der Hulst, J. M. & Sancisi, R., 1988. *Astron. J.*, **95**, 1354.
Rogstad, D. H., 1971. *Astron. Astrophys.*, **13**, 108.
Sancisi, R., 1988. In *QSO Absorption Lines: Probing the Universe*, p. 241, eds., Blades, J. C., Turnshek, D. A. & Norman, C. A., Cambridge University Press, Cambridge.
Sancisi, R., van Woerden, H., Davies, R. D. & Hart, L., 1984. *Mon. Not. R. Astron. Soc.*, **210**, 497.
Shostak, G.. & van der Kruit, P. C., 1984. *Astron. Astrophys.*, **132**, 20.
Simkin, S. M., van Gorkom, J., Hibbard, J. & Su, H.-J., 1987. *Science*, **235**, 1367.
Tacconi, L. J. & Young, J. S., 1986. *Astrophys. J.*, **308**, 600.
Tilanus, R. P. J. & Allen, R. J., 1989. *Astrophys. J. Lett.*, **339**, L57.
Tilanus, R. P. J., Allen, R. J., van der Hulst, J. M., Crane, P. C. & Kennicutt, R. C., 1988. *Astrophys. J.*, **330**, 667.
Vogel, S. N., Kulkarni, S. R. & Scoville, N. Z., 1988. *Nature*, **334**, 402.

Are the smallest galaxies optically invisible?

Claude Carignan
Université de Montréal, Canada

1 Introduction

Recent studies by Tyson (1988) and Tyson & Scalo (1988) suggest the possible existence of a large population of gas-rich dwarf irregular galaxies. Their "bursting dwarf galaxies" model would imply that a large fraction of these dwarfs remains undetected due to observational selection effects (angular diameter, surface brightness). Dekel & Silk (1986), in their cold dark matter biased galaxy formation picture, also predict that the universe is filled more uniformly with dwarf galaxies than with bright ones. Our results on DDO 154 suggest it could be a prototype gas-rich, low surface brightness, small optical diameter galaxy which happens to be relatively nearby ($\Delta \leq 4$ Mpc based on possible membership to the CVn I cloud and the magnitudes of the brightest blue stars; Carignan & Beaulieu 1989).

2 Summary of the data

DDO 154 is barely discernible on the Palomar Sky Survey. Its extrapolated central surface brightness is only $B(0) = 23.5$ mag arcsec^{-2}. The colours, however, are typical of Im galaxies with $(B - V) = 0.32$ and $(V - R) = 0.30$. Its large HI gas content and extent were discovered serendipitously by Krumm & Burstein (1984). From the VLA data, it is found that the HI extends to nearly $5D_{H_0}$ at a level $\sim 10^{19}$ cm^{-2} ($4D_{H_0}$ at a level $\sim 10^{20}$ cm^{-2}). Despite the chaotic optical appearance, the velocity field is very regular and well-defined. The analysis shows that the closing of the isovelocity contours in the outer parts is partly due to the warp of the HI disc. However, the results still suggest that the rotation curve could be truly declining for $r \geq 6$ kpc (Carignan & Beaulieu 1989), which, if confirmed, would indicate that the edge of the system may have been reached.

3 Mass model

We have deduced the distribution of mass in the galaxy, taking account of the observed light and HI distributions – a preliminary discussion is given elsewhere (Carignan & Freeman 1988). In our final best-fit model, the dark isothermal halo has a core radius $r_c = 3.0$ kpc and a central density $\rho_0 = 0.016$ M$_\odot$pc^{-3}. At the last point of the rotation curve (7.6 kpc), more than 90% of the mass is provided by the dark component. At that radius, $(M/L_B)_{total} = 80$. The so-called "disc-halo conspiracy" does not arise for this dwarf system, as V_{max}(halo) $\approx 4V_{max}$(disc). The galaxy does not lie on the Tully-Fisher relation; the maximum velocity is not determined by the luminous disc in this case.

4 Conclusion

The present results strongly suggest that galaxies must exist, even at small redshifts, where star formation has not yet turned on and that it is entirely possible that the smallest galaxies are *optically invisible*.

References

Carignan, C. & Beaulieu, S., 1989. *Astrophys. J.*, (submitted).
Carignan, C. & Freeman, K. C., 1988. *Astrophys. J.*, **332**, L33.
Dekel, A. & Silk, J. 1986. *Astrophys. J.*, **303**, 39.
Krumm, N. & Burstein, D., 1984. *Astron. J.*, **89**, 1319.
Tyson, N. D., 1988. *Astrophys. J.*, **329**, L57.
Tyson, N. D. & Scalo, J. M., 1988. *Astrophys. J.*, **329**, 618.

Can we understand the constancy of rotation curves?

Daniel Pfenniger

Observatoire de Genève, Switzerland

1 Introduction

The purpose of this study (Pfenniger 1989) is to ask whether the basic shape of such complex objects as disc galaxies can be understood by simple arguments, in a similar fashion to the modelling of stars by polytropic spheres. The now well observed extended flat rotation curves of galaxies lead to the conclusion that they contain much more matter than is indicated by their luminosity, since the inferred mass rises linearly with the radius. The notion of total mass is therefore irrelevant for understanding galactic structure, and the most suitable boundary conditions are not necessarily those of isolated objects. Apart from the problem of nature of the dark matter, the constancy of rotation curves also needs to be explained because there is no clear necessity that dark matter has to adopt precisely a profile leading to such curves.

2 Simple models

The energy content of disc galaxies is principally rotational and gravitational. As a first step, we therefore first try to understand purely rotationally supported discs and look for axisymmetric equilibria which are local energy minima.

2.1 COLLECTION OF PARTICLES IN A RIGID AXISYMMETRIC POTENTIAL

If self-gravity is negligible, a collection of particles dissipating energy but conserving angular momentum in an axisymmetric and static potential $\Phi(R, z)$, will settle to circular orbits in the plane at constant z where Φ_z vanishes. This is easily deduced by minimising the energy integral at constant angular momentum. A further condition for a minimum is that the epicyclic frequencies must be real. This trivial model leads directly to flat discs, but with no specific rotation curve since self-gravitation is neglected. Once all the particles are settled in this state, some of our initial assumptions must be modified, because *energy dissipation of circular orbits is not compatible with the detailed conservation of angular momentum.*

2.2 SELF-GRAVITATING DISC SUPPORTED PURELY BY ROTATION

If matter on circular orbits dissipates slowly with respect to the dynamical time, as tea in a cup, we can still keep an axisymmetric model, though angular momentum will no longer be perfectly conserved. In this second model, we consider a collection of self-gravitating circular orbits rotating around the z-axis.

Two extreme solutions which do not require pressure can represent an equilibrium, the infinitesimally thin disc, or the infinitely thick disc, homogeneous in z. The dissipation time-scale is assumed to be larger than the dynamical time, and we look for the asymptotic state. The total energy is $E \sim \iint dR\, dz\, R\, \rho(v_c^2 + \Phi)$ with the circular speed $v_c = (R\Phi_R)^{1/2}$, and the density ρ is linked to Φ by Poisson's equation. The integral is therefore a functional of Φ which contains up to second derivatives of Φ. In order for it to be a minimum, it must satisfy the corresponding Euler-Lagrange equation of variational calculus. Many higher order terms cancel, and we get Mestel's disc (Mestel 1963) for the thin disc solution, $\Sigma(R) \sim R^{-1}$, and a singular homogeneous wire along the z-axis for the thick disc one. *Both solutions have a constant rotation curve.*

This simple model therefore, tells us that the law of constant velocity corresponds to an energy extremum for axisymmetric, rotating, and self-gravitating systems. For Mestel's disc the second variation of E is positive, *i.e.* it is indeed an energy minimum for this kind of equilibrium. More generally, Mestel's disc is still obtained if an exterior potential is present (*e.g.* due to a central mass condensation).

2.3 DIFFERENTIALLY ROTATING, SELF-GRAVITATING POLYTROPIC DISCS

Of course the previous models can be improved by adding a pressure term. Up to now, differentially rotating polytrope models have been under-constrained by having a free rotation curve. But, by requiring that rotating equilibria also have to be energy extrema, the number of possible rotation curves is reduced, and is possibly unique. Using the same technique as above, restricting the streaming v to cylindrical rotation, the additional constraint we seek is found to be $\partial_R v^2 = \frac{\gamma-1}{\gamma-2}\partial_R v_c^2 + R\Phi_{zz}$.

3 Schematic scenario and remarks

A possible schematic evolutionary sequence of galactic discs can be proposed. Over the short time-scale of fast energy dissipation, during which angular momentum is locally well conserved, flat discs form with a mostly circular rotation, but no particular rotation curve. Subsequently, angular momentum is slowly lost from the system through instabilities, at the rate imposed by energy dissipation, causing the rotation curve to flatten. Over longer time-scales, evolution may proceed through a sequence of discs with flat rotation curves, the absolute velocity of which *increases* with time.

Since the combination of rotation, dissipation and flat rotation curves is made coherent, Newtonian gravitation is confirmed over distances as large as the most extensive observed rotation curve. The dark matter problem is then correspondingly more acute. Finally, since our arguments are rather general, other astrophysical discs may be expected to tend also toward such a property.

References

Mestel, L., 1963. *Mon. Not. R. Astron. Soc.*, **126**, 553.
Pfenniger, D., 1989. *Astrophys. J.*, in press.

How well do we know the surface density of the Galactic disc?

Thomas S. Statler
University of Colorado, USA

1 Introduction

Re-examination of the classic "Oort limit" problem has led to a controversy over the presence of substantial amounts of dark matter confined to the disc of the Galaxy. The strongest argument for its existence comes from Bahcall (1984a,b), who fits the kinematics of F stars within 200 pc of the midplane. The total integrated surface density in his preferred model is $\Sigma_0 = (67 \pm 5) M_\odot \mathrm{pc}^{-2}$, as compared with $\Sigma_{0,\mathrm{obs}} \approx 40 M_\odot \mathrm{pc}^{-2}$ for the observed luminous matter alone. He thus concludes that there must be roughly as much dark as luminous matter in the disc. On the other hand, Kuijken & Gilmore (1989a,b,c; hereafter KG), using a sample of K dwarfs extending up to $z \sim 2.5$ kpc above the midplane and a different analysis, obtain $\Sigma_{0,\mathrm{total}} = (46 \pm 9) M_\odot \mathrm{pc}^{-2}$, a value consistent with no disc dark matter at all. That these results differ by substantially more than their quoted uncertainties indicates that something is amiss. The method discussed here for obtaining Σ_0 is based on the analysis of KG, and is intended to address this problem. For a complete presentation of this material, the reader is referred to Statler (1989).

2 Modeling procedure

Central to the procedure is the representation of the Galactic potential by one of Stäckel form; in such a potential an exact third integral is known in closed form for all orbits. The distribution function (DF) of the tracer stars is assumed to be a linear combination, $f(E, I_2, I_3) = \sum_i A_i f_i$, of components of the form

$$f_i = \exp\left[\frac{E}{\sigma_{\varpi i}^2} + \frac{1}{R_0^2}\left(\frac{1}{\sigma_{\varpi i}^2} - \frac{1}{\sigma_{\theta i}^2}\right) I_2 + \frac{\Theta_{mi}}{R_0 \sigma_{\theta i}^2}\sqrt{2 I_2} \right.$$
$$\left. + \frac{1}{(R_0/z_0)^2 + 1}\left(\frac{1}{\sigma_{zi}^2} - \frac{1}{\sigma_{\varpi i}^2}\right) I_3\right],$$

where (E, I_2, I_3) are the integrals of motion, R_0 is the Solar Galactocentric distance, and z_0 is a parameter related to the tilt of the velocity ellipsoid. The individual components are tilted ellipsoidal Gaussians in velocity space; in the midplane, the tilt angle is zero, the principal dispersions are $\sigma_{\varpi i}$, $\sigma_{\theta i}$, and σ_{zi}, and the mean rotation speed is Θ_{mi}. The coefficients A_i may be negative, subject to the constraint that $f \geq 0$ everywhere in phase space.

To make a model, a library of components f_i is first chosen. The freedom to choose $\sigma_{\varpi i}$, $\sigma_{\theta i}$, and σ_{zi} for each component enables one to impose particular trends on the three-dimensional velocity structure of a given model. For a specific library, the coefficients A_i that reproduce the density law of the tracer stars are found by least squares; the likelihood of the velocity sample in the model DF is then calculated and used as the gauge of the model.

3 Is there dark matter there or not?

To test the method and measure the error bars, a choice is made for the "real" Galactic potential and tracer star DF, and random samples of 500 stars are drawn from that distribution to mimic the KG sample. A variety of models are made for both the random samples and the continuous parent distribution; see Statler (1989) for details.

The results are: firstly, that the shot noise in a sample the size of KG's is large enough to limit the uncertainty in the inferred Σ_0 to $\sim 23\%$. Secondly, there are systematic errors related to the tilt and the change of shape of the velocity ellipsoid above 1 kpc. Since there are virtually no observational constraints on these parameters, the errors are best added in quadrature and amount to about 10%. Further errors stemming from uncertainties in R_0 and the rotation curve, as estimated by KG, add another 16%. Thus the mean error (one-sigma) on KG's estimate of Σ_0 must be at least 30%, or $\pm 14 M_\odot \mathrm{pc}^{-2}$.

A statistical analysis by Gould (1988) suggests that the mean error in Bahcall's estimate of Σ_0 is probably about as large as that given above for KG's estimate. If this is the case, then the two values are marginally (1.5-σ) consistent with each other. In particular, KG are (intentionally) insensitive to ρ_0, so that their models and Bahcall's are completely consistent with a thin layer of dark matter that may be dense in the midplane but does not contribute much to Σ_0. However, if the dark matter layer is thin, then the required mass is sensitive to the distributions of the luminous components that have similar scale heights. Preliminary results from work in progress (Statler & Shull, in preparation) suggest that the inferred dynamical mass may be very sensitive to the local gas distribution, which is, unfortunately, not known quite well-enough. The question of disc dark matter thus remains, for the moment, unresolved.

References

Bahcall, J. N., 1984a. *Astrophys. J.*, **276**, 169.
Bahcall, J. N., 1984b. *Astrophys. J.*, **287**, 926.
Gould, A., 1988. preprint.
Kuijken, K. & Gilmore, G., 1988a,b,c. *Mon. Not. R. Astron. Soc.*, submitted.
Statler, T. S., 1989. *Astrophys. J.*, **344**, in press.

On the heating of the Galactic disc

P. te Lintel Hekkert[1] and Herwig Dejonghe[2]

[1] *Sterrewacht Leiden, The Netherlands*
[2] *Sterrenkundig Observatorium, Gent, Belgium*

1 Introduction

It is known from stellar kinematics in the solar neighbourhood that the velocity dispersions of old populations are significantly higher than those of young populations. This implies a heating rate which seems hard to reconcile with known heating mechanisms such as stellar encounters or interactions between stars and molecular clouds, spiral structure *etc...* Our information elsewhere in the disc comes from fairly limited and biased samples that include line of sight velocities and, in some cases, distances (Lewis & Freeman 1989). This is insufficient to infer a velocity dispersion profile reliably. In this contribution we use a dynamical model and a large database of radial velocities of OH/IR stars to deduce the velocity dispersions as function of Galactic distance.

2 The data

Our catalogue of OH/IR stars is a compilation of the 1612 MHz maser surveys by te Lintel Hekkert *et al.* (1989), Eder *et al.* (1988) and Sivagnanam & le Squeren (1986). The compilation yielded a total of 1600 positions and radial velocities. The positional information for these surveys was taken from the IRAS Point Source Catalogue (PSC).

In order to obtain an approximately constant BC[1] (on average BC = 3.4), we selected only those stars from the OH/IR catalogue with an R21[2] between 0.0 and 0.9. Although the 1612 MHz surveys have been made with different radio telescopes, the catalogue has a well defined detection limit of 3 Jy (12μm), or 8 kpc assuming a luminosity of 5000 $L_\odot$ for a given OH/IR star. We used Habing's (1986) luminosity distribution of the OH/IR stars for our modelling, recalculated for the above R21 range. Since the PSC is not complete in so-called confused areas ($|l| \leq 60° \wedge |b| \leq 2°$), these areas were excluded from our modelling.

3 Fitting procedure

A stellar system is adequately described by its distribution function F and its potential $\psi(\varpi, z)$. In this particular case, the potential bears no direct relation to the population of interest, and F and ψ are therefore quite independent.

In general, the distribution function could be a function of 3 isolating integrals of the motion. In fact, we know from the kinematics in the solar neighbourhood that this is the case. More precisely, if the Galaxy potential turned out to be Stäckel (Dejonghe & de Zeeuw 1987), the distribution functions of these local samples must depend on the third non-classical integral. On the other hand, if the Galaxy potential is not of Stäckel form, these distribution functions still must depend on an effective third integral. Since our data set is largely confined to the disc and consists of only radial velocities, there

---

[1] $L_* = BC \times \mathrm{flux}(12\mu m)$

[2] $R21 = {}^{10}\log(\mathrm{flux}(25\mu m)/\mathrm{flux}(12\mu m))$

is no direct evidence that the OH/IR population respects a third integral. We suspect, therefore, that it must be possible to construct two integral dynamical models that are compatible with the data. Since this is clearly a simpler case, both theoretically and in parameter search we will restrict ourselves to this case, and leave three integral models for future work.

Our model is a linear combination of simple Fricke-type components which are powers in $E = \psi - (v_r^2 + v_\phi^2 + v_z^2)/2$ and $L_z = \varpi v_\phi$. The coefficients are determined using a quadratic programming algorithm (Dejonghe 1989) which minimizes a χ^2 variable built on the available data. An additional constraint is that the distribution function be positive.

We subdivide the sample into rectangular bins in the plane of the sky, each bin containing approximately 20 stars. For each bin, we calculate the zeroth, first and second moment of the line-of-sight velocity distribution. This yields three observables for each bin. Errors are calculated using the normal estimates.

4 Results and discussion

The best fit to the data, with a χ^2 well within the 5% confidence limit, has a total mass of 4×10^{11} M$_\odot$ within 15 kpc. There are hardly any counter-rotating orbits, except for those with small L_z and high binding energy E, which are in the Bulge of the Galaxy.

We can use the model to calculate $\langle v_\phi \rangle = \mu_\phi$ as a function of ϖ in the plane $z = 0$ and the velocity dispersions σ_r and σ_ϕ. The asymmetrical drift is substantial: ~ 100 km s^{-1}. The velocity dispersions in ϖ and ϕ are almost equal in the centre (60 km s^{-1}), as required, but are approximately $\sigma_r = 50$ km s^{-1} and $\sigma_\phi = 70$ km s^{-1} at $\varpi = 10$ kpc.

Such dispersions are uncomfortably large to be easily explained by collisional heating effects in the galactic disc. Molecular cloud heating seems inadequate (Lacey 1984, Villumsen 1985), patchy spiral arms (Carlberg & Sellwood 1985) may be somewhat more effective – especially as we do not have much v_z information. However, as long as we do not know (a) the kind of orbits OH/IR stars are born on, and (b) how to determine ages for them reliably, it will be hard to quantify these qualitative conclusions.

References

Carlberg, R. G. & Sellwood, J. A., 1985. *Astrophys. J.*, **292**, 79.
Dejonghe, H., 1989. *Astrophys. J.*, in press.
Dejonghe, H. & de Zeeuw, P. T., 1988. *Astrophys. J.*, **329**, 720.
Eder, J., Lewis, B. M. & Terzian, Y., 1988. *Astrophys. J.*, **66**, 183.
Habing, H. J., 1986. In *Light on Dark Matter*, p. 40, ed. Israel, F. P., Reidel, Dordrecht.
Lacey, C. G., 1984. *Mon. Not. R. Astron. Soc.*, **208**, 687.
Lewis, R. J. & Freeman, K., 1989. Preprint.
te Lintel Hekkert, P., Caswell, J. L., Habing, H. J., Norris, R. P. & Haynes, R. F., 1989. In preparation.
Sivagnanam, P. & Le Squeren, A. M., 1986. *Astron. Astrophys.*, **168** 374.
Villumsen, J. V., 1985. *Astrophys. J.*, **290**, 75.

The bulge-disc interaction in galactic centres

Mark E. Bailey and **Althea Wilkinson**
University of Manchester

1 Introduction

Mass lost from stars in the central regions of galaxies may flow inwards to form both massive molecular clouds and a young, new stellar population having the form of a thick, rapidly rotating nuclear disc. Angular momentum is transferred from the clouds to the stars in the original spheroid, thereby spinning up the old bulge population within the central $\simeq 1\,\mathrm{kpc}$ and increasing the general 'boxiness' of the underlying stellar density distribution. New stars formed by the collapse of massive molecular clouds in the nuclear disc also contribute towards the general boxiness of the central nuclear bulge.

2 Bulge-disc interaction

Following previous investigations into the fate of stellar mass loss in the central regions of galaxies (*e.g.* Bailey & Clube 1978, Bailey 1980, 1982, 1985), we assume that material flows inwards ultimately to form a dense, cold nuclear disc of molecular gas. We assume a disc of radius $R_d \simeq 500\,\mathrm{pc}$ and mass $M_d \approx 10^8\,M_\odot$, embedded within a nuclear bulge of radius $R_n \simeq 1\,\mathrm{kpc}$ and mass $M_n \simeq 10^{10}\,M_\odot$, adopting a flat rotation curve. Over a Hubble time, the total stellar mass loss exceeds 10% of the original bulge mass, *i.e.* $\gtrsim 10^9\,M_\odot$.

We also assume that the nuclear disc fragments into clouds with masses $M_c \simeq 10^6\,M_\odot$, by analogy with the clouds in the disc of our Galaxy. These massive clouds, moving at the local circular velocity, suffer dynamical friction against the surrounding bulge stars, which causes them to spiral slowly into the galactic centre. The resulting angular momentum transfer depletes original polar-type loop orbits in the nuclear bulge on a timescale, t_0, comparable with the time for a cloud initially at R_0 to spiral into the galactic centre. This is given (*e.g.* Lin & Tremaine 1983) by

$$t_0 \simeq 1.4 \times 10^8 \left(\frac{M_n}{10^{10}\,M_\odot}\right)^{1/2} \left(\frac{1\,\mathrm{kpc}}{R_n}\right)^{1/2} \left(\frac{R_0}{100\,\mathrm{pc}}\right)^2 \left(\frac{10^6\,M_\odot}{M_c}\right) \mathrm{yr}.$$

The result is to reduce the surface brightness towards the poles of the bulge.

Moreover, new stars formed by the collapse of molecular clouds in the disc produce a secondary stellar population initially moving in nearly circular orbits within the nuclear disc. Interactions between these disc stars and the surrounding molecular clouds cause a steady gain in their kinetic energies perpendicular to the disc, eventually producing the appearance of a 'thick-disc' population within the bulge. The gain in stellar velocity dispersion is given (Lacey 1984) by $\sigma(t) = \sigma_0 (1 + t/t_s)^{1/4}$, where σ_0 is the original velocity dispersion and (for our nominal bulge)

$$t_s \simeq 3 \times 10^6 \frac{1}{\ln \Lambda} \left(\frac{\sigma_0}{10\,\mathrm{km\,s^{-1}}}\right)^4 \left(\frac{R_d}{500\,\mathrm{pc}}\right) \left(\frac{R}{500\,\mathrm{pc}}\right)^2 \left(\frac{10^6\,M_\odot}{M_c}\right) \left(\frac{10^8\,M_\odot}{M_d}\right) \left(\frac{200\,\mathrm{km\,s^{-1}}}{V_c}\right) \mathrm{yr},$$

where $\ln \Lambda \simeq 4$. The time for new stars to join the bulge ($\sigma \simeq 100\,\mathrm{km\,s^{-1}}$, say) is thus of order $\simeq 10^{10}\,\mathrm{yr}$. We therefore expect that much of the stellar mass loss from the original spheroid will ultimately be returned to the bulge in the form of a new, more rapidly rotating stellar population, with the orbital characteristics of a thick central nuclear disc.

We emphasize that this theory applies generally only to the central ≈ 1 kpc of spiral and elliptical galaxies, though the source of gas could in principle include much larger regions if, for example, gas flows inwards as a result of a cooling flow. In particular, we expect that it should apply to the case of our own Galactic Centre, which also appears to be an example of a compact box-bulge (Habing 1987, Harmon & Gilmore 1988).

3 Predictions

The main predictions of the model are (1) the presence of nuclear discs (including molecular clouds and evidence for star formation) in the centres of galaxies with compact box-bulges; (2) more than one component in the bulge luminosity and velocity dispersion profiles, due to the rapidly rotating thick-disc population of stars formed by inflow at earlier stages of galactic evolution; and (3) colour gradients in R and Z, due to varying ages of newly formed stars which merge into the surrounding bulge.

These predictions may be compared with those for alternative explanations of box-bulges: torquing-up a slowly rotating bulge by a massive bar in the galactic disc (*e.g.* May *et al.* 1985); or injection of new stars from outside, as a result of a suitable merger event (*e.g.* Binney & Petrou 1985). The former predicts a single bulge population and does not require either a nuclear disc or star formation; the latter would probably not be readily consistent with the absence of tails, streamers, and shells in many box-bulge galaxies. These several models are not mutually exclusive, however, though we do emphasize that the present mechanism, being based on inflowing stellar mass loss, is to some degree unavoidable.

References

Bailey, M. E., 1980. *Mon. Not. R. Astron. Soc.,* **191**, 195.
Bailey, M. E., 1982. *Mon. Not. R. Astron. Soc.,* **200**, 247.
Bailey, M. E., 1985. In *Cosmical Gas Dynamics*, p. 49, ed. Kahn, F. D., VNU Science Press, Utrecht.
Bailey, M. E. & Clube, S. V. M., 1978. *Nature,* **275**, 278.
Binney, J. & Petrou, M., 1985. *Mon. Not. R. Astron. Soc.,* **214**, 449.
Habing, H. J., 1987. In *The Galaxy*, p. 173, eds. Gilmore, G. & Carswell, B., Reidel, Dordrecht.
Harmon, R. & Gilmore, G., 1988. *Mon. Not. R. Astron. Soc.,* **235**, 1025.
Lacey, C. G., 1984. *Mon. Not. R. Astron. Soc.,* **208**, 687.
Lin, D. N. C. & Tremaine, S. D., 1983. *Astrophys. J.,* **264**, 364.
May, A., van Albada, T. S. & Norman, C. A., 1985. *Mon. Not. R. Astron. Soc.,* **214**, 131.

Dynamics of the large-scale disc in NGC 1068

J. Bland[1] and **G. N. Cecil[2]**

[1]*Rice University, Texas, USA*

[2]*Institute for Advanced Study, Princeton, USA*

1 Observations

While NGC 1068 has received much attention in recent years, little is known of the large-scale dynamics of the ionized gas in this nearby Seyfert galaxy. We have used the Hawaii Imaging Fabry-Perot Interferometer (HIFI, Bland & Tully 1989) at the CFHT to obtain detailed spectrophotometry at 65 km s^{-1} FWHM resolution for Hα and the [NII]$\lambda\lambda$6548, 6583 lines. The final maps are derived from $\sim$ 100 000 fits to spectra taken at 0.4″ increments over a 200″ field-of-view. The flux-weighted Hα + [NII] velocity field is presented in Figure 1.

Deep images of NGC 1068 reveal an outer θ-shaped ring lying roughly east-west which encompasses a visibly bright, inner disc with diameter $\sim$ 20 kpc (230″), orthogonal to the outer ring. The Hα line flux is dominated by a luminous, elliptic ring of HII regions with diameter 3 kpc and major axis 45°. The "3 kpc ring" is aligned with an oval, bar-like distortion recently discovered at $\lambda2\mu$ (Scoville *et al.* 1988). The inner disc is marked by high concentrations of atomic and molecular gas ($\sim 10^{10}$ M$_\odot$) which is thought to fuel the rapid star formation that characterizes the 3 kpc ring (Scoville *et al.* 1983). Most of the ring bolometric luminosity ($1.5 \times 10^{11} L_\circ$), which is comparable to that of the active nucleus, emerges in the far-infrared as re-radiation from dust heated by young stars (Telesco *et al.* 1984).

From Figure 1, the large-scale disc appears to undergo flat rotation with V(R$_{max}$) = 170/ sin i where R$_{max}$ $\approx$ 30″. The disc orientation is consistent with the outer ring such that $i \approx 35°$ and the position angle, $\alpha \approx 85°$. Across the inner 30″, the Z-

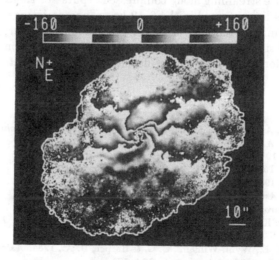

Figure 1. Hα+[NII] flux-weighted velocity map (west receding).

symmetric velocity contours are the hallmark of highly radial orbits along a bar at 45°
for the inferred disc orientation. We have identified a series of concentric, spiral "ridges"
which appear superimposed on the ordered, large-scale velocity field. These low angular
velocity features are evident in Figure 1 at 20″ east and west of the nucleus. The ridges
are continuous in velocity as they spiral across the 3 kpc ring to the north and south
and give rise to the pronounced contour stepping at larger radius. Beyond the 3 kpc
ring, these features coincide with faint beads of HII regions. The effect of these large
velocity residuals, $60/\sin i$ km s^{-1}, is seen in the CO (J=1-0) observations (Kaneko
et al. 1989), albeit at lower resolution (17″ FWHM), suggesting that the ionized and
molecular gas are coplanar and influenced by the same non-axisymmetric disturbance.

2 Discussion

From a recent high resolution study of the circum-nuclear activity, we rule out any direct
connection with the Seyfert nucleus (Cecil *et al.* 1989). Tidal distortions are discounted
only for lack of an obvious companion. We prefer an interpretation based on forcing
by the stellar bar as the outer θ−ring at a radius of $\sim 170''$ is strongly suggestive of an
outer Lindblad resonance (OLR). These are observed in 10–15% of early, barred spiral
galaxies (Athanassoula *et al.* 1982) and suggest that the $\lambda 2\mu$ bar is strong enough to
influence the large-scale dynamics of the galactic disc. We have derived a rotation
curve from the HIFI data out to a radius of 70″, and from HI measurements (Baan &
Haschick 1983) out to a radius of 5′. For an OLR at 15 kpc, the inferred pattern speed
of 10 km s^{-1} kpc^{-1} puts co-rotation at about 9 kpc; this corresponds roughly to the
outer edge of the bright, inner disc. The inner Lindblad resonances (ILR) would lie
within 2–3 kpc of the nucleus and, indeed, the ring region has been discussed in terms of
the transition from the inner to the outer ILR (Telesco & Decher 1989). Hydrodynamic
simulations of weak bars give rise to intrinsically elliptic rings near the ILR (Combes
& Gerin 1985). The ultraharmonic resonances (m=4), which occur at either side of
co-rotation, can produce streaming along compressed "spurs" or ridges (Schwarz 1981).
Since these resonances are expected in the 6–12 kpc range, it seems unlikely that they
account for the bisymmetric features in the velocity field. However, it is anticipated
that future dynamical simulations will be able to explain the large non-circular motions
and the sites of star formation in terms of density waves driven into the ISM by the
rotating stellar bar.

References

Athanassoula, E., Bosma, A., Creze, M. & Schwarz, M. P. 1982. *Astron. Astrophys.*, **107**, 101.
Baan, W. A. & Haschick, A. D., 1983. *Astron. J.*, **88**, 1088.
Bland, J. & Tully, R. B., 1989. *Astron. J., in press.*
Cecil, G. N., Bland, J. & Tully, R. B., 1989. *Astrophys. J., submitted.*
Combes, F. & Gerin, M., 1985. *Astron. Astrophys.*, **150**, 327.
Kaneko, N., *et al.*, 1989. *Astrophys. J.*, **337**, 691.
Schwarz, M. P., 1981. *Astrophys. J.*, **247**, 77 (Model A).
Scoville, N. Z., Matthews, K., Carico, D. P. & Sanders, D. B., 1988. *Astrophys. J. Lett.*, **327**, L61.
Scoville, N. Z., Young, J. S. & Lucy, L. B., 1983. *Astrophys. J.*, **270**, 443.
Telesco, C. M., Becklin, E. E., Wynn-Williams, C. G. & Harper, D. A., 1984. *Astrophys. J.*, **282**, 427.
Telesco, C. M. & Decher, R., 1988. *Astrophys. J.*, **334**, 573.

The flow of gas in barred galaxies

E. Athanassoula

Observatoire de Marseille, France

1 Technique

I have used a second-order flux-splitting code (van Albada 1985) to calculate the response of the gas in a galaxy model with a rotating bar. The model consists of one or more axisymmetric components (which one can identify with the bulge, disc or halo) and a homogeneous or inhomogeneous Ferrers-type bar (Ferrers 1877). There is one free parameter for the axisymmetric component, its central concentration, and three for the bar, its mass, axial ratio and pattern speed.

2 Results

A typical example of the gas response in the inner parts of the galaxy is given in Figure 1. The bar, which lies along the diagonal, has a length of 10 kpc. We see that the gas is very inhomogeneously distributed. There is relatively little gas over a large fraction of the surface and there are some very high concentrations in the centre and in the form of narrow strips along the leading edges of the bar. The form and location

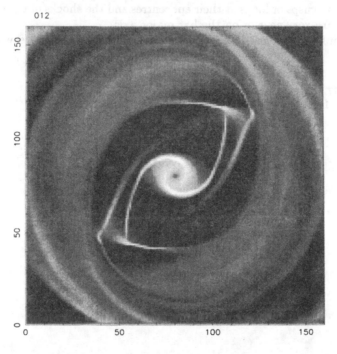

Figure 1. A typical example of the gas density response. The bar major-axis is at 45° and the side of the box is 16 kpc. The whiter areas in the grey-scale indicate higher density.

of these strips, which correspond to the shocks of the gas flow, are similar to those of the dust lanes observed in bars of galaxies of type SBb – SBc.

3 Discussion

I have made some 150 runs to cover the parameter space and have calculated the main families of periodic orbits for about one third of them. This has enabled me to determine the conditions for which the location of the shocks resembles that of the observed dust lanes.

The existence of the x_2 family is a necessary, though not sufficient, condition. For this reason models with an insufficient central concentration of the axisymmetric component do not develop offset shocks along the leading edges of the bar. This is in good agreement with the observations which show that these kind of dust lanes are not found in late type galaxies with little or no bulge.

Experiments with various values of the Lagrangian radius r_L, *i.e.* the radius of the Lagrangian points L_1 and L_2, indicate that it should be somewhat larger that the bar semi-major axis a, $r_L = (1.2 \pm 0.2)a$.

Changing the axial ratio of the bar changes the curvature of the location of the shocks. Models with thin bars can have straight shocks, as the example in Figure 1, while models with fat bars or ovals have curved shock loci, with their concave part towards the bar major axis. Certain orbits of the x_1 family for models with very eccentric bars have cusps or loops at their apocentres and the shocks of the gas responses of these models are very near or on the bar major axis.

A more complete account of these simulations will be published elsewhere.

References

Ferrers, N. M., 1877. *Quart. J. Pure Appl. Math.*, **14**, 1.
van Albada, G. D., 1985. *Astron. Astrophys.*, **142**, 491.

The warped dust lane in A1029-459

L. S. Sparke
*Scuola Normale Superiore, Pisa, Italy
and Washburn Observatory, Wisconsin, USA*
and
S. Casertano
Kapteyn Astronomical Institute, Groningen, The Netherlands

1 Introduction

In an earlier paper (Sparke & Casertano 1988), we have shown that a self-gravitating thin disc of material which is subject to the potential of a flattened spheroid can have a discrete mode of vertical oscillation. In that paper we examined the idea that the warps observed in disc galaxies might represent these discrete bending modes; there, the spheroidal component is the unseen halo. Here we consider whether the warp observed in the dust lane of the elliptical galaxy A1029-459 (Bertola, Galletta, Kotanyi & Zeilinger 1988, Bertola, Buson & Zeilinger 1988) could also be explained in terms of a warping mode; the spheroidal component is now the elliptical galaxy itself.

Fitting the warp in this dust-lane elliptical is a rather different test of the modal hypothesis from that provided by warped disc galaxies. In spiral galaxies, the disc is visible and its light distribution can be measured, while the shape of the dark halo is completely unknown. Here, by contrast, the ellipsoidal component is luminous but little is known about the disc except the ridge line of the warp. It should also be noted that this is a Type I warp (in our 1988 notation), with the plane of the dust bending *away* from the symmetry plane of the ellipsoidal component with increasing radius.

2 Model

We model the elliptical galaxy with the pseudo-Hubble profile which best matched the observations:

$$\rho(R, z) = \frac{\rho_0}{(1 + a^2/r_c^2)^{3/2}}, \qquad a^2 = R^2 + \frac{z^2}{(1 - \epsilon)^2}$$

where (R, z) are cylindrical polar coordinates, r_c specifies the halo core radius and ϵ the flattening of the density contours. The central density ρ_0 was fixed by setting the peak of the rotation curve to $260 \, \mathrm{km \, s^{-1}}$, the largest value measured in the gas disc; ϵ corresponds to the measured flattening of E3.5, and the radius r_c is set at $3''$, one-tenth of the measured de Vaucouleurs effective radius r_e.

We then guessed at the density distribution in the disc, varied the disc mass and calculated the warping mode. This shape was compared with the ridge line of the dust on the South side of the galaxy, where the warped material was better ordered, scaling our amplitude to match the observed height at $28''$ radius; Werner Zeilinger kindly supplied the measured points. The figure shows results for a ringlike density distribution similar to that observed for gas in early-type galaxies (Knapp 1987): the density rises exponentially with a scale length of $10''$ and is truncated smoothly at the edge so that the first derivatives are continuous everywhere, so as to avoid 'extra' modes caused by the sharpness of the edge (Hunter & Toomre 1969). The best-fitting curve

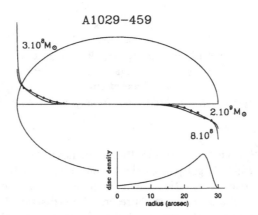

Figure 1. Maximum height above the galaxy equator for warping modes of a disc with indicated total mass in the potential specified above, compared with the observed dust lane; the best-fitting curve is drawn on both sides of the figure, with a curve corresponding to higher mass on the right side only, and to lower mass on the left side. The disc density profile is shown in the inset.

was obtained for a mass of 8.10^8 $M_\odot$; for masses below 2.10^8 $M_\odot$ there is no discrete mode and the disc cannot maintain a warp.

The disc mass required to obtain a good model depends weakly on the unknown disc parameters. We varied our model in a number of ways: making the truncation at the disc edge more or less abrupt, replacing the ringlike structure with a disc of constant density, changing the outer radius of the disc, and even reducing or increasing the core radius of the elliptical by a factor of two. Yet the best-fitting disc mass was always between 4.10^8 $M_\odot$ and 2.10^9 $M_\odot$, a fairly narrow range. Thus, our explanation of the bending as a discrete mode of oscillation of the disc in the potential of the galaxy is only viable if the mass in the disc is $\gtrsim 10^8$ $M_\odot$. Since most of the mass is likely to be in neutral hydrogen, an observation of the total HI mass in this galaxy would provide a clear-cut test of our interpretation.

References

Bertola, F., Buson, L. M. & Zeilinger, W. W., 1988. *Nature,* **335**, 705.
Bertola, F., Galletta, G., Kotanyi, C. & Zeilinger, W. W.,1988. *Mon. Not. R. Astron. Soc.,* **234**, 733.
Hunter, C. & Toomre, A., 1969. *Astrophys. J.,* **155**, 747.
Knapp, G. R., 1987. In *Structure and Dynamics of Elliptical Galaxies,* IAU Symp. **127**, p. 145, ed. de Zeeuw, T., Reidel, Dordrecht.
Sparke, L. S. & Casertano, S., 1988. *Mon. Not. R. Astron. Soc.,* **234**, 873.

Structure and evolution of dissipative non-planar galactic discs

T. Y. Steiman-Cameron[1] **and R. H. Durisen**[2]

[1] *NASA-Ames Research Center, California, USA*

[2] *Indiana University, USA*

1 Introduction

We present a simple analytic description of the evolution of a captured galactic gas disc as it settles into a preferred orientation. These discs, which often display warps and twists (*e.g.* Cen A), are widely believed to arise from the tidal capture of gas from a nearby galaxy or the accretion of a gas rich companion. The newly formed disc will initially be unstable and evolve on both the precessional and viscous time-scales. Differential precession causes a smooth continuous twist to develop. Cloud-cloud collisions within this twisted disc lead to the transport of angular momentum, causing changes in the orientations of cloud orbits and the inflow of material. Ultimately, the disc settles into a preferred orientation.

Steiman-Cameron & Durisen (1988) (hereafter SCD) derived three coupled differential equations governing the evolution of annular mass elements of a fluid disc including the effects of both orbit precession and viscosity. These equations, which describe the transport of angular momentum by Navier-Stokes stresses, describe the motion of orbit-averaged annular fluid elements. In general these equations must be solved numerically. However, analytic solutions are possible when: (1) disc precession is dominated by gravitational forces, with viscous forces being a minor perturbation. This condition is generally met. (2) The viscous timescale for inflow is much longer than the timescale for disc settling. This is true if condition four is met. (3) Settling takes place in an axisymmetric galaxy. However, if the initial disc inclination relative to a preferred orientation is small, then this condition is often approximately met even in triaxial potentials. (4) The initial inclination of the disc relative to the equatorial plane is small. However, SCD found very good agreement between analytic predictions and numerical models even for moderate initial inclinations.

2 Approximate solution

The three-dimensional geometry of a disc is provided by the inclination i and longitude of ascending node Ω of the disc as a function of radius. With the above conditions, the shape of the warp and twist in a settling disc can be determined by only two parameters. With the effective settling time $\tau_e = [(\nu/6)(d\dot{\Omega}_p/dr)^2]^{-1/3}$, where ν is the coefficient of kinematic viscosity and $\dot{\Omega}_p$ the nodal precession rate (SCD), we define the dimensionless quantities $\aleph = (\tau_p/\tau_e)$ and $T = t/\tau_p$, where $\tau_p = 2\pi/\dot{\Omega}_p$ is the precession period. If the initial disc orientation is $(i, \Omega) = (i_0, 0)$, then the time-dependent disc structure is given by (Steiman-Cameron & Durisen 1989)

$$i/i_0 = \exp[-(\aleph T)^3], \quad \text{and} \quad \Omega = 2\pi T. \tag{1}$$

Unfortunately, $\aleph$ is generally a function of radius. However, under conditions that are astrophysically interesting, $\aleph$ can be constant. For galaxies with extended halos, a useful gravitational potential is the scale-free logarithmic potential (*cf.* Sparke 1986 and references therein). Models with this potential possess similar concentric isodensity surfaces and constant circular velocity v_c with radius. The orbit-averaged nodal precession rate in this potential is

$$\dot{\Omega}_p = -(3v_c/4r)\eta \cos i, \tag{2}$$

where $\eta = [2 - 2(b^2/a^2)]/[3(b^2/a^2) + 6]$ and (b/a) is the axis ratio of galactic isodensity surfaces. The effective settling time becomes $\tau_e = (32/3)^{1/3} r^{4/3} \nu^{-1/3} |\eta v_c \cos i_0|^{-2/3}$.

For real galaxies, η will vary by less than a factor of two. Therefore, only ν is poorly determined in the expression for τ_e. Unfortunately, our limited understanding of dissipative processes in galactic discs does not permit a precise description of ν. However, the weak dependence of τ_e on ν allows reasonable approximations for ν to serve for the determination of τ_e. Assuming cloud-cloud interactions to be the dominant dissipative mechanism for settling, SCD developed a simple analytic estimate for the maximum permissible value of ν. In galaxies with flat rotation curves, this is given by $\nu_{\max} = 0.184 \, r v_{\mathrm{rms}}^2/v_c$, where v_{rms} is the root mean square cloud velocity. The use of $\nu_{\max}$ produces minimum settling times. Substituting $\nu_{\max}$ into the expression for τ_e and normalizing to the precessional period yields

$$(\tau_e/\tau_p)_{\min} = \aleph^{-1} = 0.46 \, (v_c/v_{\mathrm{rms}})^{2/3} |\eta \cos i_0|^{1/3}. \tag{3}$$

Thus $\aleph$ is constant for constant v_{rms}, while equation (2) shows that T is linear in t and inversely proportional to r. Therefore, from equations (1), looking at the disc at different times t is equivalent to looking at the disc at different radii, greatly simplifying application of these models to real galaxies. The disc configuration in the settling region, *i.e.* the region in which the inclination is rapidly changing with radius, is completely determined by $\aleph$ and i_0. The fitting parameters $\aleph$ and i_0, along with the choice of scale and determination of the viewing orientation are all that are required in principle to model observations of settling discs of small or moderate initial inclinations.

Equation (3) reveals that the settling into a preferred orientation typically takes ~ 0.5 to 2 precession periods. Using smooth particle hydrodynamic simulations of settling discs, an approach quite different from ours, Habe & Ikeuchi (1985, 1988) found settling times of ~ 1 to 3 precessional periods, for most geometries and rotation states. This agreement suggests that the timescale for settling is not strongly dependent on the dissipative mechanism, consistent with the weak dependence of τ_e on ν. The differential precession rate is the dominant determinant of settling times, as argued earlier by Simonson (1982) and by Tohline, Simonson & Caldwell (1982).

References

Habe, A. & Ikeuchi, S., 1985. *Astrophys. J.*, **289**, 540.
Habe, A. & Ikeuchi, S., 1988. *Astrophys. J.*, **326**, 84.
Simonson, G. S., 1982. *PhD thesis*, Yale University.
Sparke, L. S., 1986. *Mon. Not. R. Astron. Soc.*, **219**, 657.
Steiman-Cameron, T. Y. & Durisen, R. H., 1988. *Astrophys. J.*, **325**, 26 (SCD).
Steiman-Cameron, T. Y. & Durisen, R. H., 1989. *Astrophys. J.*, submitted.
Tohline, J. E., Simonson G. F. & Caldwell, N., 1982. *Astrophys. J.*, **252**, 92.

Non-axisymmetric magnetic fields in turbulent gas discs

K. J. Donner and **A. Brandenburg**
University of Helsinki, Finland

1 Introduction

Large-scale magnetic fields could play an important role in the dynamics of astrophysical discs. Here we report some results showing how the structure of non-axisymmetric magnetic fields is affected by differential rotation. A turbulent disc is likely to be surrounded by a gaseous corona. We shall study in particular how the field structure in the disc is affected by surrounding gas.

We are interested in the question of the origin of galactic magnetic fields. It appears that an appreciable fraction of galactic fields are bisymmetric, *i.e.* the field in alternate spiral arms is in opposite directions (Sofue *et al.* 1988). This poses a problem, since on general grounds one expects that non-axisymmetric fields should be destroyed by differential rotation on a fairly short timescale. This difficulty would be avoided if it could be shown that a non-spherically symmetric distribution of turbulent diffusivity could actually lead to dynamo generation of non-axisymmetric fields, as was suggested by Rädler (1983) and Skaley (1985). For this reason, and also in order to avoid the uncertainties involved in assuming some specific model, we have not included an α-effect in these computations. Unfortunately, we have not found any growing field modes yet, but the decaying modes are of some interest in their own right.

2 Method and results

The evolution of the magnetic field is governed by the induction equation. We solve for the eigenmodes in a system consisting of a high-conductivity disc embedded in a low-conductivity corona using the so-called Bullard-Gellman formalism. For details of the model and numerical methods, see Donner & Brandenburg (1989) and references therein.

From the point of view of mechanisms giving preference to non-axisymmetric modes our results so far are disappointing. In all cases the first bisymmetric mode decays faster than the first axisymmetric one. Increasing the rate of rotation (measured by the magnetic Reynolds number) increases the rate of decay of the bisymmetric modes and leaves the axisymmetric modes unaffected. On the other hand, within a fairly large range of magnetic Reynolds numbers the decay rate of non-axisymmetric modes is not very strongly affected. Thus, once a bisymmetric field has been generated (*e.g.* in a galaxy merger) it may survive for quite a long time on a dynamical timescale.

The structure of the field in the plane of the disc is determined by the balance between the tendency of the differential rotation to draw the frozen-in field into a tight spiral and the vertical diffusion of the field out of the disc. Thus, for a given Reynolds number in the disc, as the diffusivity in the corona becomes larger, vertical diffusion becomes faster and the spiral field in the disc becomes more open. This tendency is illustrated in Figure 1 which shows the field vectors in the disc plane for the first bisymmetric mode for two different values of the conductivity in the corona. More remarkably, there is a complete change in the geometry of the field. For a high-conductivity corona the field forms a strong trailing spiral in the differentially rotating part of the disc, whereas in the low-conductivity case it is mostly confined to the central almost rigidly rotating region. For a given coronal conductivity the type of behaviour is determined by the Reynolds number.

Clearly, the presence of a gaseous corona surrounding the disc may have a strong effect both on the decay rate and on the field structure in the plane of the disc. Observations of the magnetic fields of galaxies high above the disc plane would be of great importance in order to get some idea of the coronal conductivity.

References

Donner, K. J. & Brandenburg, A., 1989. *Geophys. Astrophys. Fluid Dyn.*, (submitted).

Rädler, K.-H., 1983. In: *Stellar and Planetary Magnetism*, p. 17, ed. Soward, A. M., Gordon and Breach, New York–London–Paris.

Skaley, D., 1985. "Sphärische Dynamos mit differentieller Rotation und ortsabhängiger Leitfähigkeit", Diploma thesis, University of Freiburg.

Sofue, Y., Fujimoto, M. & Wielebinski, R., 1988. *Annu. Rev. Astron. Astrophys.*, **24**, 459.

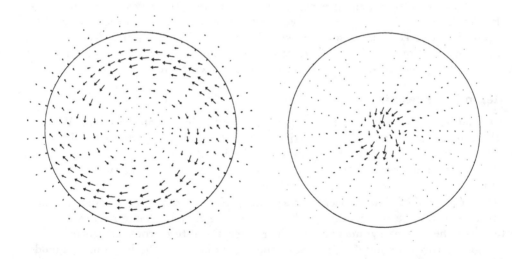

Figure 1. Field vectors in the disc plane for the first bisymmetric mode for a typical galactic rotation curve. The magnetic Reynolds number based on the central diffusivity and disc radius is 2000. The coronal diffusivity is four times (*left*) and a hundred times (*right*) larger than in the disc.

Non-axisymmetric disturbances in galactic discs

Alar Toomre

Massachusetts Institute of Technology, USA

1 Introduction

In two contexts, this talk revisited an old idea – that any concentrated mass orbiting within a shearing disc tends to create an inclined and sometimes quite intense wavelike *wake* amidst the material that flows past it. More exactly, as actually delivered, this talk began with a fairly lengthy review of such swing-amplified wake-making in the gorgeous global setting of Zang's $V =$ const disc, and only then did it proceed to my two main contexts.

2 Polarisation in the shearing sheet

One context was the idealized shearing sheet composed of thousands of identical softened mass points, with which I have been conducting a long series of numerical experiments in recent years. As I illustrated with slides and even with a homemade (!) movie, the typical impression there is one of an ever-changing *kaleidoscope of "spiral" features*, very much suggesting some sort of a recurrent strong instability in a system that really ought not to have any. Appearances aside, it turns in fact that those features are no instabilities. Instead, they are logically just superpositions of the separate wakes of the many random particles, each of which in this sense keeps acting simultaneously as both an aggressor and a victim. It is the *collection* of these wakes that forever keeps shifting in location and appearance as the individual particles constantly drift in this shear flow, but each particle in turn always carries along a wake of its own. Various extensive statistical analyses of my experiments, including two-particle correlation functions, Fourier decompositions of these streak patterns, and also a very close look at the resulting heating rates – of all of which I showed only snippets in the talk itself – leave little doubt that such wakes are indeed the bulk of the story in this local setting.

3 Edge instabilities

The other context concerned genuine instabilities, of a type to which all gravitating discs seem prone if their surface density changes abruptly enough from one value to another (usually but not necessarily to zero) as near an outer edge. I offered, and briefly defended, a simple intuitive explanation for these so-called *edge instabilities* that are logically akin to the curiously misfit "mode D" of the Gaussian disc that Kalnajs and I discovered a decade ago: Here the positive or negative extra densities that result from the almost co-rotational distortions of the edge can again be thought of as the imposed masses, and the inclined spiral waves which they create inward of themselves are their wakes, now understandably one-sided. It is the net forward pull from each such wake that supplies angular momentum to its parent edge region and thus aggravates its bulging at a very noticeable rate if (a) the edge is reasonably sharp and (b) the disc is massive and cool enough for vigorous swing amplification. Reasonable sharpness here means that the radial distance over which the disc density undergoes most of its rapid change should be no larger than about one-quarter of the axisymmetric stability length λ_{crit}, whereas the two latter conditions require that the wavelength $X \lesssim 3$ and the "temperature" $Q \lesssim 2$ in the usual swing-amplifier terminology, at least near radii where the speed of rotation is roughly constant.

Spiral instabilities in N-body simulations

J. A. Sellwood

University of Manchester

Abstract *N*-body simulations of disc galaxies that display recurrent transient spiral patterns are comparatively easy to construct, but are harder to understand. In this paper, I summarise the evidence from such experiments that the spiral patterns result from a recurrent spiral instability cycle. Each wave starts as rapidly growing, small-amplitude instability caused by a deficiency of particles at a particular angular momentum. The resulting large-amplitude wave creates, through resonant scattering, the conditions needed to precipitate a new instability.

1 Plan

The problem of spiral structure in galaxies has been worked on for many years but progress has been painfully slow. Most effort has been directed towards the development of an analytical (or at least semi-analytical) approach and many aspects of the problem have been discovered (see Sellwood 1989 for a review). Here, I collect the evidence from *N*-body simulations which indicates that the structure is continuously variable and results from a recurrent cycle of spiral instabilities.

A subsidiary purpose of this paper, is to convince the reader of the advantages of using *N*-body simulations *in tandem* with approximate analytic treatments. Without a close comparison of this nature, each separate approach is much less powerful; the limitations of the *N*-body experiments remain unquantified and the validity of the approximations in the analytic approach cannot be assessed.

The paper is divided into three distinct sections. In §2, I discuss swing-amplified noise in global simulations, and show that the behaviour in the Mestel ($V =$ const.) disc is very similar to that reported by Toomre (*e.g.* this conference) for simulations in the shearing sheet. However, other more realistic models display much larger amplitude structure and wave coherence than can be accounted for by swing-amplified noise alone.

At first sight, the instability caused by a groove in a disc, which is described in §3, is totally unrelated to the previous section. The vigorous instability provoked by such an apparently arbitrary feature leads to a large amplitude spiral pattern. I present a simplified local analytic treatment which yields a rough prediction for the eigenfrequency.

The third part of this story describes an experiment which suggests how the two previous sections might be related. The coherent waves discussed briefly in §4 appear to result from a *recurrent cycle* of groove instabilities.

2 More than particle noise

2.1 ZANG DISCS

The stability properties of the infinite Mestel (1963) disc are quite remarkable. The disc has the surface density

$$\Sigma(r) = \frac{qV_0^2}{2\pi Gr},$$

which gives an exactly flat rotation curve all the way from the axis of rotation. Here, V_0 is the orbital speed for circular motion at any radius r which arises from this mass distribution when $q = 1$. Though the surface density is singular at the centre, the mass within any radius is, of course, finite.

Zang (1976) examined the global stability properties of centrally cut-out, but otherwise full-mass, models having this density distribution and found, rather surprisingly, that the most persistently unstable modes occurred for $m = 1$. The $m = 2$ modes could be completely suppressed if the velocity dispersion were comparatively modest and the central cut-out not too abrupt (see also Toomre 1977). The central cut out he used took the form of a taper in angular momentum, J, of the distribution function: $T(J) = [1 + (r_0 V_0/J)^n]^{-1}$, where r_0 is some scale radius and the index $n \lesssim 2$ for stability. It now seems clear that the instabilities Zang found in models with $n > 2$ were edge-related modes.

In largely unpublished subsequent work, referred to in Toomre (1981), Zang and Toomre extended the study to include discs for which $q < 1$, *i.e.* embedded within a halo that did not alter the shape of the rotation curve. They found that a half-mass disc model ($q = 0.5$) could be *completely stable* to all global modes, provided the central taper was not too abrupt ($n \leq 4$). The radial velocity dispersion in their stable model $\sigma_u = 0.2835 V_0$ which gives $Q = 1.5$ for all $r \gg r_0$.

The absence of instabilities makes this model an ideal test-bed in which to study, in a global context, a variety of phenomena that have been studied in the (local) shearing sheet model. Indeed Toomre (1981) already gives a memorable illustration of swing-amplification that is far more eye-catching than any diagram produced in an analysis of the shearing sheet, even though it was first discovered in that context some fifteen years earlier (Goldreich & Lynden-Bell 1965, Julian & Toomre 1966, hereafter JT).

2.2 SWING-AMPLIFIED NOISE

I have therefore run a series of simulations of this model which illustrate swing-amplified particle noise in a global context. These experiments are the logical counterparts to Toomre's simulations in the local shearing sheet model (a preliminary report of which is given in this volume). These are essentially empirical measurements of polarisation in a disc of particles characterised by $Q = 1.5$. As quiet starts would obviously be inappropriate for such an investigation, the initial azimuthal coordinates were chosen randomly.

Unfortunately, it is necessary, for obvious computational reasons, to truncate the disc at some outer radius also. I have not yet succeeded in doing this without provoking at least a mild $m = 2$ instability at the outer edge. Though the simulations are not completely stable, therefore, the growth of the outer edge mode is slow enough that the behaviour for the first hundred dynamical times is almost completely consistent with swing-amplified particle noise, as can be seen from the following.

In order to measure the amplitude of density variations, I formed the summation

$$A(m, \gamma, t) = \frac{1}{N} \sum_{j=1}^{N} \exp[im(\theta_j + \tan \gamma \ln r_j)], \qquad (1)$$

at intervals during the simulations. Here, N is the number of particles and (r_j, θ_j) are the coordinates of the jth particle at time t. The resulting complex coefficients A are the logarithmic spiral transformation of the particle distribution and γ is the inclination angle of the spiral component to the radial direction (positive for trailing waves).

The top line of panels in Figure 1(a–c) shows the time-averaged value of $|A|$ as a function of $\tan\gamma$ for the $m = 2$, 3 & 4 components from three different experiments which span two decades in N. The time interval, in units of r_0/V_0, chosen for the average was $t = 25$ to $t = 100$ in each case – a period after the polarisation is reasonably well developed over most of the disc and before the outer edge instability had reached any significant amplitude.

The leading/trailing bias is obvious in all these plots, and is largest for $m = 2$. The horizontal dot-dash lines indicate the expectation value of $|A|(-\sqrt{\pi/4N})$ for randomly distributed particles (Sellwood & Carlberg 1984). The measured amplitude on the far leading side is completely consistent with randomly distributed particles in all three experiments. The peaks on the trailing side are higher than the leading signal by an approximately equal factor in the $N = 20K$ and $N = 200K$ experiments, but noticeably less when $N = 2K$. Spiral amplitudes in the $N = 2K$ experiment are so large as to be clearly visible which probably limits the measured bias for two reasons: peak amplitudes may well be limited by non-linear effects, and the velocity dispersion rises very quickly – making the disc less responsive. [This simulation is illustrated in Sellwood (1986).]

Following Toomre, I have estimated the bias that should be observed using the JT apparatus and found excellent agreement for the $m = 3$ & 4 components – the $m = 2$ bias in these global experiments is slightly larger than predicted by this local theory (Sellwood, in preparation).

In conclusion, results from these three experiments of this very nearly stable disc are quite consistent with theoretical expectations of swing-amplified particle noise.

2.3 MORE REALISTIC MODELS

Measurements over a comparable time interval from a parallel series of experiments with the "Sc" model of a disc galaxy used by Sellwood & Carlberg (1984, hereafter SC) are shown in the bottom row of panels in Figure 1(d–e). These experiments are not quite as comparable as would be desirable – it was necessary to use a lower mass fraction in the disc (30%) in order to prevent a rapid bar instability at the centre, and the initial velocity dispersion was set so as $Q = 1$ only. These differences shift the peak response to $m = 3$ and make the discs more responsive, respectively.

Nevertheless, the behaviour differs from that in the Zang models by more than can be accounted for by these reasons alone. The most obvious two differences are that the amplitudes on the leading side in the larger N experiments are no longer consistent with randomly distributed particles and nowhere do the signals decrease as $1/\sqrt{N}$. Though the amplitudes in the 200K particle experiment (f) are lower than those in (e), they rise throughout this time interval (1 – 4 rotation periods at the half-mass radius), attaining values no different from those in the lower N experiments by the end. This behaviour is clearly inconsistent with the hypothesis that the non-axisymmetric structure in these more realistic models can be attributed to swing-amplified particle noise.

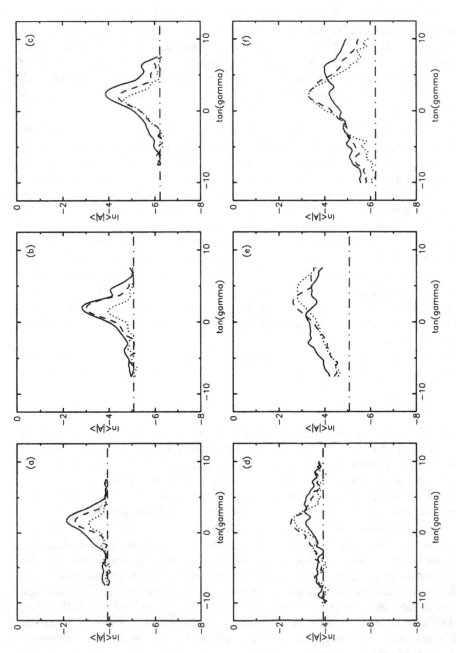

Figure 1. Time averaged values of the transform (equation 1) from two series of experiments in which the particle number is changed. The results for $m = 2$ – full drawn, $m = 3$ – dashed and $m = 4$ – dotted curves are shown in each case. The horizontal dot-dash line indicates the expectation value for N randomly distributed particles. The top row (a-c), show results from Zang models, the bottom row from Sc models, and N rises from 2K (left hand panels) through 20K (centre panels) to 200K (right hand panels).

2.4 COHERENT WAVES

As the amplitude of the spiral waves in these experiments always reached levels sufficient to heat the disc, the responsiveness and general level of activity declined as the
experiments evolved. It was therefore impossible to study structure in these uncooled
models over long periods. Therefore Carlberg and I, in further (unpublished) work with
this mass model, devised a cooling algorithm which enabled us to run a long experiment
that remained in a quasi-steady state.

Our cooling strategy in this case was to remove particles at random from the disc
and to re-insert them at some other randomly chosen position on a locally-determined
circular orbit. The distribution from which the new position was chosen was precisely
that of the initial axisymmetric disc. We devised this cooling technique, which differs
slightly from the accretion method used by SC, in order to avoid a steadily rising disc
mass. This process might be thought of as mimicking the death of a star and the formation of a new star elsewhere in the disc – the gas phase of the disc simply being omitted
from the dynamics. It is physically unrealistic in an additional important respect since
it steadily redistributes angular momentum inwards, undoing that transferred outwards
by the spirals, in order to maintain a quasi-steady distribution of angular momentum
amongst the particles. The cooling rate we adopted was also quite unrealistically high
– 15 particles per time-step, or 15% of the mass of the disc per rotation period.

We ran a $N = 20K$ model calculated according to this rule for 432 dynamical
times (one rotation period at the half-mass radius is 16 dynamical times). Starting
from an initial $Q = 1.7$, the model adjusted within the first ~ 30 dynamical times
to quasi-steady values $1.8 \lesssim Q \lesssim 2.0$ over most of the disc. The $m = 3$ component
of $|A|$, from equation (1), is contoured in Figure 2(a); the inclined stripes indicate
recurrent shearing transient spiral waves, rather similar to those reported in SC. In this
simulation, however, the amplitude of successive events soon settles to an approximately
constant value, consistent with the quasi-steady state.

Figure 2(b) shows the power spectrum of this apparently stochastic sequence of
spiral events. Narrow horizontal ridges in this diagram indicate coherent waves rotating
at constant angular frequency over a wide range of inclination angles. The narrowness
of the ridges in the frequency direction indicates that the individual waves must be very
long lived – the ridges are scarcely broader than our frequency resolution.

The spatial shapes of six of these apparently coherent waves are shown in Figure 3.
All are trailing spiral patterns which show remarkable respect for their principal resonances; the full-drawn circle marks the co-rotation resonance, the dotted circles the
inner and outer Lindblad resonances (ILR and OLR respectively). All but the most
rapidly rotating of these long-lived coherent waves possess ILRs, and only the slowest
lacks an OLR within the particle distribution. The horizontal dashed lines in Figure 2(b) mark the limits of the angular frequency range for which waves have both
Lindblad resonances within the disc of particles.

Though the waves survive for some time, they do not last indefinitely. Figure 2
shows data from the first half of the experiment – a similar analysis of the second half
(not shown) reveals other long-lived waves, but their frequencies, while in the same
range as those shown in Figure 2(b), are not identical. Each wave appears to have a
finite lifetime, which unfortunately is not easy to measure – a reasonable estimate might

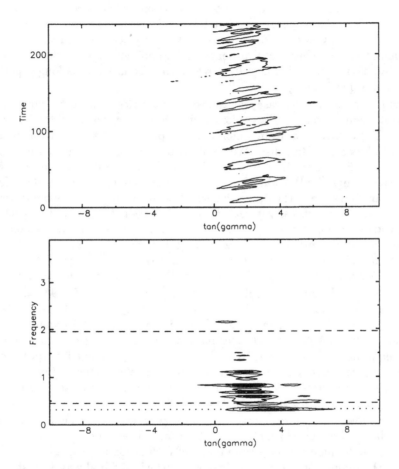

Figure 2. Results from a quasi-stationary Sc model. The rotation period at the half-mass radius in this model is 16 time units. (a) Contours of the $m = 3$ component of $|A|$ as a function of time and inclination angle. (b) A power spectrum of the same data, revealing a number of coherent waves at constant frequencies over wide ranges in pitch angles.

be ~ 160 dynamical times or ten rotations. New waves continually appear to replenish those that decay.

Long-lived coherent waves are hardly to be expected from swing-amplified particle noise. It is tempting to describe them as modes of oscillation of the disc, though the fact that the majority possess ILRs is a real surprise. Orthodox spiral mode theory (*e.g.* Toomre 1981) might be able to account for (a) with a feed-back loop and (f) as an edge mode (the horizontal dotted line in Figure 3(b) shows the circular angular frequency at the outer edge) but it could not account for most of the frequencies we observe, because such waves should be strongly damped at both Lindblad resonances.

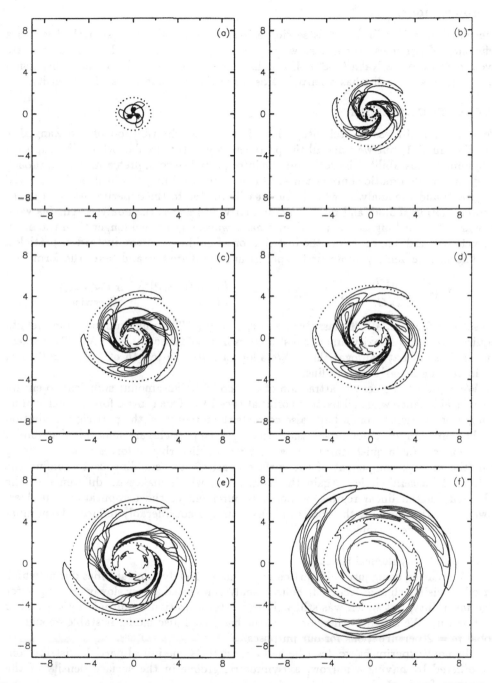

Figure 3. Six representative waves from the power spectrum in Figure 3(b). The co-rotation circle (full drawn) and the Lindblad resonances (dotted) are marked. The scales are in units of the scale radius of the Sc model (SC).

3 Groove modes

The instability described in this section arises when an otherwise smooth disc has a deficiency of particles over a narrow range of angular momenta. Further details are given in Sellwood & Kahn (1989). If the disc is cold, *i.e.* the radial velocity dispersion $\sigma_u = 0$, then such a disc has a narrow, axisymmetric groove in the surface density.

3.1 COLD DISC EXPERIMENTS

We use the half-mass Mestel disc, which is the mass distribution of the Zang disc described in §2.1, but we start all the particles on exactly circular orbits. We suppress axisymmetric instabilities by softening the inter-particle forces, preferring this approach to adding random motion only because the cause the instability can be directly observed (Figure 6) and the analysis (§3.2) is more easily related to the experiments. Discs with random motion should, and do, behave similarly. Moreover, the analysis remains valid for warm discs as long as the disturbance has a wavelength much longer than the mean epicycle size and one considers the distribution of guiding centres, instead of particles.

The surface density in our first experiment has no groove, and takes the form

$$\Sigma_0(r) = \frac{qV_0^2}{2\pi Gr}T(r), \qquad \text{with} \quad T(r) = \begin{cases} [1 + (r/r_0)^4]^{-1} & \text{if } r < r_{\max}; \\ 0 & \text{otherwise}. \end{cases}$$

We set $q = 0.5$ (*i.e.* a half-mass disc) and $r_{\max} = 9r_0$. The minimum softening length required to suppress axisymmetric instabilities in a cold disc is $\epsilon_{\min} = \lambda_{\text{crit}}/2\pi e$, where e is the base of natural logarithms. We adopt a softening length of $1.5\epsilon_{\min} \simeq 0.138r$, *i.e.* increasing linearly with radius.

We augment the central attraction of the model by an amount sufficient to ensure centrifugal balance when all particles orbit at speed V_0. This central force, which can be thought of as arising from a halo, also corrects the forces from the particle distribution for the central taper, outer cut-off and softening. The particles are smoothly distributed at the outset (*i.e.* a quiet start) and we restrict the disturbance forces to those arising from the $m = 2$ component of the surface density only, enabling us to confine the particles to a semi-circle. Again this restriction, which makes no difference to the behaviour in the linear regime, is not a requirement of the computation. It does, however, further simplify the observed behaviour and considerably reduces the number of particles we need employ.

We adopt a system of units such that $r_0 = V_0 = G = 1$. The orbital period at $r = 1$ is therefore 2π dynamical times.

A simulation using just 15K particles was run for 200 dynamical times during which period no visible change occurred. Fourier analysis revealed a very slowly growing outer edge instability, but which was still well below its saturation amplitude by the end of the test run. We conclude therefore that the basic cold disc model is stable enough to global $m = 2$ perturbations for our purposes.

The next experiment we describe had almost the same initial particle distribution, but differed by having a narrow, axisymmetric groove in the surface density of the Lorentzian form

$$\Sigma(r) = \Sigma_0(r)\left[1 - \frac{\beta w^2}{(r - r_*)^2 + w^2}\right], \tag{2}$$

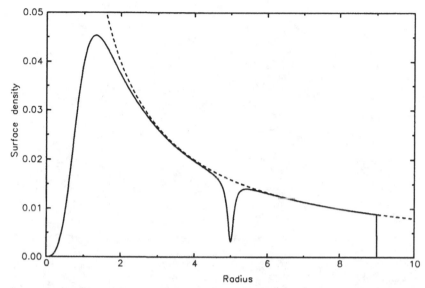

Figure 4. The axisymmetric surface density distribution in the model illustrated in Figure 5 (full drawn curve). The dashed curve shows the basic Mestel disc without an inner taper, groove or outer edge. Units are in r_0 and $V_0^2/(Gr_0)$.

as illustrated in Figure 4. We set the groove centre $r_* = 5$, its width $w = 0.1$ and fractional depth $\beta = 0.8$. Results from this simulation are shown in Figure 5 – the model is drastically unstable and develops a global, large-amplitude, spiral pattern.

The early evolution of the model is omitted from Figure 5 because no visible changes occur, but Fourier analysis can detect a rapidly growing instability from a very early stage. The Fourier coefficients are extremely well fitted by a single exponentially growing mode having the spatial form shown in Figure 6 and eigenfrequency $\omega = 0.383 + 0.072i$. The growth rate, $\Im(\omega)$, is therefore more than one third the pattern speed, $\Re(\omega)/m$, indicative of an extremely vigorous dynamical instability.

Figure 6 shows that co-rotation (the full drawn circle) is close to, but just outside the groove centre (marked by a dashed circle), and that the disturbance is very strongly localised in this region. There are weaker, trailing spiral arms extending towards the Lindblad resonances (dotted circles) on either side of the groove.

3.2 LOCAL ANALYTIC TREATMENT

The global instability is driven by the non-axisymmetric density changes that develop in a narrow range of radii around the groove. We therefore find that a local analysis of that region yields reasonable estimates of the mode frequency.

We use co-ordinates (ξ, η) to represent the forced radial and transverse displacements of a guiding centre from its circular orbit caused by the presence of wave-like disturbance potential. The origin of these displacement co-ordinates moves on a circular path at radius r with orbit speed V_0 and therefore angular velocity $\Omega(r) \equiv V_0/r$.

We consider a weak, uniformly rotating perturbing gravitational field with a potential of the form $\Phi(r)e^{i(m\theta - \omega t)}$, which is small enough for a linearised treatment of the

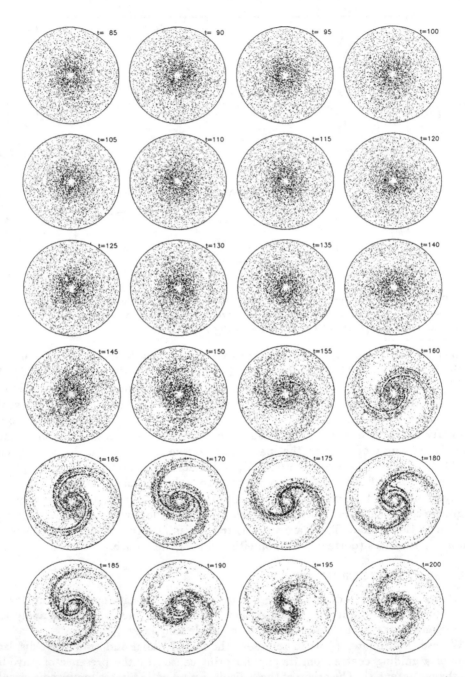

Figure 5. Evolution of a cold disc model with a deep, narrow groove. Times are in units of r_0/V_0 – a rotation period at the groove is therefore 10π. Note that the early evolution of the model, during which no visible changes occur, is omitted. The calculation was restricted so that only $m = 2$ components of the disturbance potential affected the motion of the particles.

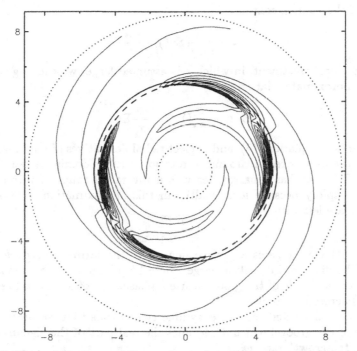

Figure 6. The best fit exponentially growing mode to the small amplitude evolution of the model shown in Figure 5. Only the positive part is shown. The full drawn circle marks co-rotation and the dotted circles the Lindblad resonances for the best-fit frequency. The groove centre is marked by the dashed circle.

forced motions to be valid.

The linearised equations of motion are

$$\ddot{\xi} + 2r\Omega\Omega'\xi - 2\Omega\dot{\eta} = \frac{\partial\Phi}{\partial r}$$

$$\ddot{\eta} + 2\Omega\dot{\xi} = \frac{im\Phi}{r},$$

with the exponential factor understood. We have used a dot to denote differentiation with respect to time (following the path of the guiding centre) and a dash denotes differentiation with respect to r. These equations can be readily integrated; setting $\kappa^2 = 2\Omega^2$ (for a flat rotation curve) we obtain

$$\xi \sim -\frac{m}{V_0\kappa\nu}\Phi \qquad \text{and} \qquad \eta \sim \frac{im}{r\kappa^2\nu^2}\Phi, \tag{3}$$

where $\nu = (\omega - m\Omega)/\kappa$, a dimensionless forcing frequency that has a small real part near co-rotation. We have therefore neglected terms containing $\partial\Phi/\partial r$, since these enter with lower power of ν in the denominator and are considerably smaller than the terms we retain.

The equation of continuity relates the disturbance surface density, Σ_1, to the displacements of the guiding centres; *viz.*

$$\Sigma_1 = -\frac{1}{r}\frac{\partial}{\partial r}(\Sigma r \xi) - \frac{im}{r}\Sigma \eta.$$

In our current local treatment, in which r is assumed large, we can neglect curvature and use the approximate relation

$$\Sigma_1 \simeq -\frac{\partial}{\partial r}(\Sigma \xi) - \frac{im}{r}\Sigma \eta.$$

We substitute from equations (3), and neglect radial derivatives of κ and Φ, which vary slowly across the narrow region around co-rotation; *i.e.* the dominant radial derivatives are those of Σ (which has a narrow feature in the radial range of interest) and $1/\nu$ (which varies rapidly because ν is small). In this approximation, two of the terms cancel and we are left with

$$\Sigma_1 \sim \Sigma'\frac{m}{V_0 \kappa \nu}\Phi. \tag{4}$$

Equation (4) clearly gives a surface density of the form observed for the mode (Figure 6). The disturbed density is large where Σ' is large, but the opposite signs of this gradient on each side of the groove cause a phase shift of almost π (*i.e.* a 90° shift in real space) across the groove.

Since we have neglected curvature, symmetry demands that co-rotation should be at the centre of the groove, which determines the real part of the eigenfrequency. We now estimate the growth rate, ω_i.

Because our softening length is large compared with the width of the groove, we may compute the potential of the entire disturbance as if it were a single sinusoidal wave on the circle at $r = r_*$ having an amplitude

$$C = \int \Sigma_1 dr \simeq \int \Sigma'\frac{m}{V_0 \kappa \nu}\Phi dr, \tag{5}$$

where the range of integration includes all stars whose orbits are substantially perturbed.

We can write the surface density (2) in the vicinity of the groove in the form

$$\Sigma(x) = \Sigma_0(r)\left[1 - \frac{\beta w^2}{(r - r_*)^2 + w^2}\right] \equiv \frac{\Sigma_0(r)}{2}\left[2 + i\beta w\left(\frac{1}{x - iw} - \frac{1}{x + iw}\right)\right],$$

where $x = r - r_*$. Using this in (5), we find

$$C = \frac{m}{V_0}\Phi \int_{-\infty}^{\infty} \frac{-i\beta w \Sigma_0}{2}\left[\frac{1}{(x - iw)^2} - \frac{1}{(x + iw)^2}\right]\frac{1}{\omega - m\Omega}dx.$$

Writing $\omega - m\Omega = mV_0(x - \lambda)/r_*^2$, where $\lambda = r_*(1 - \omega r_*/mV_0)$, we may evaluate the integral by closing off the contour in the lower half of the complex plane (since $\Im(\lambda) < 0$), and obtain

$$C = \frac{m\Phi}{V_0} \cdot \frac{-i\beta w \Sigma_0 r_*^2}{2mV_0} \cdot \frac{-2\pi i}{(\lambda - iw)^2}.$$

Because co-rotation is at r_*, $\lambda = -i\omega_i r_*^2/mV_0$ (*i.e.* purely imaginary). The above expression therefore simplifies to

$$C = \frac{\pi m^2 \beta w \Sigma_0 r_*^2 \Phi}{(\omega_i r_*^2 + mV_0 w)^2}. \tag{6}$$

The potential in the groove arising from this sinusoidally varying disturbance with mass per unit length C is

$$\Phi = 2GCK_0\left(\frac{m\epsilon}{r_*}\right),$$

where K_0 is the Bessel function of imaginary argument and ϵ is the softening length. Substituting this into (6), and setting $\Sigma_0 = qV_0^2/2\pi Gr$, we obtain the desired expression for the growth-rate of the mode

$$\omega_i = \frac{mV_0 w}{r_*^2}\left\{\left[\frac{q\beta r_*}{w}K_0\left(\frac{m\epsilon}{r_*}\right)\right]^{1/2} - 1\right\}. \tag{7}$$

This expression indicates that we expect instability only for grooves in which the fractional depth

$$\beta > \frac{1}{qK_0}\frac{w}{r_*}, \tag{8}$$

i.e. not for infinitesimally shallow grooves. The critical depth is very small, however; *e.g.* in our experiments, $q = 0.5$, $w = 0.02r_*$ and $K_0 \simeq 1.45$ (for $m = 2$), so we should expect an instability whenever the fractional depth exceeds merely 2.8%! Notice also that the critical depth required increases with the groove width (though our analysis assumes this to be small); it would seem therefore, that the instability requires a critical density *gradient* on the sides of the groove.

We may compare the predictions of this local theory with the empirical results from our experiments. The eigenfrequency expected is $\omega_r = 0.4$ (for co-rotation at the groove centre) and $\omega_i = 0.035$ [from (7)]. That observed in the simulation is $\omega = 0.383+0.072i$. The real parts are in reasonable agreement, but our prediction considerably underestimates the observed growth rate. Two approximations in our analysis are largely responsible for this poor agreement, the principal being the additional contribution to Φ which comes from the supporting response of the disc. Our neglect of curvature appears to be of lesser importance; in other experiments, in which we raised the azimuthal mode number to $m = 3$ and 4, where it is more justifiable to neglect curvature, predictions for the growth rate from equation (7) are somewhat closer to the observed values (and co-rotation approaches the groove centre more closely still).

But most of the growth rate discrepancy is removed only in an improved treatment which includes the supporting response from the background disc (Sellwood & Kahn 1989). The spiral response of the disc on either side of the groove can be thought of as the polarisation response, or wake, induced by the growing non-axisymmetric disturbance in the groove. Discussions of wakes usually focus on the steady response to a large co-orbiting disturbing mass (*e.g.* JT); in this case, we have an exponentially growing disturbance that induces an exponentially growing response.

The instability produces a growing sinusoidal distortion to each side of the groove, and we must expect it to saturate once the distortions exceed the groove width, *i.e.* when further perturbations to the orbits of particles no longer produce corresponding increases in disturbance forces. At this stage, the mass distribution at the radius of the original groove has large non-axisymmetric variations – to a good approximation, it will consist of two blobs (for the $m = 2$ component). These large amplitude blobs will dissolve only slowly and each will continue to induce a spiral wake for some considerable time after the mode has saturated, as can be seen in Figure 5.

As the clearest possible example of how untested theory can be completely misleading, I note that the analysis provided here implies that a ridge, rather than a groove, would not provoke instabilities. As a ridge would be characterised by a negative value for β, equation (8) indicates that no instability should be expected. A simple experimental test of this prediction showed that it was completely wrong! A ridge provokes instabilities just as fierce as those from a groove. We were able to understand ridge modes only when we extended the analytic treatment presented here to consider wavelike perturbations travelling around the ridge. This more sophisticated treatment is presented in Sellwood & Kahn (1989)

4 A recurrent instability cycle

In this section I argue that an experiment from my recent collaboration with D. N. C. Lin provides a connecting link between the two previous sections. Our project was begun in the hope that N-body simulations of low mass particle discs around a central point mass might reveal non-axisymmetric instabilities of relevance to accretion discs. The experiments did reveal non-axisymmetric instabilities, but which are, however, likely to be of much greater significance to spiral structure in galaxies than to instabilities in accretion discs!

The simulation, which is discussed in detail elsewhere (Sellwood & Lin 1989), revealed a recurrent instability cycle of the following form: An instability somewhere in the disc causes a non-axisymmetric wave to grow to large amplitude. Particles at the Lindblad resonances of the wave are strongly scattered and change their angular momenta. As the resonances are narrow, the distribution function is severely depleted over a narrow range of angular momentum, which leads to a new groove-type instability with co-rotation at the radius of the Lindblad resonance of the previous wave.

Figure 7 shows how the distribution function changed in the experiment between the start and a moment 267 dynamical times later. The particle distribution is shown in action space: the abscissa of each is its angular momentum and the ordinate its radial action. (Particles on more eccentric orbits have larger radial actions.) The arrows point to the principal resonances (co-rotation, inner and outer Lindblad resonances) for the five waves observed in the experiment up to the moment illustrated. The resonant scattering that has occurred has produced substantial heating (the particles have much larger radial actions) and the distribution function is also much less smooth than at the start. Each tongue of particles reaching to large radial action was produced by scattering at a Lindblad resonance.

But the most interesting aspect of this revealing diagram is that the most recent wave has left a clear deficiency of particles having angular momenta ~ 1.35. This

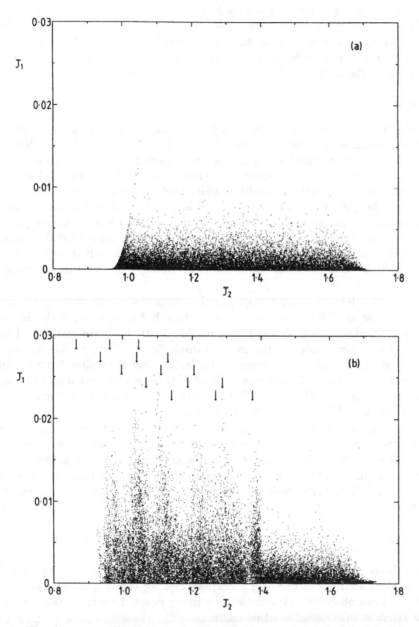

Figure 7. The phase space distribution of particles in the experiment described in Sellwood & Lin (1989) (a) Shows the situation at the start and (b) after 267 dynamical times. The spread in J_2 (angular momentum) reflects the radial extent of the annulus and those particles with large J_1 (radial action) are on more eccentric orbits. Only one fifth of the 100K particles used in the simulation are plotted in each panel. The arrows in (b) point to the locations of co-rotation and the two Lindblad resonances for the five waves observed in the model by time 267.

deficiency drives a new instability in the disc which begins its linear growth at about the time illustrated. The new instability is quite clearly a groove-type mode – corotation lies at precisely the angular momentum of the deficiency in the distribution function. Note that in this case we have a "groove" in the distribution of guiding centres; because particles have random motion, there is no noticeable groove in the surface density of the disc.

5 Discussion

In this paper I have collected the evidence indicating that the recurrent transient spirals seen in many simulations results from spiral *modes* that grow rapidly and decay only to be replaced by new instabilities. This picture is radically different from two other current views of spiral structure. Bertin *et al.* (1989, and references therein) argue that spiral patterns result from mild instabilities which lead to quasi-steady waves once nonlinear effects in the gas are taken into account. It is not at all clear that the models they propose could avoid the far more vigorous recurrent cycle of groove modes presented here. Toomre, on the other hand, stresses the role of local spiral "streaks" resulting from density fluctuations in the disc (*e.g.* this conference). Though these undoubtedly occur, it is far from clear that real galaxy discs are "lumpy" enough for strong spirals to be produced in this way.

The original evidence for something more than noise came from the experiments of SC. The discussion of §2 bolsters that case considerably by contrasting the Sc models with the almost perfect illustration of swing-amplified particle noise manifested by the Zang discs – spiral amplitudes in the more realistic Sc disc hardly changed with N. Moreover, the recurring transient patterns in cooled Sc models resulted from the superposition of a few long-lived coherent waves. Most of these waves, which could be picked out by Fourier analysis, had two Lindblad resonances within the disc and therefore could not be accounted for by orthodox mode theory.

In the two subsequent sections I demonstrate that a groove-like feature in the angular momentum distribution provokes a fierce instability and show how modes of this type can recur in an instability cycle. Resonant scattering by a large amplitude wave, created by the saturation of one mode, carves a groove that precipitates a new linear instability. As the instability cycle also heats the disc, making it less susceptible to further instabilities, activity must fade after comparatively few cycles unless the disc is cooled. The presence of a dynamically significant fraction of gas in the galaxy is therefore essential if the galaxy is to continue to display spiral structure over a long period.

Though the connection still needs to demonstrated in massive discs, this recurrent groove mode cycle seems at last to offer an explanation for the very intriguing spiral activity that had been observed in simulations for many years. Experiments are currently in hand to attempt to substantiate this claim.

This paper has also brought out the importance of relating results from simulations to the theory, and *vice-versa*. Only through the quantitative comparison of the results from the Zang model experiments with the predictions of swing-amplified particle noise, could the more vigorous structure in the Sc models be clearly identified as something more than noise. On the other hand, theory, unfettered by cross checks with experimental results, can easily go astray. An excellent example is provided by the beautifully straightforward linear analysis, which gives a reasonable description of groove modes, but which incorrectly predicts that axisymmetric density ridges in a disc are not destabilising!

Much of the work reported in §2 was done in collaboration with Ray Carlberg and has yet to be published. Franz Kahn was my collaborator for work described in §3. The author acknowledges the support of an SERC Advanced Fellowship.

References

Bertin, G., Lin, C. C., Lowe, S. A. & Thurstans, R. P., 1989. *Astrophys. J.,* **338**, pp. 78 & 104.

Goldreich, P. & Lynden-Bell, D., 1965. *Mon. Not. R. Astron. Soc.,* **130**, 124.

Julian, W. H. & Toomre, A., 1966. *Astrophys. J.,* **146**, 810 (JT).

Mestel, L., 1963. *Mon. Not. R. Astron. Soc.,* **126**, 553.

Sellwood, J. A., 1986. In *The Use of Supercomputers in Stellar Dynamics*, Lecture Notes in Physics **267**, p. 5, eds. Hut, P. & McMillan, S., Springer-Verlag, New York.

Sellwood, J. A., 1989. In *Nonlinear Phenomena in Vlasov Plasmas*, p. 87, ed. Doveil, F., Éditions de Physique, Orsay.

Sellwood, J. A. & Carlberg, R. G., 1984. *Astrophys. J.,* **282**, 61 (SC).

Sellwood, J. A. & Lin, D. N. C., 1989. *Mon. Not. R. Astron. Soc.,* in press.

Sellwood, J. A. & Kahn, F. D., 1989. *Mon. Not. R. Astron. Soc.,* in preparation.

Toomre, A., 1977. *Annu. Rev. Astron. Astrophys.,* **15**, 437.

Toomre, A., 1981. In *Structure and Evolution of Normal Galaxies*, p. 111, eds. Fall, S. M. & Lynden-Bell, D., Cambridge University Press, Cambridge.

Zang, T. A., 1976. *PhD thesis*, MIT.

Long-lived spiral waves in N-body simulations

Neil F. Comins and **Michael C. Schroeder**
University of Maine, USA

1 Description of simulation

In an effort to better understand disc galaxies, we have developed a Cartesian, 2-D, N-body and hydrodynamic computer code. The results presented here use only the N-body portion of the code. To accommodate the variable time-step length required by the Courant condition for hydrodynamic flows, we use a second order predictor-corrector integration scheme (Schroeder & Comins 1989) with the same accuracy as the more familiar time-centred leap frog scheme.

The particles are distributed as a Kuz'min disc, and we add a fixed 'halo', having between 65% and 75% of the total gravitational potential, to stabilize the system against bar-mode instabilities. Tangential and radial velocity dispersions establish an initial Toomre Q of 1.0 over the disc. The resulting disc appears to be stable to non-axisymmetric perturbations.

We add a rotating, logarithmic, two-armed, spiral perturbation to the potential. The amplitude of this spiral is ramped up and down as a Gaussian (Toomre 1981). This spiral perturbation grows from 2% of its maximum amplitude to full strength in 1/2 a rotation period and then decays in the same manner. Both trailing arm spirals (TASs) and leading arm spirals (LASs) are used with varieties of pitch angles and pattern speeds. All such perturbations lead to strong non-axisymmetric responses in the disc. Unless indicated otherwise, the pattern speed of the perturbation is 1/2 the co-rotation speed of the particles at the half-mass radius. Final Qs range between 1.1 in the interior and 3.5 near the disc edge.

2 Runs and results

We ran both LAS and TAS spiral perturbations on a 50×50 grid with $10\,000$ stars. LAS perturbations transferred their energy to TASs via swing amplification. TAS perturbations led to TAS arms. In both cases the spiral arms persisted for more than 5 rotation periods, but the maximum amplitude 2-armed spirals were generated by LAS perturbations (probably because they were amplified by the swing amplifier).

Simulations using $5\,000$, $20\,000$ and $40\,000$ particles were also run. The resulting spirals had peak amplitudes within 6% of the $10\,000$ particle case. As shown in Figure 1, the pattern speed of the spiral eventually settled, in all cases, just above the angular velocity of $\Omega - \kappa/2$ for all but the outer quarter of the disc. (As usual, Ω is the angular frequency of circular motion and κ is the Lindblad epicyclic frequency.) As a result, we conclude that these spiral arms are primarily generated and maintained as kinematic spiral arms (Toomre 1977). When perturbation strengths of 0.5% and 2% were used, the overall patterns were similar to the 1% case. The maximum amplitude of the resulting SDW scales linearly with the perturbation amplitude.

We made runs with perturbations at 1/4 the co-rotation speed at the 1/2 mass radius, and with the full co-rotation speed to complement the 1/2 co-rotation speed

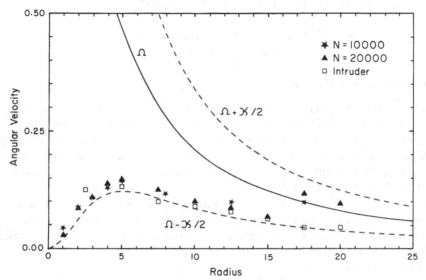

Figure 1. Measured pattern speeds for two runs with different numbers of particles and for a tidal intruder run. The curves show relevant angular velocities for this disc.

runs described above. The pattern speeds of the resulting waves were found to be independent of that of the forced wave.

We reran prior interacting galaxy experiments (Sundelius *et al.* 1987) with the Kuz'min disc. Tidal forces induced spiral arms as a companion with 30% of the test galaxy's mass passed by in a parabolic, prograde orbit with a perigalactic distance equal to the disc radius. This creates waves with the same pattern speed as the perturbations described above (Figure 1).

3 Conclusions

The persistence of the trailing arm spiral waves appears to be caused by the kinematic spiral arm mechanism described by Kalnajs (1973). The spiral arms wrap up over many rotation periods because this disc's ILR has angular velocity gradients producing TAS waves for $R > 5$ and LAS waves for $R < 5$. A more detailed paper on these results is in preparation.

MCS was supported for this work by NASA Graduate Student Fellowship NGT-50098.

References

Kalnajs, A. J., 1973. *Proc. Astron. Soc. Aust.*, **2**, 174.
Schroeder, M. C. & Comins, N. F., 1989. *Astrophys. J.*, to appear.
Sundelius, B., Thomsson, M., Valtonen, M. J. & Byrd, G. G., 1987. *Astron. Astrophys.*, **174**, 67.
Toomre, A., 1977. *Annu. Rev. Astron. Astrophys.*, **15**, 437.
Toomre, A., 1981. In *Structure and Evolution of Normal Galaxies*, p. 111, eds Fall, S. M. & Lynden-Bell, D., Cambridge University Press, Cambridge.

Overstable modes in stellar disc systems

P. L. Palmer, **J. C. B. Papaloizou** and **A. J. Allen**

Queen Mary College, London

1 Introduction

In this article we describe some results from a more general study on overstability in stellar systems (Palmer *et al.* 1989). Our analysis is based on a linear normal mode analysis of equilibria, as first discussed by Kalnajs (1977). The modes we shall discuss are *global* modes, and we do not employ the usual WKB approximation.

2 Unstable modes

We perturb an equilibrium described by a distribution function f with the potential given by

$$\Phi_1(r, \phi, t) = \psi(r)e^{-ik\phi}e^{i\omega t} \tag{1}$$

and insist on total energy conservation. The eigenfunctions $\psi(r)$ are then given by

$$r\Sigma_1(\psi) = \int_0^R K(r, r', \omega)\psi(r') \, dr' \tag{2}$$

where Σ_1 is the perturbed surface density which corresponds to (1) through the Poisson equation. We may consider this as an unbounded, positive definite, self-adjoint operator on ψ. Equation (2) is an eigenvalue problem for the eigenvalue ω. The kernel K is square integrable and self-adjoint. It involves an integration of the distribution function over that part of phase space for which motion passes through both r and r'. For stars moving on nearly circular orbits, this can be replaced by integration over epicycle amplitudes and guiding centre radii.

Let the epicycle frequency be κ and the angular frequency be Ω. We now assume that $2\Omega - \kappa$ has a maximum value in the disc, and let this maximum value be ω_0. This is true for any reasonable non-singular mass distribution. We then evaluate the kernel K in the neighbourhood of the disc where $2\Omega - \kappa = \omega_0$. One can then show that

$$K(r, r', \omega) \simeq \frac{K_1(r, r')}{\sqrt{\omega - \omega_0}}.$$

For ω close to ω_0 the integral in (2) is dominated by this part of the disc. Although some care has to be taken as ω approaches ω_0, one can show that there is a spectrum of real frequencies with $\omega > \omega_0$ (see Palmer *et al.* 1989 for further details) which are short wave modes. The analysis requires $\omega - \omega_0 \geq O(A^2)$ where A is the epicycle amplitude.

It is clear that these neutral modes are negative energy modes, in the WKB sense, since the pattern rotates more slowly than the stars (Lynden-Bell & Kalnajs 1972). Any energy loss mechanism will then destabilise them, leading to an overstable spiral mode. One can easily show that the presence of an outer Lindblad resonance will destabilise these modes, as will all resonances of higher order in the epicycle amplitude (*e.g.* $\omega = 2\Omega + 2\kappa$). The co-rotation resonance does not always destabilise, depending upon the gradient of the vortensity (Goldreich & Tremaine 1979):

$$\frac{d}{dr}\left(\frac{\Sigma\Omega}{\kappa^2}\right).$$

3 Implications for spiral structure

Spiral structure in disc galaxies depends as much upon turning the gas into stars as on forming spiral patterns in the stellar disc (Sellwood & Carlberg 1984). The origin of spiral structure may be due to normal modes, as a mode has fixed frequency and so does not shear out. The spiral pattern is maintained because structure on scales smaller than the epicycle amplitudes of the stellar motions cannot be sheared out. Although spiral structure in galaxies does not look like grand design, we note that the spectrum of neutral modes discussed above is very dense. This leads to mode interaction and beats. Indeed beats have been found in simulations of the corresponding unstable spherical systems (Palmer *et al.* 1989). We expect this interference to disrupt any grand design, leading to a much more blotchy appearance to the spiral structure.

The densely packed spectrum is also important for simulations of $m = 2$ instabilities in discs (Ostriker & Peebles 1973, Sellwood 1985, Toomre 1977, Athanassoula & Sellwood 1986). The neutral modes may be destabilised at co-rotation, which operates over a very narrow band of radii for a given mode frequency. As a result the instability could be switched off, or greatly enhanced (see Sellwood's article in this volume) by a small redistribution around the co-rotation radius.

The importance of the co-rotation resonance has often been emphasised (Toomre 1981). We point out, however, that instability will arise through any energy loss mechanism from the neutral modes. In particular, unstable spherical systems display an extra Lindblad resonance with $\omega = \kappa$. Since disc galaxies are believed to be embedded in a dark halo, this same resonance could couple halo material with these neutral modes leading to instability. In this case co-rotation may lie beyond the outer edge of the disc. Observations of the dust lanes in the spiral arms of at least some disc galaxies do seem to indicate this (Petrou & Papayannopoulos 1986).

References

Athanassoula, E. & Sellwood, J. A., 1986. *Mon. Not. R. Astron. Soc.*, **221**, 213.
Goldreich, P. & Tremaine, S., 1979. *Astrophys. J.*, **233**, 857.
Kalnajs, A. J., 1977. *Astrophys. J.*, **212**, 637.
Lynden-Bell, D. & Kalnajs, A. J., 1972. *Mon. Not. R. Astron. Soc.*, **157**, 1.
Ostriker, J. P. & Peebles, P. J. E., 1973. *Astrophys. J.*, **186**, 467.
Palmer, P. L., Papaloizou, J. C. B. & Allen, A. J., 1989. *Mon. Not. R. Astron. Soc.*, to appear.
Petrou, M. & Papayannopoulos, T., 1986. *Mon. Not. R. Astron. Soc.*, **219**, 157.
Sellwood, J. A., 1985. *Mon. Not. R. Astron. Soc.*, **217**, 127.
Sellwood, J. A. & Carlberg, R. G., 1984. *Astrophys. J.*, **282**, 61.
Toomre, A., 1977. *Annu. Rev. Astron. Astrophys.*, **15**, 437.
Toomre, A., 1981. In *Structure and Evolution of Normal Galaxies*, p. 111, eds. Fall, S. M. & Lynden-Bell, D., Cambridge University Press, Cambridge.

Galactic seismological approach
to the spiral galaxy NGC 3198

M. Noguchi[1], T. Hasegawa[2] and M. Iye[1]
[1]*National Astronomical Observatory, Tokyo, Japan*
[2]*University of Tokyo, Japan*

1 Motivation and technique

We construct a series of N-body simulations to model NGC 3198. Each model is composed of an exponential disc of 40K particles and a rigid spherical halo. The disc scale-length was set equal to the value derived in the photometry by Wevers (1986) and the mass-to-luminosity ratio, M/L, of the disc component, which cannot be derived directly from the observational data, is the only free parameter of the model. The halo mass distribution was adjusted so that the rotational velocity due to the disc and halo reproduces the rotation curve observed by Bottema (1988). The dynamical evolution of the disc is investigated using the particle-mesh N-body code of Noguchi (1987).

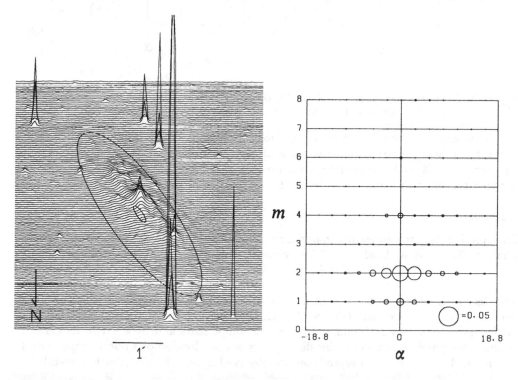

Figure 1. (a) V-band image of NGC 3198 taken with the 105cm Schmidt Telescope at Kiso Observatory. The luminosity distribution in the annular region bounded by the two ellipses was transformed to give the power spectrum indicated in (b), where the radius of circle is proportional to the amplitude of each component relative to the axisymmetric component. The abscissa is the radial wave number α defined in Iye *et al.* (1982) and m is the azimuthal wave number.

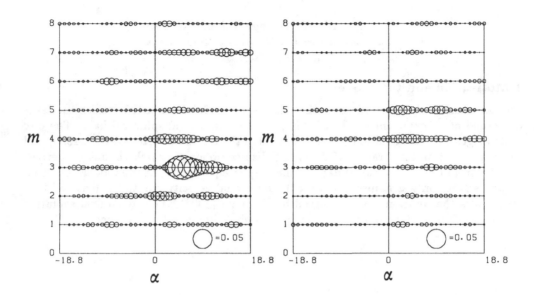

Figure 2. Power spectrum of model spirals. (a) Disc with $M/L = 2.4$ at $T = 3.77$. (b) Disc with $M/L = 1.8$ at $T = 6.28$. One rotation period at the disc edge is 6.28.

We monitored the evolution of morphological patterns in the simulations by calculating the power spectrum representation of the spiral structure (*cf.* Iye *et al.* 1982), and compared it with the structure observed in NGC 3198 (Figure 1) in the hope of deriving the most probable value for M/L.

2 Results

The spiral structures appearing in our three numerical models were not rigorously stationary. The disc with $M/L = 3.6$ is violently unstable and turns into a persistent bar quite unlike NGC 3198. The disc with $M/L = 2.4$ generates a spiral structure the amplitude of which exceeds the observed one within a half rotation period (Figure 2a). The disc with $M/L = 1.8$ develops a spiral structure the amplitude of which is comparable to the observed one (Figure 2b). However, a discrepancy exists such that the model spiral structure is dominated by $m = 2$ components and by a bar in its central region. One possible reason for this discrepancy is the adoption of a purely exponential disc instead of a disc which reproduces the observed luminosity distribution exactly.

References

Bottema, R., 1988. *Astron. Astrophys.*, **197**, 105.
Iye, M., Okamura, S., Hamabe, M. & Watanabe, K., 1982. *Astrophys. J.*, **256**, 103.
Noguchi, M., 1987. *Mon. Not. R. Astron. Soc.*, **228**, 635.
Wevers, B. M. H. R., 1986. *Astrophys. J. Suppl. Ser.*, **66**, 505.

Characteristics of bars from 3-D simulations

D. Friedli and **D. Pfenniger**
Observatoire de Genève, Switzerland

1 Introduction

The formation and the evolution of bars, the mutual influence of the bulge and the bar and the effects of vertical resonances, have been studied in a series of N-body simulations ($10^5 \leq N \leq 5 \times 10^5$) over long time scales ($T \geq 2000$ Myr). A PM method is used with a 3-D polar grid having an exponential spacing in R (the central resolution is 0.2 kpc or less), and a linear spacing in ϕ and z. More detail will be given in a future paper. The particles are distributed into a bulge and a disc components in hydrostatic equilibrium (Satoh & Miyamoto 1976). The dimensionless parameters are the bulge-to-disc mass ratio $\mu = M_b/M_d$, and the bulge-to-disc scale length ratio $\beta = b/a$.

2 Some results

2.1 VELOCITY DISPERSION

At $T = 0$ the velocity ellipsoid is isotropic by construction, but it becomes rapidly anisotropic, such that $\sigma_R \geq \sigma_\phi \geq \sigma_z$ everywhere, and its size decreases with R. Due to heating by time-dependent perturbations, σ_R and σ_ϕ grow considerably, *as well as σ_z in the bar region since vertical resonances exist.* This large scale heating is more efficient than that induced by local perturbations.

2.2 BULGE EVOLUTION

The disc, and subsequently the bar, flattens the initially spherical bulge, which aligns itself with the bar, *i.e.* the bulge co-rotates with the bar. For example with $\mu = 0.18$ and $\beta = 1/12$, the bulge axis ratios are $1 : 0.84 : 0.82$ ($T = 2000$ Myr).

2.3 BAR SHAPE

The horizontal and the vertical bar ellipticity, ϵ_h and ϵ_v, are measured at the most eccentric contour line of the projected density. With time, ϵ_v increases or oscillates,

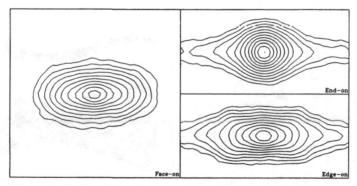

Figure 1. Projected density of a stable, prolate and box-shaped bar.

while ϵ_h decreases steadily before reaching an asymptotic value. With M_b, ϵ_h decreases and ϵ_v increases. The extreme axis ratios found are $1 : 0.35 : 0.34$, $(\beta = 1/3, \mu = 0.18)$, and $1 : 0.69 : 0.11$, $(\beta = 1/16, \mu = 0.33)$.

2.4 BOXINESS AND VERTICAL RESONANCES

A slow growth of vertical structures is observed and the bar is typically box- or peanut-shaped when viewed edge-on (Figure 1). Inside the bar there is a strong and extended $1/1$ resonance between the horizontal and the vertical epicyclic frequencies κ and ν. Thus, every horizontal resonance between κ, the rotation frequency Ω, and the pattern speed Ω_p coincides with the analogous vertical resonance involving ν. Vertical resonances allow particles to leave the ($z = 0$) plane, and are favourable for feeding box-shaped orbits found in barred potentials (Pfenniger 1984). Simulations with Cartesian grids have also produced box-shaped figures (Hohl *et al.* 1979; Combes & Sanders 1981). Since this shape is equally well obtained with our polar grid, *boxiness is confirmed to be naturally associated with fast rotating bars*.

2.5 ORBITAL STRUCTURE

As for a smooth potential (Pfenniger 1984), we have determined the main families of periodic orbits inside the exact box-shaped bar potential resulting from a simulation. The horizontal bar shape is supported by the main direct family and by $4/1$ orbits in the $z = 0$ plane (Figure 2). The retrograde family exists but is *vertically unstable* all the way, and is also partly horizontally unstable, indeed Ω_p is such that $\Omega + \Omega_p \approx \nu \approx \kappa$. The box-shape is produced by the general 3-D shape of the direct regular orbits, in particular by *the 3-D orbits trapped by the main plane direct family* (Figure 3). Fully 3-D families of periodic orbits supporting the box-shape exist too.

References

Combes, F. & Sanders, R. H., 1981. *Astron. Astrophys.,* **96**, 164.
Hohl F., Zang T. A. & Miller, J. B., 1979. *NASA Ref. Publ.* **1037**.
Pfenniger, D., 1984. *Astron. Astrophys.,* **134**, 373.
Satoh, C. & Miyamoto, M., 1976. *Publ. Astron. Soc. Jpn*, **28**, 599.

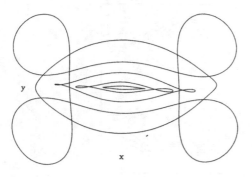

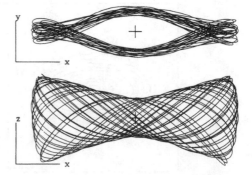

Figure 2. Periodic orbits supporting the bar.   **Figure 3.** A "regular" orbit supporting the box.

Spirals and bars in linear theory

J.F. Sygnet[1], **M. Tagger**[2] and **R. Pellat**[3]

[1] *Institut d'Astrophysique, Paris, France*
[2] *Service d'Astrophysique, Saclay, France*
[3] *Ecole Polytechnique, Palaiseau, France*

1 Introduction

We have recently shown (Tagger *et al.* 1989a,b and references therein) that the linear theory of density waves in a flat self-gravitating disc contains, in addition to the usual tightly wound waves, another type of perturbation which is essentially bar-like and which dominates the mode structure in the vicinity of the co-rotation radius. We briefly discuss this analytical result, which can be illustrated by numerical calculations, and by its relationship to the disc response to external forcing.

Two different descriptions of density waves have been used in the past: steady waves and shearing perturbations. The difficulty of a unique description stems from the flat disc geometry where the solution of the Poisson equation in the vertical dimension involves an integral operator. The WKBJ approximation, in practice the assumption of tightly wound spirals, allows us to calculate waves with well defined physical properties and has the important advantage of incorporating a properly defined boundary condition at infinity, but it cannot be used to describe the efficient swing amplification mechanism.

2 Spirals and Bars

Swing amplification is most simply described in the shearing sheet model, where the relevant equations can be Fourier transformed very easily. The solution $\hat{\phi}(k)$ can be easily computed for "large" radial wavenumber k, but we found that a problem arises when one transforms back to real space. It has already been noted that when one computes the inverse Fourier transform the integrand oscillates rapidly at large k, except at saddle points K_j. The contributions of these saddle points $C_j \exp(ik_j x)$ to $\phi(x)$ can be identified with the usual short and long, leading and trailing spiral waves (Goldreich & Tremaine 1978).

At this point we note that if the only contribution to the integral at large k comes from the saddle points, it does not rule out other contributions in the neighbourhood of the origin ($\mid k \mid \sim \eta \equiv$ azimuthal wavenumber) where the asymptotic expansions do not hold. To obtain the correct solution it is best to follow contours of steepest descent from the saddle points and try to connect them in order to obtain a new contour along which the inverse Fourier transform can be integrated, which allows us to identify all possible contributions to the potential. (An example is shown in Figure 1 for a region where the waves propagate.) Apart from the above mentioned saddle points, there remains the cut $[\pm i\eta, \pm i\infty]$, introduced by the square root of $k^2 + \eta^2$ in Poisson's equation, due to the thin disc nature of the galaxy, which produces a contribution along the cut.

The $\Re(k)$ part of this new contribution is neither positive nor negative, and is therefore a purely radial feature, *i.e.* a bar. We find that the "spiral" contributions are larger than the "bar" one in the propagation region, whereas, in the forbidden region

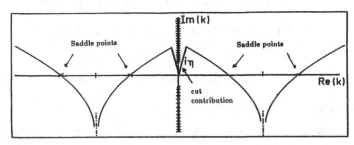

Figure 1. Integration contour for the inverse Fourier integral.

close to co-rotation, the saddle points leave the real axis, making the amplitude of the spiral contribution exponentially smaller, while the "bar" contribution retains its magnitude and becomes the dominant one by far!

We emphasize that the "bar" contribution we find does not have the radial propagation properties of a wave; we see it rather as the effect of direct long-range interaction between particles modified by their collective response, *i.e.* an effect of a "gravitational wake".

3 Numerical Results

Figure 2 shows the results of a direct numerical solution in Fourier space, followed by an inverse Fourier transform. One sees waves propagating far from co-rotation (a combination of leading and trailing waves in the RHS and in the LHS the transmitted trailing wave propagating outward) but also a cental feature with a practically constant phase.

The case of an external forcing (*e.g.* by a nearby satellite in the case of a planetary ring) must be reconsidered in view of these results. Since spiral waves are evanescent in the forbidden region, only long trailing waves are exited near the Lindblad resonances. Here however, another contribution is also found at $|k| \sim \eta$, corresponding to our bar-like feature (Tagger *et al.* 1989b). We find trailing spirals close to both Lindblad resonances, plus a bar-like perturbation which is dominant close to co-rotation: one clearly sees the bar extending throughout the forbidden region, connected to trailing waves further out.

References

Goldreich, P. & Tremaine, S., 1978. *Astrophys. J.*, **222**, 850.

Tagger, M., Sygnet, J. F. & Pellat, R., 1989a. *Astrophys. J. Lett.*, **337**, L9.

Tagger, M., Sygnet, J. F. & Pellat, R., 1989b. In *Proceedings of a course on Plasma Astrophysics in Varenna (Italy)*, p. 335, ESA SP-285, Noordwijk.

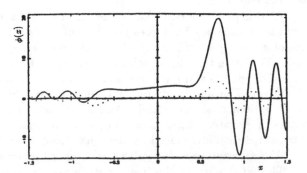

Figure 2. Mode structure from direct numerical integration.

Stellar hydrodynamical solutions for Eddington discs

N. W. Evans

Queen Mary College, London

1 Introduction

For any astrophysical disc, the most immediate puzzle to solve is what holds it up against gravitational collapse. Here, we shall solve the equations of stellar hydrodynamics for Eddington discs – stellar dynamical discs whose gravitational potential is separable in elliptic coordinates (*e.g.* Lynden-Bell 1962). In other words, we shall take the potential, and therefore the density, as given, and try to find the components of stress that support the disc. For an assumed mean streaming field, this gives the velocity dispersions.

2 Apparatus

Let (x, y) be Cartesian coordinates. Then elliptic coordinates (λ, μ) are defined as the roots for τ of

$$\frac{x^2}{\tau + \alpha} + \frac{y^2}{\tau + \beta} = 1 \tag{1}$$

where α and β are constants. The separable, or Stäckel, potential has the form

$$\psi = \frac{f(\lambda) - f(\mu)}{\lambda - \mu} \tag{2}$$

where f denotes an arbitrary function. The equations of stellar hydrodynamics are

$$\frac{\partial}{\partial \lambda} \rho \sigma_\lambda^2 + \frac{\rho(\sigma_\lambda^2 - \sigma_\mu^2)}{2(\lambda - \mu)} = \frac{\rho \partial \psi}{\partial \lambda} \tag{3}$$

$$\frac{\partial}{\partial \mu} \rho \sigma_\mu^2 + \frac{\rho(\sigma_\mu^2 - \sigma_\lambda^2)}{2(\mu - \lambda)} = \frac{\rho \partial \psi}{\partial \mu} \tag{4}$$

where $(\sigma_\lambda^2, \sigma_\mu^2)$ are components of the stress tensor referred to an orthonormal axis set. On transferring to a new set of coordinates (s, t) defined by

$$s = \frac{(\lambda + \mu)}{2}, \qquad t = \frac{(\lambda - \mu)}{2} \tag{5}$$

a separable equation for $\Delta = \sigma_\lambda^2 - \sigma_\mu^2$ may be constructed. This can be used (Evans & Lynden-Bell 1989) to give a complete formal solution to (3) and (4) by explicit construction of the Green's function. The components of the stress tensor are everywhere determined by specifying as boundary conditions the anisotropy parameter

$$\beta(\lambda, \mu) = 1 - \frac{\sigma_\mu^2}{\sigma_\lambda^2} \tag{6}$$

on $\mu = -\alpha$, that is on the y axis above the focus of the coordinate system.

184

Evans

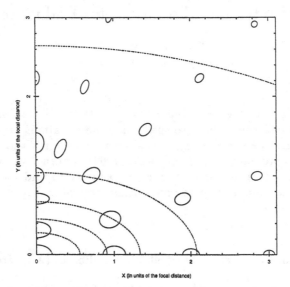

Y (in units of the focal distance)

X (in units of the focal distance)

Figure 1. The velocity ellipse for a hydrodynamical solution for the elliptic disc.

3 Results for the perfect elliptic disc

The model chosen for investigation here is the perfect elliptic disc (de Zeeuw 1985). The surface density of the disc is

$$\Sigma = \frac{\Sigma_0}{(1 + m^2)^{\frac{3}{2}}}, \qquad m^2 = \frac{x^2}{a^2} + \frac{y^2}{b^2} \tag{7}$$

where Σ_0 is the central surface density. The density is stratified on concentric ellipses with semi-axes ma and mb.

Numerical techniques for solution of the stellar hydrodynamical equations are discussed in Evans (1989). In Figure 1, the velocity ellipse is plotted correctly oriented for a physical (*i.e.* everywhere positive) solution for the elliptic disc with $\frac{b}{a} = 0.5$. Contours of equal density are drawn in broken lines. The squares of the velocity dispersions are normalised to take values in the range [0,10] by measuring in units of tenths of the central potential and the unit of length is taken as the focal distance. The solution shows how the disc is held up by stars moving on orbits with large radial excursions. It is this radially anisotropic dispersion tensor that supports the disc against gravitational collapse.

I gratefully acknowledge support from the U.K. Science and Engineering Research Council. I wish to thank Prof. D. Lynden-Bell for much useful advice and encouragement.

References

Evans, N. W., 1989. *Inter. Journ. Comp. Math.*, in press.
Evans, N. W. & Lynden-Bell, D., 1989. *Mon. Not. R. Astron. Soc.*, **236**, 801.
Lynden-Bell, D., 1962. *Mon. Not. R. Astron. Soc.*, **124**, 95.
de Zeeuw, P. T., 1985. *Mon. Not. R. Astron. Soc.*, **216**, 273.

Theory of gradient instabilities of the gaseous Galactic disc and rotating shallow water

A. M. Fridman

Astronomical Council of the Academy of Sciences of the USSR

Abstract We develop the theory of instabilities in a rotating gaseous disc and in shallow water for the case where there is a break in the surface density and sound velocities, as well as the rotation rate, at a particular radius. Different instabilities of sub-sonic and supersonic flows have been investigated. We also prove the identity of the linearised dynamical equations for the gaseous disc of the Galaxy and for our rotating shallow water experiments.

1 Introduction

The present paper pursues two aims: (1) to prove that gradient instabilities can lead to spiral structure in galaxies, and (2) to give the theory of gradient instabilities in rotating shallow water, when viscosity effects can be neglected. The behaviour in rotating shallow water has been investigated in an experiment known as "Spiral" at the Plasma Physics Department of the Institute of Atomic Energy.

It is natural to ask why such different subjects as galactic discs and shallow water are combined in this one paper. The reason is that the dynamical behaviour of a gaseous galactic disc and rotating shallow water are described by one and the same set of differential equations. Clearly, shallow water may be considered as a 2-D gaseous dynamical system (Landau & Lifshitz 1986) similar to the gaseous disc of our galaxy. However, viscosity effects near the bottom in the experimental set-up are absent in galaxies and the latter contain forces of self-gravitation that are absent in shallow water.

We here demonstrate that viscosity in our shallow water experiments with the "Spiral" apparatus has no significant influence on the perturbations that develop (§2), and that the linearised dynamical equations of the gaseous galactic disc, including self-gravitation, may be transformed to the shallow water equations (§4).

The situation in the Galaxy is more complex still, however, for besides the gaseous disc, it contains a number of stellar sub-systems having various morphological peculiarities. The unperturbed gravitational potential in the gaseous disc is determined by the total distribution of all the sub-systems of the Galaxy, so in this sense, the gaseous disc rotates in an external potential, much as the rotation in the "Spiral" experiment is caused by the vessel.[1] Fortunately, the perturbed gravitational potential in the gaseous disc is determined mainly by the perturbed gas density – the perturbed density of stars is smaller (§3). So the role of the "external" field of the stellar sub-systems is limited to creation of the rotation curve.

---

[1] A possible exception is the part of the curve from 400 pc to $\gtrsim 1$ kpc, where a massive molecular disc of radius ~ 700 pc and having a sharp edge in the central galaxy region, seems to contribute remarkably to the potential of the region. It is also located where there is a strong negative gradient in the Galaxy rotation curve, which is responsible for spiral instabilities (Fridman 1989a).

As mentioned already, self-gravitation does not change the structure of the equations if one uses, instead of the sound velocity, c_s, the typical velocity spread of perturbations in the gaseous self-gravitating medium, c_g (§4). We note that a similar redetermination can also be made for shallow water (Landau & Lifshitz 1986), in which the sound velocity is replaced by a typical velocity of the spread of surface density perturbations in the water, the latter being much slower than the former. We also find $c_g \ll c_s$ in regions of the gaseous disc far from the centre of the Galaxy (*e.g.* the Solar neighbourhood). In the centre of the Galaxy, where we think there is a spiral generator, the self-gravitation force is comparable to the pressure force $c_g \sim c_s$.

Having established the equivalence of the two sets of linearised equations describing shallow water dynamics and the gaseous disc in the Galaxy, we then solve these analytically in §5 for the case where there is a tangential discontinuity in the rotation velocity, density and typical gravitating sound velocity. We obtain the dispersion relation and examine the conditions under which *gradient instabilities* might be expected. We discuss the nature of these instabilities and relate them to other well known instabilities in §§6 & 7.

The stabilising role of the "negative" density gradient revealed in earlier papers (Morozov *et al.* 1976a, b) not only produces a noticeable decrease in the growth-rate of the shear instability, but also an essential decrease in the pattern speed when compared to the calculations by Morozov (1977, 1979), who neglected the density gradient. The new results obtained here are therefore in much better agreement with the results from the "Spiral" experiments (Morozov *et al.* 1984, 1985, Fridman *et al.* 1985) and also correspond better with the view (*e.g.* Marochnik & Suchkov 1984) that co-rotation should be far out in the Galaxy. Our conclusions are summarised in §8.

2 The influence of molecular viscosity on the shallow water experiments

The estimates given in this section concern only the first "Spiral" experiments described in Morozov *et al.* (1984, 1985) and Fridman *et al.* (1985).

The relative importance of viscosity to the behaviour in the shallow water experiments depends upon the values of a number of dimensionless parameters. The first of these is the Ekman number (Pedlosky 1984)

$$E_\nu \equiv \frac{\nu}{\Omega_0 H^2},$$

where ν is the coefficient of kinematic molecular viscosity (Landau & Lifshitz 1986), Ω_0 is the angular velocity of the vessel and H is the depth of the liquid. In our experiments, H varied from 0.2 to 0.4 cm and (for water) $\nu = 0.01$ cm^2 s^{-1}; thus $E_{\nu,\max} \simeq 0.25[\Omega_0(r)]_{\min}^{-1}$. In the "rotating periphery" experiments $[\Omega_0(r)]_{\min} \simeq 2$ s^{-1}, therefore $E_{\nu,\max} \simeq 1/8$, *i.e.* viscosity effects are seen to be small.

The Ekman number can also be defined as $E_\nu \equiv \delta^2/H^2$, where $\delta = (\nu/\Omega)^{1/2}$ is the depth of the Ekman layer. Despite its relatively small value ($\delta^2 \ll H^2$), the entire depth of water H can be brought into uniform rotation in some typical time τ_{sp}, the spin down, or spin up time. When we made an abrupt change to the rotation velocity, Ω_1, we observed an abrupt "jump" in the mode number – the number of arms changed from m_1 to m_2. The time to establish a new steady regime is τ_{sp}, the value of which is estimated as follows. The radial equation of motion, in order of magnitude, is

$$\frac{\partial v_r}{\partial t} \sim \nu \frac{\partial}{\partial z}\left(\frac{\partial v_r}{\partial z}\right). \tag{1}$$

Since v_r varies with z only over the typical interval $(0, \delta)$, we may approximate $\partial v_r/\partial z \sim (\partial v_r/\partial z)\Theta(\delta - z)$, where $z > 0$ and $\Theta(\delta - z)$ is the Heaviside step function.

Integrating (1) over z from 0 to H we obtain:

$$\frac{\partial}{\partial t}v_r H \sim \nu\left.\frac{\partial v_r}{\partial z}\right|_0^\delta \sim \nu\frac{v_r}{\delta}.$$

Multiplying the last quantity by the density ρ and the area of the base of the vessel, S, we obtain the equation for the rate of change of radial impulse, $\mathcal{P}_r = \rho S H v_r$,

$$\frac{\partial \mathcal{P}_r}{\partial t} \sim \frac{\mathcal{P}_r}{\tau_{\rm sp}}, \qquad \tau_{\rm sp} \sim \frac{H\delta}{\nu} \sim \frac{H}{\sqrt{\nu\Omega}}.$$

We can now estimate the time to establish a stationary regime, *e.g.* for $m = 2$, and the time taken to "jump" to a new mode when we suddenly change the angular velocity by $\Delta\Omega$. In the first case, this time will be much longer than $\tau_{\rm sp,min} \simeq 0.47$ s (for $\Omega = 18$ s^{-1} and $H > 0.2$ cm). In the second case, for $\Delta\Omega \simeq 2$ s^{-1} we have $\tau_{\rm sp} \simeq 1.4$ s. The time for instability, $\tau_{\rm in}$, which depends upon the smeared width of the velocity kink, is much less. If this smearing is caused by molecular viscosity (*i.e.* a laminar Ekman layer) then $\delta_{\rm lam} \sim \sqrt{\nu/\Omega_1} \simeq 0.024$ cm is much less than the radial wavelength λ ($\lambda \approx 6$ cm for $m = 2$). In this case $\gamma \simeq \Omega_1(\simeq 18$ s$^{-1})$ and $\tau_{\rm in} \sim 1/\gamma \simeq 0.05$ s.

These arguments indicate that molecular viscosity scarcely affects the instabilities in our experiments with water. Moreover, Nezlin *et al.* (1987) found that an increase in the viscosity of the liquid by a factor of ten did not change the qualitative appearance of the spiral pattern. The spiral structure disappears only when liquids of still higher viscosity are used.

Antipov *et al.* (1988) were able to give a negative answer to the question of the influence of turbulent viscosity on the structure observed in the experiments. They showed that the "viscous lifetime" of structures is fully determined by laminar viscosity.

3 Disturbance gravitational potential in the Galactic disc

The aim of this section is to show that

$$\frac{\tilde{\sigma}_g}{\tilde{\sigma}_*} \gg 1, \tag{2}$$

where $\tilde{\sigma}_*$ is the perturbation density in the stars associated with spiral wave in the Galaxy and $\tilde{\sigma}_g$ is that in the gas. If condition (2) were true throughout the entire Galaxy, then the perturbation potential $\tilde{\psi}$, given by

$$\tilde{\psi} = -2\pi G(\tilde{\sigma}_g + \tilde{\sigma}_*)/|k|$$

depends only on $\tilde{\sigma}_g$.

The inequality (2) follows from the hydrodynamical conception of spiral density wave generation (Fridman 1978) and from observational data, as we now show. Let the observed sharp gradient in the rotation velocity at a radius ~ 1 kpc cause an instability in the shear flow. The kink in the rotational velocity is clearly seen in the gaseous disc (Haud 1979) but is not seen in the stellar component – because the dispersion of stellar velocities is so high, the gradient is strongly smeared out. Thus, if an instability also develops in the shear flow of the stellar "gas", then its growth-rate, γ_*, would be much less than the growth-rate, γ_g, in the gas. A large amplitude perturbation will therefore develop much more quickly in the gas. As a result, the perturbation in the gravitational potential will be (Polyachenko & Fridman 1976, English translation: Fridman & Polyachenko 1984, hereafter PF)

$$\tilde{\psi} = -\frac{2\pi G \tilde{\sigma}_g \mathcal{R}_g}{|k|}, \tag{3}$$

where k is the wavenumber of the perturbation and $\mathcal{R}_g$ is a "reduction factor" allowing for the thickness, h_g, of the gas layer having the approximate form

$$\mathcal{R}_g \simeq \frac{1}{1 + |k| h_g/2}. \tag{4}$$

More accurate estimates of this function are plotted in Shu (1968), Vandervoort (1970) and also in PF and Polyachenko & Fridman (1981).

Perturbations in the gravitational field in turn cause perturbations to the stellar density. Since $\tilde{\sigma}_*/\sigma_* \ll 1$, the perturbations are well described by linear theory. Using results from Lin, Yuan & Shu (1969), we have

$$\frac{\tilde{\sigma}_*}{\sigma_{0,*}} = -\frac{k^2 \tilde{\psi}}{\kappa^2} \mathcal{R}_\nu(\chi), \tag{5}$$

where

$$\mathcal{R}_\nu(\chi) \equiv \frac{1}{\chi} \left[1 - I_0(\chi)e^{-\chi} - \sum_{n=1}^{\infty} \frac{(-1)^n I_n(\chi)e^{-\chi}}{(\nu/n)^2 - 1} \right] \mathcal{R}_*. \tag{6}$$

We have used standard notation in this expression, namely

$$\chi \equiv k^2 c_{r,*}^2/\kappa^2, \qquad \nu^2 \equiv (\omega - m\Omega_0)^2/\kappa^2, \tag{7}$$

with Ω_0 and κ being the angular frequency of circular motion and the epicyclic frequency respectively and ω is the eigenfrequency of the perturbation. In (6), I_n is the Bessel function of imaginary argument and $\mathcal{R}_*$ is the "reduction factor" which allows for the thickness of the stellar disc. This last factor, which was omitted by Lin *et al.*, has the same form as for the gas (equation 4).

The Galaxy rotation curve is almost Keplerian over the radial range $0.4 < r < 1$ kpc (Haud 1979, Rohlfs & Kreitschman 1988), *i.e.* $\Omega_0 \sim r^{-3/2}$, but is more nearly $\Omega_0 \sim r^{-1}$ over an extended region including the Solar vicinity. We adopt the values[2] given in Table 1 (see Sanders, Solomon & Scoville 1984, and references therein) and

$$\kappa^2(r_c) = \Omega_0^2(r_c), \qquad \kappa^2(r_\odot) = 2\Omega_0^2(r_\odot).$$

---

[2] Spiral waves can be seen now at $r \gtrsim 2 - 3$ kpc, so the value $\lambda_c \simeq 1.2$ kpc at $r = 1$ kpc should be considered conditional as yet.

Table 1

	Inner Galaxy $r = r_c$	Solar neighbourhood $r = r_\odot$
r	1 kpc	10 kpc
V_0	200 km s^{-1}	220 km s^{-1}
λ	~ 1.2 kpc	~ 2 kpc
$c_{r,*}$	100 km s^{-1}	50 km s^{-1}
σ_*	300 M$_\odot$ pc^{-2}	80 M$_\odot$ pc^{-2}
$\mathcal{R}_*$	0.25	0.45
c_s	20 km s^{-1}	8 km s^{-1}
h_g	0.06 kpc	0.14 kpc
σ_g	6 M$_\odot$ pc^{-2}	8 M$_\odot$ pc^{-2}
$\mathcal{R}_g$	0.86	0.76
$\mathcal{R}_*/\mathcal{R}_g$	0.29	0.59

Using these data and formulae (7) we find that $\chi_\odot > \chi_c \simeq \pi^2 \gg 1$.

The asymptotic expansion of the Bessel function $I_n(\chi)$ for $\chi \gg 1$ is

$$I_n(\chi) \simeq \frac{e^\chi}{\sqrt{2\pi\chi}} \left[1 + O\left(\frac{1}{\chi}\right)\right].$$

Therefore the terms containing Bessel functions in (6) are small in comparison to unity.[3]

Substituting equations (3) and (6) into (5), using the asymptotic form for the Bessel function and assuming $\nu \sim 1$, we finally obtain

$$\left(\frac{\tilde\sigma_*}{\tilde\sigma_g}\right)_{max} \approx \frac{G\sigma_* \lambda}{c_{r,*}^2} \frac{\mathcal{R}_*}{\mathcal{R}_g}.$$

Using values from Table 1, we estimate

$$\left(\frac{\tilde\sigma_*}{\tilde\sigma_g}\right)_{max} \approx 0.04 \quad \text{at } r = r_c; \qquad \left(\frac{\tilde\sigma_*}{\tilde\sigma_g}\right)_{max} \approx 0.166 \quad \text{at } r = r_\odot.$$

In the light of these values, we shall henceforth neglect the potential of the perturbed stellar component when considering the development of spiral perturbations.

4 Proof of the equivalence of the linearised dynamical equations

Having proved $\tilde\psi$ to be determined only by $\tilde\sigma_g$, we can write down the system of linearised dynamical equations for the gaseous disc in the Galaxy:

$$\frac{\partial v_r}{\partial t} + \Omega_0 \frac{\partial v_r}{\partial \phi} - 2\Omega_0 v_\phi = -\frac{\partial}{\partial r}\left(c_{g,0}^2 \eta\right), \tag{8}$$

---

[3] The asymptotic formula given does not take into account the fact that the function $I_n(\chi)$ is a monotonously decreasing function of n (*e.g.* Yanke, Edme & Lesh 1984) for a fixed $\chi \gg 1$. Therefore, the alternating series converges and its sum, in accordance with the convergence test, must be less than its first term in the series.

$$\frac{\partial v_\phi}{\partial t} + \Omega_0 \frac{\partial v_\phi}{\partial \phi} + \frac{\kappa^2}{2\Omega_0} v_r = -\frac{1}{r}\frac{\partial}{\partial \phi}\left(c_{g,0}^2 \eta\right), \tag{9}$$

$$\frac{\partial \eta}{\partial t} + \Omega_0 \frac{\partial \eta}{\partial \phi} + \frac{\partial v_r}{\partial r} + (1 + r\ln' \sigma_0)\frac{v_r}{r} + \frac{1}{r}\frac{\partial v_\phi}{\partial \phi} = 0. \tag{10}$$

Here, a prime denotes differentiation with respect to r and we have introduced the following designations

$$c_{g,0}^2 \equiv c_{s,0}^2 - \frac{2\pi G\sigma_0}{|k|\mathcal{R}_g}, \qquad c_{s,0}^2 \equiv \frac{dP_o}{d\sigma_0}, \qquad \eta \equiv \frac{\sigma}{\sigma_0}, \tag{11}$$

As before, we denote stationary values by a subscript 0 and we omit the tilde over perturbed quantities, using it later only to denote magnitudes of perturbed quantities. In writing these equations, we have assumed a linearised equation of state

$$\frac{P}{\sigma_0} = c_{s,0}^2 \eta.$$

If we make the substitution

$$c_{g,0}^2 = c_{w,0}^2 \equiv gH_0, \qquad \eta = \frac{H}{H_0},$$

in equations (8) – (10), we obtain the system of linearised equations for rotating shallow water (Batchelor 1973). This is the system of equations used by Fridman (1989b) to describe perturbations in the "Spiral" experiments on f-plane.

5 Solution of equations for the case of tangential discontinuities

Since the coefficients in equations (8) – (10) do not depend on ψ or t, we look for solutions of the form
$$A(r, \phi, t) = \tilde{A}(r)\exp[i(m\phi - \omega t)].$$

The equations of motion (8) and (9) may then be reduced to one equation

$$\frac{\partial}{\partial r}(c_{g,0}^2 \tilde{\eta}) = \frac{2m\Omega_0}{r\hat{\omega}}c_{g,0}^2 \tilde{\eta} + (\hat{\omega}^2 - \kappa^2)\tilde{\xi}, \tag{12}$$

and the continuity equation (10) may be written as

$$\frac{d}{dr}(r\sigma_0\tilde{\xi}) = -r\sigma_0\left[2\frac{m\Omega_0}{\hat{\omega}}\frac{\tilde{\xi}}{r} + \left(1 - \frac{m^2}{r^2}\frac{c_{g,0}^2}{\hat{\omega}^2}\right)\tilde{\eta}\right]. \tag{13}$$

The new quantities in these equations are defined as

$$\hat{\omega} \equiv \omega - m\Omega_0, \qquad v_r \equiv \frac{d\xi}{dt} = \left(\frac{\partial}{\partial t} + \frac{v_{0,\phi}}{r}\frac{\partial}{\partial r}\right)\xi = -i\hat{\omega}\xi.$$

Observations indicate (Haud 1979, and references therein) a sharp fall of the rotation curve of the gaseous component over the radial range $0.3 < r < 1.1$ kpc. The position of the centre of this drop (*i.e.* at $r = R = 0.7$ kpc) is remarkable because it coincides with the edge of the central gaseous disc whose surface density $\sigma_{g,1}$ exceeds by two orders of magnitude the gas surface density $\sigma_{g,2}$ for $r > R$. Furthermore, the order of magnitude of the gas surface density does not change for all $r > R$.

In the light of this, we model the angular rotation velocity, $\Omega_0(r)$, sound velocity, $c_{g,0}(r)$, and gas surface density, $\sigma_0(r)$, by a discontinuity at the distance $r = R$. *i.e.* for $r < R$

$$\Omega_0(r) = \Omega_1 \equiv \text{const.}, \qquad \sigma_0(r) = \sigma_1 \equiv \text{const.}, \qquad c_{g,0}(r) = c_{g,1} \equiv \text{const.};$$

and for $r > R$

$$\Omega_0(r) = \Omega_2 \equiv \text{const.}, \qquad \sigma_0(r) = \sigma_2 \equiv \text{const.}, \qquad c_{g,0}(r) = c_{g,2} \equiv \text{const.}.$$

If we integrate equations (12) and (13) over the radial range $R - \epsilon < r < R + \epsilon$ and then let $\epsilon \to 0$ we obtain the following "matching" conditions

$$\left[\tilde{\eta} c_{g,0}^2 + R\Omega_0^2 \tilde{\xi} \right]_{R-0}^{R+0} = 0, \qquad \left[\tilde{\xi} \sigma_0 \right]_{R-0}^{R+0} = 0. \tag{14}$$

We now reduce the system of two ordinary differential equations of first order, (12) and (13), to one ordinary differential equation of second order. The solution of the latter will contain two arbitrary constants which we will obtain from the "matching" conditions (14). The desired equation has constant coefficients on the two different sides of the discontinuity at $r = R$.

From (12), we have

$$\tilde{\xi} = \frac{c_{g,0}^2}{\hat{\omega}^2 - \kappa^2} \left(\frac{d\tilde{\eta}}{dr} - \frac{2m\Omega_0}{r\hat{\omega}} \tilde{\eta} \right), \tag{15}$$

which we differentiate with respect to r, taking into account that all the coefficients are constants. We use the resulting expression in (13) and by simple manipulations we obtain the differential equation for cylinder functions of imaginary argument

$$\tilde{\eta}'' + \frac{1}{r}\tilde{\eta}' - \left(k^2 + \frac{m^2}{r^2} \right) \tilde{\eta} = 0, \qquad k^2 \equiv \frac{4\Omega_0^2 - \hat{\omega}^2}{c_{g,0}^2}. \tag{16}$$

The general solution of (16) is (Yanke *et al.* 1984)

$$\tilde{\eta} = \mathcal{Z}_m(ikr) = C_1 \mathcal{I}_m(kr) + C_2 \mathcal{K}_m(kr).$$

Since $\mathcal{I}_m(x) \to \infty$ as $x \to \infty$, and $\mathcal{K}_m(x) \to \infty$ as $x \to 0$ we have the following solution on the two different sides of the discontinuity

$$\tilde{\eta}_1 = C_1 \mathcal{I}_m(k_1 r); \qquad \tilde{\eta}_2 = C_2 \mathcal{K}_m(k_2 r), \tag{17}$$

where

$$k_{1,2} \equiv \frac{4\Omega_{1,2}^2 - (\omega - m\Omega_{1,2})^2}{c_{g,1,2}^2}.$$

Using the conditions (14) to "match" the two solutions (17) at $r = R$ we obtain a system of two transcendental equations for the the two unknown coefficients in (17). For this system to have a non-trivial solution, the determinant is required to be zero. Thus we come to the following dispersion relation (see Appendix)

$$k_1^2 \alpha_2 - k_2^2 \alpha_1 Q \mu^2 + \frac{M^2}{R^2} \alpha_1 \alpha_2 (1 - Qq^2) = 0, \qquad (18)$$

where

$$\alpha_1 \equiv \frac{2m}{x - m} - k_1 R \frac{\mathcal{I}_m'(k_1 R)}{\mathcal{I}_m(k_1 R)},$$

$$\alpha_2 \equiv \frac{2mq}{x - mq} - k_2 R \frac{\mathcal{K}_m'(k_2 R)}{\mathcal{K}_m(k_2 R)},$$

$$M \equiv \frac{R\Omega_1}{c_{g,1}}, \quad q \equiv \frac{\Omega_1}{\Omega_2}, \quad Q \equiv \frac{\sigma_1}{\sigma_2}, \quad \mu \equiv \frac{c_{g,1}}{c_{g,2}} \quad x = \frac{\omega}{\Omega_1},$$

and

$$k_1 = \frac{M}{R}\sqrt{4 - (x - m)^2}, \qquad k_2 = \frac{M}{\mu R}\sqrt{4q^2 - (x - mq)^2}. \qquad (19)$$

The primes denote differentiation with respect to the argument of the cylinder function. The parameter M has the sense of a Mach number in the region of the discontinuity (calculated at the value of the 'interior" velocity, $r = R-0$), and the parameters q, Q^{-1} and μ characterise the ratios of the physical variables on either side of the discontinuity.

The dispersion relation (18) transforms into that given by Morozov (1977, 1979) when $Q = \mu = 1$, because in that paper the surface density, σ_0, and gas velocity dispersion, $c_{g,0}$, were considered constant throughout the disc and self-gravitation effects were neglected. As we shall see below (see also Fridman 1989a), the changes in surface density and gas velocity dispersion introduce qualitatively new effects.

6 Gradient instabilities for small Mach numbers

From (19) we see that $k_1 R \sim M$, which we here assume to be small, and $k_2 R \sim Mq\mu$. For small enough Mach numbers, $M \ll \mu/q$, $k_2 R \ll 1$. Expanding the cylinder functions $\mathcal{I}_m(k_1 R)$ and $\mathcal{K}_m(k_2 R)$ in a series for small arguments gives

$$\frac{\mathcal{I}_m'(k_1 R)}{\mathcal{I}_m(k_1 R)} \sim \frac{m}{k_1 R}, \qquad \frac{\mathcal{K}_m'(k_2 R)}{\mathcal{K}_m(k_2 R)} \sim -\frac{m}{k_2 R}.$$

We therefore find

$$\alpha_1 = -m\left(1 - \frac{2\Omega_1}{\hat{\omega}_1}\right), \quad \alpha_2 = m\left(1 + \frac{2\Omega_2}{\hat{\omega}_2}\right), \quad \hat{\omega}_{1,2} = \omega - m\Omega_{1,2}.$$

Using these expressions for α_1 and α_2, the dispersion relation (18) takes the form

$$(1 + Q)x^2 - 2[(m - 1) + (m + 1)qQ]x + m[(m - 1) + q^2 Q(m + 1)] = 0,$$

which has the solution

$$x_{1,2} = (1 + Q)^{-1}\Big\{ m(1 + qQ) + (qQ - 1)$$
$$\pm i[m^2 Q(1 - q)^2 - (1 - qQ)^2 - m(Q - 1)(1 + q^2 Q)]^{1/2}\Big\} \qquad (20)$$

This expression does not contain μ because we have neglected all terms containing M – the approximation $M \ll 1$ is equivalent to $c_g \to \infty$, so it cannot matter how many times one infinity is larger than another. If we were to include subsequent terms in the expansion for small M, the equation would contain μ^2, but we shall confine our attention to the expression given, which is zeroth order in M.

When we set $Q = 1$ in (20) we, of course, obtain the solution found by Morozov (1977, 1979)

$$x_{1,2} = \frac{1}{2}\left\{m(1+q) + (q-1) \pm i[(m^2-1)(1-q^2)]^{1/2}\right\}, \tag{21}$$

which demonstrates that instabilities arise at any value of $q \neq 1$. This is exactly the *Kelvin-Helmholtz instability* (KHI), which arises irrespective of whether the inside rotates faster or slower than the outside. The physics of the KHI is discussed in detail in PF, but can be briefly summarised as follows. The limit of $M \ll 1$ corresponds to an incompressible fluid, $c_g \to \infty$. We can therefore use the Bernoulli equation (Landau & Lifshitz 1986) $v^2/2 + P/\sigma = $ const. As the functions fall exponentially on either side of the discontinuity (Landau 1944), the perturbed flow takes place as in a tube-like narrow region near $r = R$. Let region I be inside the circle of radius R and region II outside. Above a "hump" in region II, the velocity is higher than in the neighbouring regions, the flow in the tube must be preserved. Thus the pressure above the "hump" is lower (Bernoulli's equation) and the "hump" grows.

We can also consider the case of solid body rotation of the whole system, *i.e.* $q = 1$, but which has a density discontinuity, $Q \neq 1$. In this case (20) becomes

$$x_{1,2} = (1+Q)^{-1}\left\{m(1+Q) + (Q-1) \pm i[-(1+Q)^2 + m(1-Q)^2]^{1/2}\right\}$$

We therefore expect a *flute instability* (FI) when $Q < 1$: the higher density region compresses the region of lower density.

Thus the solution (20) when $q \neq 1$ and $Q < 1$ contains two instabilities: KHI and FI. By varying these two parameters we may trade-off one against the other. For example, one can stabilise the FI by counter-acting with additional centrifugal force, *i.e.* by raising q. Similarly, it is possible to stabilise a KHI by a negative density gradient. Examples of such stabilisation are given below.

Strictly speaking, in the general case one ought not to speak about the two forms of instability separately. The solution (20) describes the generation of a *shear-flute instability* (SHFI) or *gradient instability* (GI) when $M \ll 1$.

6.1 SPECIAL CASE

We now give an example of how a KHI could be stabilised by a density decrease, and we will assume $Q \gg 1$ so that the effects are large. We adopt the values $q \simeq 0.1$ and $Q \simeq 100$ as in our Galaxy, though it should be obvious that the results obtained will have no relation to our Galaxy (for which $M \gg 1$) or to any other spiral galaxy. (The author knows of no galaxy for which $M \ll 1$.)

Using the above suggested values, equation (20) becomes

$$x_{1,2} = q(m+1) \pm iQ^{-1/2}[m^2(1-q)^2 - (m-1)q^2Q - m]^{1/2}, \tag{22}$$

which gives the instability condition

$$m^2(1-q)^2 > (m+1)q^2Q + m,$$

or, when $q \ll 1$

$$q^2Q < m\frac{m-1}{m+1}.$$

It follows that no $m = 0$ or $m = 1$ instabilities are possible. An $m = 2$ instability develops when $q^2Q < 2/3$, *etc.* Thus a large negative density gradient can stabilise the system.

From (22) we find the pattern speed of the perturbation

$$\Omega_p \equiv \frac{\omega}{m} = \frac{m+1}{m}\Omega_2, \qquad (\text{if } Q \gg 1 \text{ and } Qq \gg 1).$$

For comparison, Morozov (1977, 1979) found for $Q = 1$

$$\Omega_p = \frac{\Omega_1 + \Omega_2}{2} - \frac{\Omega_1 - \Omega_2}{2m}.$$

The difference in pattern speeds between the two cases is not large when m is small. *e.g.* for $m = 2$ the ratio, $(\Omega_p)_{Q \gg 1}/(\Omega_p)_{Q=1} \equiv \alpha_p \simeq 6q$, or 0.6 when $q = 0.1$.

However, as we know from theoretical (Morozov 1977, 1979) and experimental (Morozov *et al.* 1984, 1985, Fridman *et al.* 1985) results, when $M \ll 1$, modes of large m are generated. The differences in pattern speeds at large m between the two cases ($Q \gg 1$ and $Q \sim 1$) is much greater – about a factor of 5 when $q = 0.1$. Thus, we expect disturbances to have a lower azimuthal velocity, and larger co-rotation radius, when $Q \gg 1$, than they would have if $Q = 1$.

7 Gradient instabilities at large Mach number

Writing x as $x_1 + iMx_2$, we find from (19)

$$k_1 \simeq \frac{M^2x_2}{R} - i\frac{M}{R}(x_1 - m),$$

$$k_2 \simeq \frac{M^2x_2}{\mu R} - i\frac{M}{\mu R}(x_1 - mq). \qquad (23)$$

Using these in (18), we obtain

$$x_1 = \frac{m(1 + Q\mu q)}{1 + Q\mu}, \qquad x_2 = \frac{1 - Qq^2}{1 + Q\mu},$$

or

$$x = (1 + Q\mu)^{-1}\left[m(1 + Q\mu q) + iM(1 - Qq^2)\right]. \qquad (24)$$

Again we can compare this solution with that obtained by Morozov (1977, 1979) by setting $Q = \mu = 1$ to find

$$x = \frac{1}{2}[m(1 + q) + iM(1 - q^2)].$$

We note that this solution, which predicts instability only if $q < 1$, differs from (21), which indicates instability for all $q \neq 1$. We now see that when $M \gg 1$, an instability develops only when the interior rotates more rapidly than the exterior and we denoted this a *centrifugal instability* (CI) (Morozov *et al.* 1984, 1985, Fridman *et al.* 1985). The physics of this instability is similar to that of FI and differs essentially from that of KHI.[4]

For the special case of uniform rotation of the whole system ($q = 1$), we have from (24)

$$x = (1 + Q\mu)^{-1} \left[m(1 + Q\mu) + iM(1 - Q) \right]$$

which describes FI of supersonic flow. As before, the condition for this instability is independent of M.

As for the $M \ll 1$ case, the solution (24) for $M \gg 1$ does not describe some definite instability developing when $q > 1$ or when $Q < 1$, but some *centrifugal-flute instability* (CFI) or again GI. The condition for this instability for $M \gg 1$ is that $Qq^2 < 1$.

The pattern speed in the general case given by (24) is

$$\Omega_p \equiv \frac{\Re(\omega)}{m} = \frac{1 + Q\mu q}{1 + Q\mu} \Omega_1.$$

which reduces to $\Omega_p = \frac{1}{2}(1 + q)\Omega_1$ in the special case $Q = \mu = 1$ (Morozov 1977, 1979). In the Galaxy, $q \simeq 0.1$, $Q \simeq 100$ and $\mu \sim 1$, and we predict $\Omega_p \simeq 0.36\Omega_1/2$[5]. This has a larger co-rotation radius than was predicted by Morozov (1977, 1979), who considered only the $Q = 1$ case and obtained $\Omega_p = \Omega_1/2$, *i.e.* about a factor three higher. The revised theory given here yields estimates closer to the values observed in the experiments (Morozov *et al.* 1984, 1985, Fridman *et al.* 1985) and also corresponds better with modern ideas (*e.g.* Marochnik & Suchkov 1984) of the pattern speed in the Galaxy.

Finally we consider the form of the spiral pattern generated in the outer part of the disc by the GI. The perturbed surface density, σ, is given by (17) which, using the wavenumber from equation (23), may be approximated as

$$\sigma \sim \mathcal{K}_m(k_r r)e^{im\phi} \sim r^{1/2}\exp\left\{ -\frac{\Omega_1^2 Rr}{c_{g,1}c_{g,2}}\frac{(1 - Qq^2)}{(1 + Q\mu)} + im\left[\phi + \frac{\Omega_1 r}{c_{g,2}}\frac{(1 - q)}{(1 + Q\mu)} \right] \right\}. \quad (25)$$

We may draw two conclusions from this expression:

---

[4] Landau (1944) proved that KHI for 2-D perturbations stabilises for $M > 2\sqrt{2}$ as can be readily seen. Referring back to the discussion in §6, but now for $M \gg 1$, the "hump" in region II is perceived by the flow to make the supersonic nozzle narrower, so the velocity does not grow (as in the strongly sub-sonic case), but decreases. As a result the pressure above the "hump" increases, pushing it back into region I.

[5] Q and μ are not independent. At $r = R$ ($= 0.7$ kpc) $c_{g,0}^2 \approx c_{s,0}^2$ and μ is fully determined by the equation of state. For a polytropic model $P/\sigma^\gamma = $ const, $\mu = Q^{(1-\gamma)/2} = Q^{-1/2}$ if $\gamma = 2$ (Pasha & Fridman 1989). Thus $\mu = 0.1$ if $Q = 100$ which corresponds to a turbulent velocity of ~ 80 km s^{-1} in the molecular disc. If we use this value, rather than the inconsistent value of ~ 20 km s^{-1} assumed above, we obtain $\Omega_p \approx 0.26\Omega_1/2$, which is not significantly different.

(1) The necessary condition for finiteness of the solution coincides with the condition for development of instabilities when $M \gg 1$.
(2) Density waves trail only when the angular velocity decreases with radius *i.e.* for $q < 1$.

This last condition is also necessary for the development of GI in the system. Equation (25) also indicates that the radial wavelength is

$$\lambda_r = \frac{2\pi}{k_r} = 2\pi \frac{c_{g,2}}{m\Omega_1} \frac{(1+Q\mu)}{(1-q)}.$$

8 Conclusions

(1) We have shown that the perturbed gravitational potential of the Galaxy in the plane $z = 0$ is mainly determined by the perturbed surface density of the gaseous disc. The contribution from the perturbed density in the stellar disc is negligible.
(2) By introducing the gravitating sound velocity, c_g, the linearised dynamical equations for the gaseous gravitating disc are reduced to the similar equations for rotating shallow water.
(3) Viscosity effects near the bottom of the vessel have little influence on the generation of density waves in the "Spiral" experiments. These experiments may therefore be considered as modelling the process of spiral arm formation in the gaseous disc of the Galaxy.
(4) We have investigated the instabilities that arise when the 2-D gaseous gravitating disc (or rotating shallow water) has a break in the surface density, sound velocity and angular rotation rate at a radius $r = R$. The instability has been called a *gradient instability* (GI), but can also be termed a *shear-flute instability* (SHFI) for $M \ll 1$ or a *centrifugal-flute instability* (CFI) for $M \gg 1$.
(5) In the special case of a uniform density medium ($Q = 1$), SHFI is a Kelvin-Helmholtz instability (KHI) and the CFI is a CI (Morozov 1977, 1979, Morozov *et al.* 1984, 1985, Fridman *et al.* 1985).
(6) In the special case of uniform rotational velocity on both sides of the discontinuity ($q = 1$), both SHFI and CFI become FI with the condition for instability independent of the Mach number M.
(7) The pattern speed of the perturbations, Ω_p, is determined by q, Q and μ. The more general treatment given here predicts pattern speeds close to those observed in the experiments (Morozov *et al.* 1984, 1985, Fridman *et al.* 1985).
(8) Perturbations have the form of trailing spirals (in the case where $M \gg 1$) if the angular velocity in the disc decreases with radius.

I express my gratitude to Profs. A. G. Morozov and M. V. Nezlin for fruitful discussions.

References

Antipov, S. V., *et al.* 1988. *Physics of Plasmas*, **14**, 1104.
Batchelor, J., 1973. *An Introduction for Fluid Dynamics*, Mir, Moscow.
Fridman, A. M., 1978. *Usp. Fiz. Nauk*, **125**, 352.
Fridman, A. M., 1989a. *Pis'ma Astron. Zh.*

Fridman, A. M., 1989b. *Dokl. Akad. Nauk SSSR.*, in press.

Fridman, A. M., Morozov, A. G., Nezlin, M. V. & Snezhkin, E. N. 1985. *Phys. Lett. A*, **109**, 228.

Haud, U. A., 1979. *Pis'ma Astron. Zh.*, **5**, 124.

Landau, L. D., 1944. *Dokl. Akad. Nauk SSSR*, **44**, 151.

Landau, L. D. & Lifshitz, E. M., 1986. *Hydrodynamics*, Nauka, Moscow.

Lin, C. C., Yuan, C. & Shu, F. H., 1969, *Astrophys. J.*, **155**, 721.

Marochnik, L. S. & Suchkov, A. A., 1984. *Galaxy*, Nauka, Moscow.

Morozov, A. G., 1977. *Pis'ma Astron. Zh.*, **3**, 195.

Morozov, A. G., 1979. *Astron. Zh.*, **56**, 498.

Morozov, A. G., Fainshtein, V. G. & Fridman, A. M., 1976a. *Dokl. Akad. Nauk SSSR*, **231**, 588.

Morozov, A. G., Fainshtein, V. G., Polyachenko, V.L. & Fridman, A. M., 1976b. *Astron. Zh.*, **53**, 946.

Morozov, A. G., Nezlin, M. V., Snezhkin, E. N. & Fridman, A. M., 1984. *Pis'ma Zh. Ehksp. Teor. Fiz.*, **39**, 504.

Morozov, A. G., Nezlin, M. V., Snezhkin, E. N. & Fridman, A. M., 1985. *Usp. Fiz. Nauk*, **145**, 161.

Nezlin, M. V., Rylov, A. Yu., Snezhkin, E. N. & Trubnikov, A. S. 1987. *Zh. Ehksp. Teor. Fiz.*, **92**, 3.

Pasha, I. I. & Fridman, A. M., 1989. *Zh. Ehksp. Teor. Fiz.*, in press.

Pedlosky, G., 1984. *Geophysical Hydrodynamics*, vols 1-2, Mir, Moscow.

Polyachenko, V. L. & Fridman, A. M., 1976. *Equilibrium and Stability of Gravitating Systems*, Moscow: Nauka. English translation: Fridman, A. M. & Polyachenko, V. L., 1976. *Physics of Gravitating Systems*, 2 vols, Springer-Verlag, New York.

Polyachenko, V. L. & Fridman, A. M., 1981. *Pis'ma Astron. Zh.*, **7**, 136.

Rohlfs, K. & Kreitschman, G., 1988. *Astron. Astrophys.*, **201**, 51.

Sanders, D. B., Solomon, P. M. & Scoville, N. Z., *Astrophys. J.*, **276**, 182.

Shu, F. H., 1968. *PhD thesis*, Harvard University.

Vandervoort, P. O., 1970. *Astrophys. J.*, **161**, 67.

Yanke, E., Edme, F. & Lesh, F. 1984. *Special Functions*, Mir, Moscow.

Appendix

Substituting in (15) from (17) we obtain the following expressions for ξ_1 and ξ_2 in the two different regions

$$\xi_1 = C_1 \frac{\alpha_1 \mathcal{I}_m}{k_1^2 R}, \qquad \xi_2 = C_2 \frac{\alpha_2 \mathcal{K}_m}{k_2^2 R}. \tag{A1}$$

Using the second "matching" condition (14), we obtain

$$C_1 \frac{\sigma_1 \alpha_1}{k_1^2} \mathcal{I}_m = C_2 \frac{\sigma_2 \alpha_2}{k_2^2} \mathcal{K}_m. \tag{A2}$$

and from the first "matching" condition we find

$$\tilde{\eta}_1 + \frac{M^2}{R} \tilde{\xi}_1 = \mu^2 \tilde{\eta}_2 + \frac{M^2 q^2}{R} \tilde{\xi}_2,$$

or with the help of (17) and (A1) we have

$$\left(1 + M^2 \frac{\alpha_1}{k_1^2 R^2}\right) \mathcal{I}_m C_1 = \left(\mu^2 + M^2 q^2 \frac{\alpha_2}{k_2^2 R^2}\right) \mathcal{K}_m C_2. \tag{A3}$$

The system of homogeneous transcendental equations (A2) and (A3) has non-trivial solutions for C_1 and C_2 when the determinant of the system is zero. This condition gives us the dispersion relation (18) we seek.

Stability criteria for gravitating discs

Valerij L. Polyachenko

Astronomical Council of the Academy of Sciences of the USSR

Abstract The local dispersion relation and corresponding marginal stability condition for perturbations of an arbitrary degree of axial asymmetry are derived within the frame of the hydrodynamical model with plane pressure. This is just the condition to which Lin & Lau (1979), Bertin & Lin (1988), Morozov (1985) and others aspired to obtain (without success). The most curious fact is that the result has been, in essence, available since 1965, though in a slightly masked form. From the marginal stability condition, we find the sufficient conditions which guarantee lack of quasi-exponentially increasing (even temporary) perturbations. In particular, for the flat rotation curve ($v = \Omega r = $ const), the condition is reduced to $Q > Q_m = \sqrt{3}$ where Q is the Toomre stability parameter. At the beginning of the paper, we make some remarks concerning stability studies of gravitating discs.

1 Some General Remarks

I first give a brief historical background to the problem of the stability gravitating discs. I restrict myself to analytical results only and do not touch upon those obtained by computer simulations.

It is well known that Toomre (1964) derived the stability criterion for local axisymmetric perturbations of stellar discs.

The next substantial theoretical (analytical) result was obtained independently (by different methods) by Kalnajs (1972) and by Polyachenko & Shukhman (1972, 1973) (also Polyachenko 1972). We studied linear perturbations of uniformly rotating discs with the equilibrium surface density $\sigma_0(r) = \sigma_0\sqrt{1 - r^2/R^2}$, a quadratic potential $\phi_0(r) = \frac{1}{2}\Omega_0^2 r^2$, and the distribution function, in the frame of reference associated with the disc,

$$f_0 = \frac{\sigma_0}{2\pi\sqrt{1-\gamma^2}}\left[(1-\gamma^2)(1-r^2) - v_r^2 - v_\varphi^2\right]^{-1/2}, \qquad |\gamma| \leq 1, \tag{1}$$

where γ is the angular velocity of the disc as a whole in units of Ω_0 (*i.e.* it is assumed that $\Omega_0 = 1$, $R = 1$). This equilibrium phase density was given by Bisnovaty-Kogan & Zel'dovich (1970). They also noticed that the distribution function (1) permits the essential generalization $f_{0\gamma} \to f_0 = \int_{-1}^{1} A(\gamma) f'_{0\gamma} d\gamma$, where $A(\gamma)$ is an arbitrary function, $f'_{0\gamma}$ is the original distribution function (1), but written, for all γ, in the inertial frame of reference. In particular, for a certain choice of $A(\gamma)$, one can obtain the distribution function

$$f_0 = \frac{\sigma_0}{\pi}\theta\left[(1-r^2)(1-v_\varphi^2) - v_r^2\right], \qquad \left(\theta(x) = \begin{cases} 1, & x > 0 \\ 0, & x < 0 \end{cases}\right). \tag{2}$$

In these papers, we derived the exact spectra of eigenfrequencies for both the radial (axisymmetric) and non-radial modes, for all these models. Further considerations were reported by Morozov *et al.* (1973) and the most detailed analysis is given in my monograph with Fridman (1976) (English translation: Fridman & Polyachenko 1984). It may be worth noting that the most elegant form of the dispersion equation for this

problem was later given by Antonov (1976) who independently considered the model (2);
he reduced the dispersion relation to a very simple form and then proved its stability. I
note also that the derivation of disc spectra in the papers of the author and Shukhman
and of Antonov was only a part of a more general programme (now mainly realized)
of studying the linear perturbations of collisionless systems with quadratic potentials
(homogeneous layer, sphere and ellipsoid). Details can be found in the monograph by
Polyachenko & Fridman (1976, 1984).

We drew attention to the fact that stabilization of non-axisymmetric perturbations
in the disc (in particular, the bar mode) requires considerably larger velocity dispersions
than the critical dispersion for radial perturbations (given by the Toomre criterion).

Predominance of the largest scale, non-axisymmetric disc modes (first of all, the bar
mode) has allowed us (*e.g.* Polyachenko & Fridman 1976) to clarify the old hypothesis
of QSSS (quasi-stationary spiral structure) by Lin & Shu (1964). With reasonable
assumptions about the magnitude of stellar velocity dispersions, the central region
remains unstable to large-scale non-axisymmetric modes (in the linear approximation).
A bar-like standing wave is produced there, the frequency of which is defined by the
equilibrium parameters of this region. In turn, the bar excites a trailing spiral density
wave – mainly in the flattest and coldest subsystems of the galaxy. The established
stationary amplitude of the spirals is defined, for instance, by the non-linear effects
which stabilize the bar mode. It seems that just this approach (or one very similar) is
assumed in the modern version of the gravitational theory of galactic spiral arms; *i.e.*
the modern theory of global modes by Lin, Lau, Mark, Bertin and others, who usually
refer to the old QSSS-hypothesis by Lin & Shu (1964). However, in that paper, Lin &
Shu meant something quite different from the bar mode instability.

2 Local Stability Criteria

Unfortunately, exact studies are possible only for quite idealized models. They may be
considered only as the known limits for further computer calculations of more reasonable
models. But the promising program of such calculations, which was begun in the series
of papers by Kalnajs (1976), is not so far completed. (It is interesting that the similar
program for spherical collisionless systems initiated by the author and Shukhman (1981)
is now in a rather developed state.)

So I turn now to studies of stability criteria for localized non-axisymmetric perturbations which can be also performed analytically. Even in this limited field, there is
a rather large number of papers. I shall restrict myself mainly to the papers dealing
with the often-used hydrodynamical model with plane pressure: *e.g.* Lin & Lau (1979),
Bertin & Lin (1988) and Morozov (1985). The older paper by Goldreich & Lynden-
Bell (1965), which could be added to this list, will be considered separately for reasons
explained somewhat later.

In the first three papers, the study of linear stability was carried out by the standard
method of WKB-type, which assumed that $k_r \gg k_\varphi$. Here k_r and k_φ are the radial and
azimuthal components of the wave vector. This leads to the following relation for the
stability boundary:

$$Q^2 = 4(\bar{\lambda} - \bar{\lambda}^2 + g^2\bar{\lambda}^4),$$ (3)

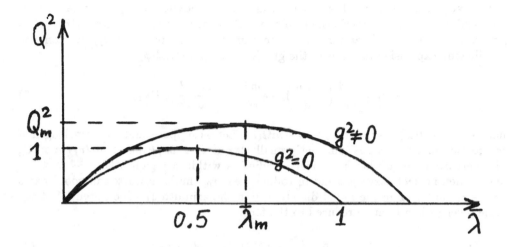

Figure 1. Dependence of Q^2 on $\bar{\lambda}$ according to equation (3) for small values of g^2.

where $Q = \kappa_0 c_s / \pi G \sigma_0$ is the Toomre stability parameter, $\bar{\lambda} = \lambda / \lambda_c$, λ is the wavelength of perturbation, $\lambda_c = 4\pi^2 G \sigma_0 / \kappa_0^2$ is the critical Toomre wavelength and $g^2 = 8r|(\Omega_0^2)'|k_\varphi^2 (\pi G \sigma_0)^2 / \kappa_0^6$ is the dimensionless parameter taking into account that perturbations are slightly non-axisymmetric: $g^2 \ll 1$. I should also note that the form of equation (3) does not coincide exactly with that given by Lin & Lau – if we expand their equation in the small parameter $\epsilon = g^2 \bar{\lambda}^2 \ll 1$ and omit the unnecessary terms which exceed the accuracy adopted, we obtain equation (3).

For radial perturbations ($m = g^2 = 0$), equation (3) gives the well-known formula $Q^2 = 4(\bar{\lambda} - \bar{\lambda}^2)$, which is shown by the lower curve in Figure 1. Points under the curve are unstable, those above are stable. As this curve has the maximum $Q_m^2 = 1$ at $\bar{\lambda} = 0.5$, discs with $Q > Q_m(= 1)$ are stable with respect to axisymmetric perturbations for arbitrary wavelengths. Non-axial symmetry ($0 < g^2 \ll 1$) leads to slightly larger values of Q_m:

$$Q_m^2(g^2) = 1 + \frac{1}{4}g^2, \qquad (\bar{\lambda}_m = \frac{1}{2} + \frac{1}{4}g^2). \qquad (4)$$

This means that non-axisymmetric perturbations are more unstable.

These results, which were obtained within the framework of perturbation theory in the small parameter g^2, indicate that the suppression of instability for larger g^2 requires larger values of Q. In such a situation, it is evident that the answer to the main question of the stability theory – what conditions are necessary for complete stability? – lies beyond such a perturbation theory. It can be said that this theory itself pushes us out of its frame.

Localized perturbations with arbitrary values of the parameter g^2 (or the parameter $s = k_\varphi / k_r$), *i.e.* without the usual WKB approximations, were considered in our recent papers (Polyachenko 1987, Polyachenko & Strel'nikov 1988). Let us begin from the cold case, where the velocity dispersion (sound speed) $c_s = 0$. The generalization for hot discs is rather simple.

By excluding velocities from the linearized hydrodynamical pressureless equations, and considering perturbations to the surface density of the form $\sigma_1(r,\varphi,t) = \sigma_1(r)\exp(-i\omega t + im\varphi)$ (r and φ – polar coordinates in the plane (x,y)), we come to the following expression in terms of the gravitational potential ϕ_1:

$$-\sigma_1 = \frac{1}{r}\frac{d}{dr}\left(r\epsilon\frac{d\phi_1}{dr}\right) - \epsilon\frac{m^2}{r^2}\phi_1 - \frac{2m}{r\omega_*}\frac{d}{dr}(\epsilon\Omega)\phi_1, \tag{5}$$

where $\epsilon \equiv \sigma_0(r)/(\omega_*^2 - \kappa^2)$ and $\omega_* \equiv \omega - m\Omega(r)$. Let us suppose that the wavelength of perturbation in azimuth φ is sufficiently small: $2\pi r \gg l_\varphi \equiv 2\pi/k_\varphi$, $k_\varphi \equiv m/r$, $(m \gg 1)$, while perturbations are localized near a radius r_0 within a region of size $\sim \Delta \ll r_0$. We can then neglect curvature and introduce, near r_0, the local Cartesian coordinates $x = r - r_0$ and y (x in the radial direction, y in the azimuthal). The potential of the simple layer ϕ_1 can then be reduced to the form

$$\phi_1(x) = -2G\int_{-\infty}^{\infty} dx' K_0(k_\varphi|x - x'|)\sigma_1(x'), \tag{6}$$

where K_0 is the MacDonald function. Thus, in the general case, the problem is reduced to the integro-differential equation

$$\phi_1(x) + 2G\int_{-\infty}^{\infty} dx' K_0(k_\varphi|x - x'|)\sigma_1[\phi_1(x')] = 0, \tag{7}$$

where $\sigma_1[\phi(x)]$ is determined by the formula (5). Let us introduce the natural dimensionless coordinate $x = (r - r_0)/\Delta$ where Δ is the distance between co-rotation r_0 and Lindblad resonances. Expanding the unperturbed quantities in powers of x, and omitting terms which are small in the ratio Δ/r_0, we obtain from (5):

$$-\sigma_1(x) = \frac{\sigma_0}{\kappa_0^2\Delta^2}\left\{\frac{1}{z^2-1}\phi_1'' - \frac{2z}{(z^2-1)^2}\phi_1' + \left[-\frac{k^2}{z^2-1} + \frac{4k\Omega_0/\kappa_0}{(z^2-1)^2}\right]\phi_1\right\}, \tag{8}$$

where $\sigma_0 \equiv \sigma_0(r_0)$, $z = x + iy$, $\nu = \gamma/\kappa_0$ is the dimensionless growth-rate ($\nu \ll 1$) and $k \equiv k_\varphi\Delta$.

The equation which results from substitution of equation (8) into equation (7) can be studied as follows. On introducing Fourier transformations into equation (7) (which is natural because it contains a convolution), we are led to an equivalent fourth-order differential equation of a rather complicated form. It is remarkable (and apparently no accident) that this equation happens to be factorizable and reduces to the form (now for a hot disc):

$$\left[\left(\frac{d}{dq} + \nu\right)^2 + A_1\left(\frac{d}{dq} + \nu\right) + B_1\right]\left[\left(\frac{d}{dq} + \nu\right)^2 + B_3\right]\phi_q = 0, \tag{9}$$

where ϕ_q denotes the Fourier transformation of the perturbed potential $\phi_1(x)$,

$$A_1 = \frac{4q}{(q^2 + k^2)},$$

$$B_1 = 1 + \frac{3k^2}{(q^2 + k)^2} + \frac{2\alpha^2}{(q^2 + k^2)}, \qquad (10)$$

$$B_3 = 1 - \frac{2\alpha^2}{(q^2 + k^2)} + \frac{3k^2}{(q^2 + k^2)^2} - \frac{2\sqrt{q^2 + k^2}}{c}(1 - a\sqrt{q^2 + k^2}),$$

with

$$\alpha^2 = \frac{2\Omega_0}{r|\Omega_0'|}, \qquad k^2 = \alpha^4 - \alpha^2, \qquad c = \frac{\kappa_0^2 \Delta}{\pi G \sigma_0}, \qquad a = \frac{c_s^2}{2\pi G \sigma_0 \Delta}.$$

There are three parameters: α^2, a and c. By substituting $\phi_q = e^{\nu q}\psi_q$ one can eliminate ν from equation (9).

Let us restrict ourselves to only one of two simpler equations, contained in the factorized equation, namely:

$$\frac{d^2 \psi_q}{dq^2} + B_3(q)\psi_q \doteq 0. \qquad (11)$$

Firstly, we could consider this equation as a Schrödinger equation with the potential $U(q) = -B_3(q)$. But proper solutions of this equation, satisfying the necessary boundary conditions, exist only for the cold case. Even these solutions are scarcely meaningful as they exist only simultaneously with the strongly unstable solutions characteristic of cold systems.

We note that after the substitution $k_r \Delta = q = -k_\varphi \tau$, $(\tau = r|\Omega_0'|t - k_{r_0}/k_{\varphi_0})$, equation (11) takes the form

$$\frac{d^2 \psi}{d\tau^2} + \left[k_\varphi^2 \Delta^2 - \frac{2\alpha^2}{1 + \tau^2} + \frac{3}{(1 + \tau^2)^2} - \frac{2\sqrt{1 + \tau^2}(k_\varphi \Delta)^3}{c}(1 - a\sqrt{1 + \tau^2}k_\varphi \Delta) \right] \psi = 0. \quad (12)$$

The first three terms in the bracket coincide with the corresponding terms in the equation derived by Goldreich & Lynden-Bell (1965) for the Cauchy problem of the evolution of perturbations proportional to $\exp[-ik_\varphi \tau x + ik_\varphi y]$ (in a local Cartesian reference system). The last two terms in equation (12) are different from those in the equation by Goldreich & Lynden-Bell, because they did not consider an infinitesimally thin disc. Equation (12) may be written as

$$\frac{d^2 \psi}{dt^2} + 4k^2 A^2 B_3[q(t)]\psi = 0, \qquad (A = 1/2r|\Omega_0'|). \qquad (13)$$

As we see, the behaviour of the solution depends on the sign of B_3: it oscillates when $B_3 > 0$ and has a growing mode when $B_3 < 0$. Substituting $d/dt^2 \to -(\omega - m\Omega_0)^2$ in equation (13) leads to a local dispersion relation, and the equation $B_3(q) = 0$ marks the condition of marginal stability, namely

$$Q^2 = 4\left(\bar{\lambda} - \bar{\lambda}^2 + g^2\bar{\lambda}^4 - \frac{3\kappa_0^2}{16\Omega_0^2}g^4\bar{\lambda}^6 \right). \qquad (14)$$

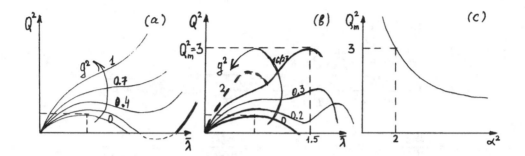

Figure 2. Curves of marginal stability for the hydrodynamical model with plane pressure according to (a) Lin & Lau (1979) or Bertin & Lin (1988), (b) Polyachenko & Strel'nikov (1988) and (c) qualitative dependence of $Q_m^2(\alpha^2)$ for the maximum of the function $Q_{\alpha^2}^2(\bar{\lambda}, g^2)$ as a function of two variables.

Equation (14) can be properly related to an analogous equation by Lin & Lau (1979)

$$Q^2 = 4 \left(\bar{\lambda} - \frac{\bar{\lambda}^2}{1 + g^2 \bar{\lambda}^2} \right). \tag{15}$$

For this once, the equation is written out in its original form, without any expansions. Note, however, that equation (15) was obtained for small g^2 and $\epsilon = g^2 \bar{\lambda}^2$, while it was employed by Lin & Lau even for $g^2 \sim 1$. As expected, equation (14) and that which results from equation (15) for small ϵ coincided, but are quite different when $\epsilon \sim 1$. The expression used by Lin & Lau is plotted in Figure 2a; it is monotonic and so is incorrect. The correct curves have maximum values for all α^2 and g^2. Moreover, at a given value of α^2 these maxima first increase with g^2 and then decrease (Figure 2b). Therefore, the function $Q_{\alpha^2}^2(\bar{\lambda}, g^2)$ has a maximum $Q_m^2(\alpha^2)$ as a function of two variables; when $\alpha^2 = 2$, *i.e.* for the flat rotation curve ($v_0 = r\Omega = $ const), $Q_m^2 = 3$. This maximum is at $g^2 = 16/27$, $\bar{\lambda} = 1.5$, *i.e.* for $m = 2v_0/3c_s \approx 4$ (four-armed azimuthal symmetry) for $v_0 = 250$ km/s, $c_s = 45$ km/s (typical values for the Solar vicinity in the Galaxy). I think this result indicates that the symmetry of a response to excitation by a bar (for instance) is not certainly two-armed. Qualitative dependence of $Q_m^2(\alpha^2)$ is shown in Figure 2c.

One should bear in mind that evolution of each mode takes place at every given moment in accordance with an instantaneous value of the radial wavenumber that changes linearly with time. The time of crossing the region of unstable values of q, as well as the degree of amplification, are restricted (are small about $Q_{\max}$). When $Q > Q_{\max}$, the region of amplification disappears.

3 Discussion

The theory of linear perturbations can be developed in two ways: by studying either the initial-value problem or the eigenvalue problem – *viz.* searching for solutions with exponential time dependence ($\sim e^{-i\omega t}$). One usually prefers the second approach when considering a stability problem because one believes that it is the only way to obtain the complete solution (in particular, the determination of the stability boundary). Besides, it is often much the simpler method.

Goldreich & Lynden-Bell (1965) (henceforth GLB) adopted the first (Cauchy) approach to the problem of local stability of a gravitating disc, within the framework of the hydrodynamical model with plane pressure; the second approach was followed by Lin & Lau (1979) and by Morozov (1985). In both cases, they sought solutions localized near a radius r_0 and, moreover, represented the angular velocity as $\Omega(r) = \Omega(r_0) + \Omega'(r_0)(r - r_0)$, the remaining terms of the Taylor series being omitted.

In the case of a plane-parallel flow, $v = v_x = v_x(y)$, such an approximation is analogous to a Couette approximation when the linear law $v_x(y) = v_x(y_0) + v_x'(y_0)(y - y_0)$ is adopted. As is well-known for the incompressible Couette flow, the linearized problem reduces to the Rayleigh equation. In the case of perturbations that conserve the unperturbed vortex $((\mathrm{rot}_z \vec{v}_1) = 0)$, it turns out to be impossible to satisfy the necessary conditions at the flow boundaries (or at $y \to \pm\infty$) (*e.g.* Case 1960). Localized perturbations in incompressible rotating flow were considered (using the GLB method) in the paper by Lominadze *et al.* (1988). It can be easily seen that this problem also reduces to the Rayleigh equation in the x-representation. Hence it follows at once that there are no vortex free eigensolutions that vanish far from the localised perturbation, because the boundary conditions cannot be satisfied.

The dispersion equation of the usual WKB type (used by Lin, Lau, Bertin and others) is manifestly insufficient because it indicates that ever larger velocity dispersions are required for stability as the azimuthal wavenumber of the perturbation is increased. (Within the approximation of WKB theory, it is assumed to be small.) Thus, if we want to find sufficient stability criteria for arbitrary localized perturbations, we must go beyond the limitations of WKB theory.

We have tried to generalize the studies by Lin & Lau, and others to the case when perturbations had arbitrary degree of axial asymmetry. It turns out that vortex free perturbations (or to be more exact, the perturbations conserving generalized vortex $(\mathrm{rot}_z \vec{v}/\sigma)_1 = 0$) also cannot be proper eigensolutions (satisfying the necessary boundary conditions) in the approximation analogous to the Couette one, *i.e.* for the linear shear of the angular velocity. Thus, in such an approximation (and for perturbations of the type under consideration) the evolutionary solutions of GLB-type are the only possible ones. It turned out that the GLB-type[1] equation was automatically contained (though it was not evident beforehand) in the much more complicated equation which initially appeared when studying the stability problem by the eigenfunction method ($\sim e^{-i\omega t}$): this simpler equation can be singled out because the more general equation factorizes.

As is well known, eigensolutions of the Rayleigh equation can exist provided that $v_0'' = 0$ somewhere within the flow region. In the rotating flow of gravitating compressible fluid under consideration, it is possible that the existence of global modes (*i.e.* satisfying the boundary conditions) requires some analogous condition.

---

[1] The difference is connected with the infinitesimally thin disc approximation we have used directly; of course, our equation can be also derived from the GLB-approach for this case.

We have formulated the equation derived in terms of the local dispersion relation $\omega_*^2 = \omega_*^2(k_r, k_\varphi)$ and, in particular, obtained the marginal stability curve. It is evident that Lin, Lau, Bertin, Morozov, and others aspired to obtain just the latter relation – but without success because they restricted themselves by only the WKB approximation with $k_r \gg k_\varphi$. Attempts to extend the results of WKB approximation to the systems and perturbations with $k_r \sim k_\varphi$, $g^2 \sim 1$, $\epsilon \sim 1$ do not stand up to criticism. Moreover, they lead to results which are qualitatively incorrect (*cf.* Figures 2a and 2b). The interpretation of the local dispersion relation as well as the marginal stability condition is given at the end of §2. Of course, it differs from the usual interpretation, for the homogeneous or quasi-homogeneous systems. (The case under consideration is substantially inhomogeneous.)

We have succeeded in obtaining a sufficient stability criterion (see Figure 2c) which guarantees lack of any (even small) time-periods of quasi-exponentially increasing perturbations. It is clear that this stability criterion is sufficient with some margin since, near $Q = Q_{\text{max}}$, the perturbations do not practically increase. At the same time, under the usual interpretation of the marginal curve as the boundary of exponential instability, it is supposed that even unstable perturbations near the stability boundary should grow arbitrarily strongly for a sufficiently long time. In reality, it is not the case due to the drift of the radial wavenumber, which takes the system out of the instability region. So it would be more relevant to determine the conditions for perturbations to amplify by a certain factor (Polyachenko & Strel'nikov 1989).

In principle, the local stability of collisionless (stellar) systems can be studied in a similar way (Polyachenko & Strel'nikov 1989). In this case, we have the integral equation by Julian & Toomre (1966), instead of the differential equation by GLB. Using this integral equation like the GLB equation above, we have derived the local dispersion relation and corresponding marginal stability curves $Q^2 = Q_{\alpha^2}^2(\bar{\lambda}, g^2)$. By studying the marginal curves, we could show that the function $Q_{\alpha^2}^2(\bar{\lambda}, g^2)$ has a maximum $Q_m^2(\alpha^2)$ as a function of two variables. This is quite similar to the hydrodynamical case. Moreover, when $\alpha^2 = 2$ (*i.e.* for the flat rotation curve) $Q_m^2 \simeq 3$ in the case under consideration. (Recall that $Q_m^2 = 3$, for the same rotation law, in the hydrodynamics).

References

Antonov, V. A., 1976. *Uchen. zap. Leningrad Univ. (Sov.)*, **32**, 79.

Bertin, G. & Lin, C.C., 1988. In *Evolution of Galaxies*, p. 255, (Proceedings of the 10th IAU Regional Astronomy Meeting). Ed. J. Palouš, Publ. Astron. Inst. Czech. Acad. Sci. **69**.

Bisnovaty-Kogan, G. S. & Zel'dovich, Ya. B., 1970. *Astrophysica (Sov.)*, **6**, 149.

Case, K. M., 1960. *Phys. Fluids*, **3**, 149.

Fridman, A. M. & Polyachenko, V. L., 1984. *Physics of Gravitating Systems*, Springer-Verlag, New York.

Goldreich, P. & Lynden-Bell, D., 1965. *Mon. Not. R. Astron. Soc.*, **130**, 125.

Julian, W. H. & Toomre, A., 1966. *Astrophys. J.*, **146**, 810.

Kalnajs, A. J., 1972. *Astrophys. J.*, **175**, 63.

Kalnajs, A. J., 1976. *Astrophys. J.*, **205**, 745, 751.

Lin, C. C. & Lau, 1979. *Stud. Appl. Math.*, **60**, 97.

Lin, C. C. & Shu, F. H., 1964. *Astrophys. J.*, **140**, 646.

Lominadze, J. G., Chagelashvily, G. D. & Chanashvili, R. G., 1988. *Letters to Astron. J. (Sov.)*, **14**, 856.

Morozov, A. G., Polyachenko, V. L. & Shukhman, I. G., 1974. Preprint of SibIZMIRAN, 5-74, Irkutsk.

Morozov, A. G., 1985. *Astron. J. (Sov.)*, **62**, 805.

Polyachenko, V. L., 1972. Dissertation, Leningrad Univ.

Polyachenko, V. L., 1987. *Astron. Circ. (Sov.)*, No.1490.

Polyachenko, V. L. & Fridman, A. M., 1976. *Equilibrium and Stability of Gravitating Systems*, (in Russian), Nauka, Moscow.

Polyachenko, V. L. & Shukhman, I. G., 1972. Preprint of SibIZMIRAN, 1-2, Irkutsk.

Polyachenko, V. L. & Shukhman, I. G., 1981. *Astron. J. (Sov.)*, **58**, 933.

Polyachenko, V. L. & Strel'nikov, A. V., 1988. *Astron. Circ. (Sov.)*, No.1529.

Polyachenko, V. L. & Strel'nikov, A. V., 1989. To be published.

Toomre, A., 1964. *Astrophys. J.*, **139**, 1217.

Stability of two-component galactic discs

Alessandro B. Romeo

Scuola Internazionale Superiore di Studi Avanzati, Trieste, Italy

1 Local stability

The properties of the marginal stability curve are summarized in Figure 1, where we have adopted the following scaling and parametrisation:

$$\bar{\lambda} \equiv \frac{k_H}{|k|}, \quad k_H \equiv \frac{\kappa^2}{2\pi G \sigma_H}; \quad \alpha \equiv \frac{\sigma_C}{\sigma_H}, \quad \beta \equiv \frac{c_C^2}{c_H^2}; \quad Q_H \equiv \frac{c_H \kappa}{\pi G \sigma_H}.$$

In these formulae the subscripts "H" and "C" denote the stars of the active disc and the cold interstellar gas, respectively. For more information and definition of the notation used see Romeo (1985, 1987, 1988), Bertin and Romeo (1988).

2 Discrete global spiral modes

2.1 STAR-DOMINATED EQUILIBRIUM MODELS

We use two-component equilibrium models which incorporate the essential features of the cold interstellar gas, as suggested by some recent observational surveys. Appreciable modifications to the structure of the modes, with respect to the corresponding one-component cases, are present only when a peaked distribution of molecular hydrogen is simulated (see Figure 3b).

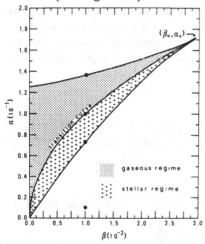

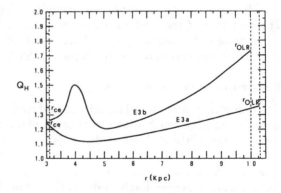

Figure 1. Inside the *two-phase region* of the (β, α) plane the marginal stability curve in the $(\bar{\lambda}, Q_H^2)$ plane exhibits two maxima. In the *stellar (gaseous) regime* the stellar (gaseous) peak occurring at intermediate (short) wavelengths is dominant; the curve $\alpha = \sqrt{\beta}$ corresponds to the *transition* between these two regimes.

Figure 2. The *local stability parameter* Q_H for the two-component equilibrium models E3a, simulating the neutral atomic hydrogen distribution, and E3b, simulating the presence of a ring of molecular hydrogen. These profiles have been derived consistently with a mechanism of *self-regulation*.

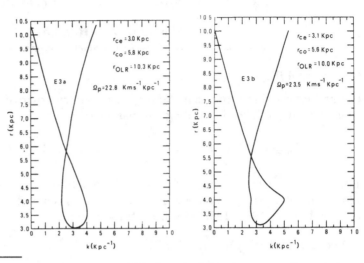

Figure 3. Propagation diagrams relative to the $(m = 2, n = 0)$ mode for the two-component equilibrium models (a) E3a and (b) E3b. We observe a shift of the mode towards a *more tightly wound spiral structure* with a smaller co-rotation radius than in the corresponding one-component case.

However, even in the cases where no qualitative modifications are present (see Figure 3a), the basic states which support these modes are characterized by relatively high stellar velocity dispersions, *i.e.* by values of the stability parameter $Q_H > 1$ (see Figure 2). For more information see Romeo (1985, 1988), Bertin and Romeo (1988).

2.2 GAS-DOMINATED EQUILIBRIUM MODELS: QUALITATIVE PREDICTIONS

We expect to have "double propagation diagrams" characterized by other turning points in addition to r_{ce} and r_{co}; thus, more complicated wave channels and cycles are available. The structure of the modes is expected to be considerably modified; in particular, the presence of the gaseous peak at short wavelengths suggests a *high degree of winding*. More comments are reported by Romeo (1985, 1988).

3 Astrophysical implications

Gas-rich early normal spiral galaxies should exhibit relatively high planar stellar velocity dispersions, consistently with a mechanism of self-regulation; non-monotonic profiles are expected for relatively high H_2 densities.

A *new regime of extremely tightly wound spirals* is expected when the stellar component is much hotter than the gaseous component in gas-rich early-type galaxies (*e.g.* NGC 488?); correspondingly, a discontinuity in the behaviour of the planar stellar velocity dispersion profile may be observed.

For a more complete discussion see Romeo (1985, 1987).

References

Bertin, G. & Romeo, A. B., 1988. *Astron. Astrophys.*, **195**, 105.
Romeo, A. B., 1985. *Tesi di Laurea*, Università di Pisa, Italy.
Romeo, A. B., 1987. *M. Phil. Thesis*, SISSA, Trieste, Italy.
Romeo, A. B., 1988. *Preprint, SISSA Astro. 167*, Trieste, Italy.

The smoothed particle hydrodynamics of galactic discs

Pawel Artymowicz[1,2] and **Stephen H. Lubow**[1,3]

[1] *Space Telescope Science Institute, Baltimore, USA*

[3] *The Johns Hopkins University, Baltimore, USA*

1 Introduction

We apply the technique of smoothed particle hydrodynamics (SPH), a gas-dynamical Lagrangian numerical scheme, to analyse the non-linear response of a gaseous disc to an imposed potential. For the first time, we compare SPH and semi-analytical results for a galaxy model with small spiral pitch angles. Density amplitudes, phases and general density profile shapes, including the subtle effects of the $n = 2$ ultra-harmonic resonance, are in good agreement throughout the disc. We therefore establish the applicability of SPH to wide range of problems involving density waves in galactic discs.

The SPH scheme, which is based on kernel estimates of the physical parameters of the gas, has been described elsewhere (*e.g.* Gingold & Monaghan 1982, Monaghan & Lattanzio 1985). In SPH, gradients of physical quantities, such as pressure, are expressed as gradients of the kernel alone through an integration by parts. The resulting SPH fluid equations contain bulk flow parameters, such as the gas sound speed and viscosity, which are explicitly specified. This helps constrain the problem and permits a stringent test of the technique. SPH has some major advantages over the familiar grid-based schemes: first, there is no need to solve the continuity equation separately; second, the convective terms in the momentum equations are represented exactly; third, the mesh is adaptive – it becomes dense wherever the gas density becomes large – which is important in the highly non-linear flows often found in galaxies. SPH pays for these advantages with its procedure for evaluation of Eulerian derivatives (used to calculate pressure and, especially, viscosity) which cannot easily be justified *a priori*.

2 Choice of problem

Neither SPH, nor other particle simulation methods, have previously been tested quantitatively against semi-analytical results for realistic galactic flow problems. In order to assess the applicability of SPH to spiral arm problems, we compute the gas disc response to external driving by the potential of a tightly wrapped spiral. Modelling such a system presents a demanding test for any numerical code.

The galaxy potential we adopt is that used by Contopoulos & Grosbøl (1986) to model the Sc(s) galaxy NGC 5247. We investigated several choices for the value of their parameter A, which determines the amplitude of the perturbation, and values of the pitch angle i_0 between 5° and 15°. We choose the viscosity to correspond to a mean free path ~ 100 pc and a sound speed of 8 km sec^{-1}.

Lubow, Balbus & Cowie (1986) have solved the steady state, non-linear, viscous fluid equations directly to find the gas response to small pitch angle spiral forcing. We have computed the density profiles in 6 radial bins extending from the centre to

---

[2] On leave from N. Copernicus Astronomical Centre, Warsaw, Poland

the co-rotation radius in the model galaxy with the same parameters as in the SPH calculations.

3 Results

We used about 7 500 particles in the 2-D simulation and allowed the smoothing length, h, to vary spatially in order to give better spatial resolution in the high-density regions; we choose h such that each particle interacted with a given number (usually ~ 15) of its nearest neighbours. We used linked lists of particles belonging to large square cells, each containing many particles, in order to speed up the neighbour finding process. The simulation reached a steady state in 3 to 6 orbital periods of the pattern, ($\sim 10^4$ time-steps).

We find very good agreement between the density maps of the models computed by the two different methods. Both predict the existence of rather narrow shocks in the inner part of the disc and a more linear response towards co-rotation; the density contrast at each radius closely coincided throughout the disc. The shapes of the induced shocks agree as well. For instance, in the intermediate radius range in a model with a 7° pitch angle, there is a sharply rising density on the inner side of the arm where the gas is shocked, and a gradual density decrease on the other side of the arm as the gas re-expands.

The comparison furnishes an impressive confirmation of the ability of SPH to reproduce fairly subtle effects, such as the ultra-harmonic resonance located between the inner Lindblad resonance and co-rotation. Both methods produce a secondary bump in the wave profile near the ultra-harmonic $n = 2$ resonance, or equivalently two fragmentary arms between the main arms. It is important to note that any sizeable difference in the parameter values between the two methods (*e.g.* if the sound speed differs by 20%) considerably degrades the agreement of the shock profiles.

We conclude that SPH is a reliable scheme for steady-flow calculations, as our test problem. We expect it perform equally well in explicitly time-dependent studies.

References

Contopoulos, G. & Grosbøl, P., 1986. *Astron. Astrophys.*, **155**, 11.
Gingold, R. A. & Monaghan, J. J., 1982. *J. Comput. Phys.*, **46**, 429.
Lubow, S. H., Balbus, S. A. & Cowie, L. L., 1986. *Astrophys. J.*, **309**, 494.
Monaghan, J. J. & Lattanzio, J. C., 1985. *Astron. Astrophys.*, **149**, 135.

Tidal generation of active disc galaxies by rich clusters

Gene G. Byrd[1] and Mauri Valtonen[2]
[1] *University of Alabama, USA*
[2] *University of Turku, Finland*

1 Seyferts with companions

Galaxy activity is correlated with companions (Keel *et al.* 1985, van der Hulst *et al.* 1986). Using matched samples of Seyferts and controls, Dahari (1984) searched for companions, measuring the galaxy-companion separations and their sizes. He measure the tidal perturbation strength by a parameter $P = $ (companion mass)/(separation)^3 in units of the galaxy mass and radius. Dahari found that more Seyferts (37%) have companions than do normal spirals (21%), and that Seyferts with companions are perturbed more strongly. Selection effects cause companions of higher redshift Seyferts to be missed and Byrd *et al.* (1987) estimate that 75% to 90% of Dahari's Seyferts have companions.

Byrd *et al.* (1986) tested the correlation using computer models of tidally perturbed spiral galaxies. Observations require a gas mass inflow rate of > 0.5 M$_\odot$ yr^{-1} for Seyfert activity. We used a self-gravitating 60 000 particle disc and inert "halo" perturbed by a companion on a parabolic orbit. Tidal perturbation of the disc throws gas clouds into nucleus-crossing orbits to fuel activity. The experiments demonstrated that the inflow rate exceeded the required value at perturbation levels matching those where Dahari finds many more Seyferts than normals. We therefore conclude that observed companions of Seyferts do have tidal fields sufficient to trigger activity.

2 Seyferts in rich clusters

If individual gravitational encounters are responsible for activity, the incidence of activity should correlate with the enounter rate. Gavazzi & Jaffe (1987) argue that individual encounters should be less important in rich clusters than in groups. Merritt (1984) discusses how tidal stripping by the cluster gravitational field quickly reduces the importance of galaxy-galaxy tidal encounters after the cluster has formed. However, Gavazzi & Jaffe (1987) find that late type spiral galaxies in rich clusters are radio sources ten times more powerful than spirals outside clusters. They propose that ram pressure of the cluster medium on disc clouds induces massive star formation, supernovae, *etc.* Ram pressure is thought to strip cluster spirals of gas making SO's. But, Gisler (1980) and Dressler (1980) observe that in rich clusters, late type Sc's (least susceptible to ram pressure stripping of gas) are reduced in relative abundance while the more susceptible Sa's have the same relative abundance. Cluster medium ram pressure stripping has problems.

We propose and test an alternative, tidal triggering by the cluster gravitational field. Our computer model is the same except the tidal field is symmetric in a fixed direction. This matches a spinning disc galaxy falling into a rich cluster. We have run 25 simulations to explore various levels of P and halo/disc ratios. For halo/disc of 0 to 1, the required level is $\log P > -2.2$; more massive halos raise this to $\log P > -1$. Conservatively, we require $P > 0.1$ for activity and calculate the radius at which the cluster tidal field is > 0.1. We use a King model of Coma (Merritt 1984) with a core radius 250 kpc and velocity dispersion 100 km/s along with a 225 km/s rotation and 25 kpc radius for the disc galaxy. Our results indicate tidal triggering occurs at < 3 radii core or 750 kpc. A simple 6×10^{14} solar mass point mass cluster model gives about the same result. Most of cluster galaxies are inside this radius.

3 Galaxy evolution in clusters

If frequent enough, cloud-cloud collision pressures are much greater than cluster medium ram pressure in tidally affected galaxies because the cluster medium is so thin. Here pressure is proportional to number density $\times$ velocity^2. For the cluster medium, 10^{-3} atom/cm$^3 \times (10^3$ km/s$)^2 = 10^3$. Cloud-cloud collisions even in our galaxy exert pressures of 10^3 atom/cm$^3 \times (7$ km/s$)^2 = 10^4$. We find typical collision speeds of 100 km/s between disc gas clouds in our tidally perturbed models, which should generate bursts of star formation. Collisions in a tidally disturbed galaxy exert $10^3 \times (100$ km/s$)^2 = 10^7$.

The reduced abundance of Sc's in rich clusters can be explained by the following "tidal action sequence". Tidally initiated disc collisions of molecular clouds in spiral arms and nuclear inflow result in bursts of star formation. Supernovae, *etc.* eject gas from the disc and nucleus (with some subsequent help from ram pressure). Late type spirals are thus preferentially transformed to anemic spirals and SO's. Metal enrichment of the cluster medium is explained also. Our simulations and those of Noguchi (1987) indicate bars (or bar-like structure) result from very strong tidal perturbations, explaining the excess of barred spirals in the central regions of the Coma Cluster (Thompson 1981). Bothun & Dressler (1986) found HI-poor, blue disc galaxies in the central portions of the Coma Cluster which show star formation. We interpret these galaxies as part way down the tidal action sequence.

References

Bothun, G. D. & Dressler, A., 1986. *Astrophys. J.*, **301**, 57.

Byrd, G. G., Valtonen, M., Sundelius, B. & Valtaoja, L., 1986. *Astron. Astrophys.*, **166**, 75.

Byrd, G. G., Sundelius, B. & Valtonen, M., 1987. *Astron. Astrophys.*, **171**, 16.

Dahari, O., 1984. *Astron. J.*, **89**, 966.

Dressler, A., 1980. *Astrophys. J.*, **236**, 351.

Gavazzi, G. & Jaffe, W., 1987. *Astrophys. J.*, **310**, 53.

Gisler, G. R., 1980. *Astron. J.*, **85**, 623.

Keel, W. C., Kennicutt, R. C., Hummel, E. & van der Hulst, J. M., 1985. *Astron. J.*, **90**, 708.

Merritt, D., 1984. *Astrophys. J.*, **276**, 26.

Noguchi, M., 1987. *Astron. Astrophys.*, **203**, 259.

Thompson, L. A., 1981. *Astrophys. J. Lett.*, **244**, L43.

van der Hulst, J. M., Hummel, E., Keel, W. C. & Kennicutt, R. C., 1986. In *Spectral Evolution of Galaxies*, p. 103, eds. Chiosi, C. & Renzini, A., Reidel, Dordrecht.

Formation of spiral arms in the early stages of galaxy interaction

Maria Sundin

Chalmers University of Technology, Sweden

1 Simulation

A videomovie, showing a simulation of a disc galaxy perturbed by a companion, has been made. It is easy to see some time-dependent phenomena in the movie that would have been hard to discover on paper plots.

The videomovie shows results from a 2-dimensional N-body simulation with 60 000 particles using a polar coordinate code. The disc is self-gravitating and surrounded by an inert spherical halo. It has Mestel's density distribution, which gives a flat rotation curve, and the companion is represented by a point mass which passes on an initially parabolic orbit.

The initially stable disc evolves as follows: A very sharply defined material arm quickly appears. After some time a counter arm is formed due to self-gravitation. Soon after the appearance of the counter arm, a third arm forms from the remains of the first material arm. This third arm has a higher radial velocity than the counter arm which makes it pass, or expand, through the latter. Some time later, after the passage of the companion, a density wave pattern, much more long lived than the material arms, is formed. The density waves first appear between the material arms giving the impression of a fork in the arm, such as has been observed in several galaxies. As the material arms are decaying when this happens, a fork would be a short lived phenomenon.

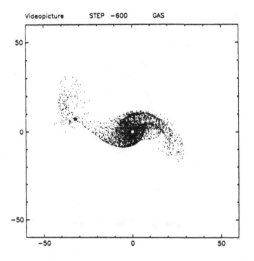

Figure 1. Material spiral arms form when the companion is close to the disc.

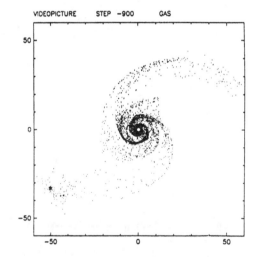

Figure 2. Density waves appear some time after the passage of the companion.

Formation of leading spiral arms
in retrograde galaxy encounters

M. Thomasson[1], K. J. Donner[2,3], B. Sundelius[1,4]
G. G. Byrd[5], T.-Y. Huang[6,7] and M. J. Valtonen[6]
[1] *Onsala Rymdobservatorium, Sweden*
[2] *Nordita, Denmark*
[3] *University of Helsinki, Finland*
[4] *Chalmers University of Technology, Sweden*
[5] *University of Alabama, USA*
[6] *University of Turku, Finland*
[7] *University of Nanjing, China*

1 Introduction

Observations indicate that tidal interactions are important for the formation of large scale spiral structures (Elmegreen & Elmegreen 1982, 1983, 1987; Kormendy & Norman 1979). In our earlier numerical experiments we have found that a two-armed trailing spiral pattern is generated in a direct (prograde) encounter (Sundelius *et al.* 1987). In this paper we investigate the effect of a retrograde transient encounter on a disc galaxy (see also Thomasson *et al.* 1989).

2 Theory and numerical experiments

Using a resonant perturbation theory we have studied stellar orbits in a disc perturbed by a point mass in a retrograde orbit acting for a finite time. The perturbation produces slowly precessing stellar orbits that align to form a single, leading arm. (A leading arm has its tip pointing in the direction of rotation.)

We tested the theory with a two-dimensional N-body code (see *e.g.* Figure 1), and found two necessary conditions for a leading arm to dominate the pattern in a disc:

(i) The perturbation from the retrograde companion must be large enough.
(ii) The disc must be stabilized by a halo at least as massive as the disc.

3 Comparison with observations

Two factors determine the fraction of leading spirals in a sample of interacting galaxies: the relative lifetimes of leading and trailing arms, and the range of orbital inclinations of the companion that produce leading arms.

We have not found any difference in lifetime between the two kinds of patterns in our N-body experiments. The fraction of leading arms formed during encounters should be less than 50%, since trailing patterns are more easily formed. Based on three-body experiments by Saslaw *et al.* (1974), we believe that something like 25% of all spirals with a large companion should have a leading spiral pattern.

Pasha (1985) examined 189 galaxies and found four leading arm candidates. Having examined these four carefully, we think that not more than two of them really are leading spirals. This means that of the tidally perturbed galaxies in the sample, only approximately 5% have leading arms.

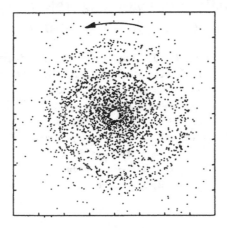

Figure 1. An example of a leading spiral in one of our experiments

4 Discussion

Why are there so few leading spirals? Trailing arms are created by the powerful swing amplifier (Toomre 1981), a mechanism that, however, is killed by a large halo. The theory for leading arm formation does not include self-gravity, *i.e.* a leading pattern can exist in a galaxy with a large halo. We have also seen in the *N*-body experiments that a halo is necessary for a leading pattern to dominate in a disc. Since so few leading spirals are observed, we conclude that very few spiral galaxies have halos with masses larger than the disc mass.

References

Elmegreen, B. G. & Elmegreen, D. M., 1983. *Astrophys. J.,* **267**, 31.

Elmegreen, D. M. & Elmegreen, B. G., 1982. *Mon. Not. R. Astron. Soc.,* **201**, 1021.

Elmegreen, D. M. & Elmegreen, B. G., 1987. *Astrophys. J.,* **314**, 3.

Kormendy, J. & Norman, C. A., 1979. *Astrophys. J.,* **233**, 539.

Pasha, I. I., 1985. *Sov. Astron. Lett,* **11**, 1.

Saslaw, W. C., Valtonen, M. J. & Aarseth, S.J., 1974. *Astrophys. J.,* **190**, 253.

Sundelius, B., Thomasson, M., Valtonen, M. J. & Byrd, G. G., 1987. *Astron. Astrophys.,* **174**, 67.

Thomasson, M., Donner, K. J., Sundelius, B., Byrd, G. G., Huang, T.-Y. & Valtonen, M. J., 1989. *Astron. Astrophys.,* **211**, 25.

Toomre, A., 1981. In *Structure and Evolution of Normal Galaxies,* p. 111, eds Fall, S. M. & Lynden-Bell, D., Cambridge University Press, Cambridge.

The influence of galaxy interactions on stellar bars

M. Gerin[1,2], F. Combes[1,2] and E. Athanassoula[3]
[1] *École Normale Supèrieur, France*
[2] *Observatoire de Meudon, France*
[3] *Observatoire de Marseille, France*

1 Introduction

Several statistical studies have emphasized that both bars and companions can generate prominent structures in spiral galaxies and that bars are more commonly found in non-isolated galaxies (*e.g.* Elmegreen & Elmegreen 1982). Numerical simulations made by Noguchi (1987) show that a galaxy perturbed by a massive companion develops a bar rapidly after the passage at pericentre. But little is known so far of the interplay between the effect of a companion and the evolution of a bar. We therefore ran a series of encounters between a barred spiral galaxy and a less massive companion ($M_C/M_G = 0.5$). The methods and more results are given in Gerin *et al.* (1989).

2 Variation of the bar strength and pattern speed

For strong interactions there is a short lived increase of the bar strength during the passage, followed by a decrease to the reference level (Figure 1). The relative position of the bar and companion at pericentre determines the amplitude of this transient effect. Figure 2 presents the relative variation in the maximum amplitude of the bar, $\Delta P_2/P_2$, and in its angular velocity $\Delta\Omega_b/\Omega_b$, as function of the angle α between the bar and companion at pericentre ($\alpha = \phi_{\text{bar}} - \phi_{\text{com}}$). An *increase* of the bar strength and a *decrease* of its pattern speed occur when α is *positive*, *i.e.* when the companion lags behind the bar. The opposite is true for negative α.

We give more details from two runs which maximize these effects. We selected particles in a ring around co-rotation for the bar to study variations of energy and angular momentum during the run. These variations are displayed in Figure 3. When

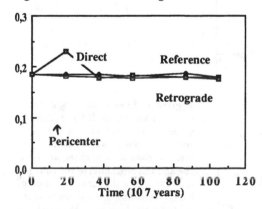

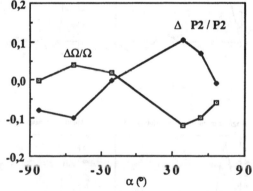

Figure 1. Maximal bar strength $P_{2,\text{max}}$ for a direct, retrograde and reference model with a bar.

Figure 2. Variation of the maximum bar strength, and bar pattern speed as a function of the phase angle α at pericentre.

$\alpha = 39°$, particles at the end of the bar *lose energy* while those at right angles gain energy. Particles just in phase with the companion (and some at 180°) escape. When $\alpha = -53°$, particles at the end of the bar *gain energy* due to the interaction with the companion, while those at right angles lose energy and angular momentum.

The effect of the companion is maximum when the phase lag at pericentre is $\pm45°$ since, when the bar symmetry axis is at 0° or 90° to the companion axis, the torque exerted by the companion on the bar (considered as a rigid body) is zero. This torque is negative when the companion lags the bar (at pericentre), and positive when it leads the bar. Thus, when $\alpha = 39°$, particles at the end of the bar may settle on more elongated orbits and contribute more strongly to the bar. The bar length may increase also, and its pattern speed decrease due to the displacement of the end of the bar towards the exterior of the galaxy, where the angular velocity is lower. On the contrary, when $\alpha = -53°$, the companion will extract some particles from the bar. Since the resulting bar rotates more rapidly, it will have to be shorter in order to continue to be accommodated within co-rotation.

These modifications of the bar parameters occur without major changes in the axisymmetric galaxy potential: the angular velocity and epicyclic frequency the particle motions do not change much in all runs. The location of the principal resonances will therefore change due to the interaction. In SB galaxies undergoing an interaction, we may observe a short-lived starburst phenomenon if the orbits are favourable (*e.g.* the case of NGC 1097; Gerin *et al.* 1989).

Retrograde passages do not perturb the bar.

References

Elmegreen, D. M. & Elmegreen, B. G., 1982. *Mon. Not. R. Astron. Soc.*, **201**, 1021.
Gerin, M., Combes, F. & Athanassoula E., 1989. *Astron. Astrophys.*, Submitted.
Noguchi, M., 1987. *Mon. Not. R. Astron. Soc.*, **228**, 635.

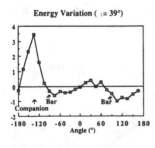
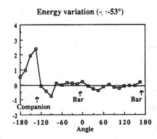

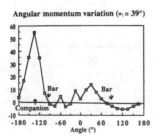
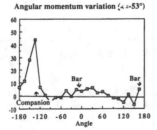

Figure 3. Energy and angular momentum variations for particles located at the end of the bar, versus their phase angle at pericentre. Variations are taken between the and the beginning of the run. (a) $\alpha = 39°$. Note the large variation of energy and angular momentum for particles in phase with the companion (formation of a tidal bridge), and for particles at 180° (formation of a tidal tail). (b) $\alpha = -53°$.

Disc galaxies - work in progress in Gothenburg

Stefan Engström and **Björn Sundelius**
Chalmers University, Sweden
and
Magnus Thomasson
Onsala Space Observatory, Sweden

1 Problem under study

We study interactions between disc galaxies, in particular the case of a main system and a small perturber. Here we look at the behaviour using a responsive disturber in contrast to the more common approximation of a rigid mass distribution. It is possible to isolate the effect of the perturber's dynamics by studying the difference between simulations with and without internal motion in the disturber. One facet of extended systems is the possible stripping of cool gas from the smaller galaxy in the case when the systems do not merge. Another line of study is cloud-cloud collisions, their impact on cooling the gas and keeping the velocity dispersion down.

2 Numerical model

A particle-mesh code is used to simulate these systems. It is possible to evolve a 2-dimensional Cartesian grid (512 × 512 maximum) with 200K particles in a reasonable time (approximately 15s/time-step). The code is quite flexible since it does not need any description of the free populations other than their mass density and desired velocity dispersion. An option is to use a rigid potential in order to mimic a hot (spherical) component. The rigid components are moved along with the centre of mass for the respective system.

3 Experiments

The experiment displayed in the poster session had two galaxies with a mass ratio of 5:1. The orbit was initially parabolic and direct, as seen from the larger system's point of view. The larger galaxy was the same in all experiments: 50% (by mass) rigid halo, 40% warm (Q=1) exponential disc, 10% cold exponential disc. The Plummer halo was concentrated towards the centre with the ratio of length scales Plummer:Exponential disc = 1:2. Simulations were carried out with three different perturbers:

1 Main system perturbed by a point mass.
2 Disturber identical to main system except for scaling and mass, direct encounter.
3 Same as 2, except for a retrograde encounter for the smaller disc.

4 Results

The main system showed a strong reaction, mostly due to an instability because of insufficient softening for this system. The smaller disc did not suffer from this effect and our softening was small enough (1/200 of the computational box) to achieve a reasonable resolution.

One can see how dynamical friction is enhanced when going from a point mass to retrograde to direct encounter (experiments 1 to 3 to 2).

The direct encounter perturber was totally disrupted while the retrograde variant was almost unaffected by the collision.

We observed a streak of initially warm particles scattered over the lower half of the larger system in simulation 2, though this was not seen in the retrograde run.

5 Cloud-cloud collisions in disc galaxies

In a real galaxy, gas clouds undergo dissipative collisions with other gas clouds. This lowers the velocity dispersion of the cloud population and affects the general appearance of the spiral pattern. We study this in a new polar grid program in which clouds are allowed to collide inelastically. This is done by introducing a second (Cartesian) grid. Two particles in a box in this grid that approach each other collide and their relative velocity is reduced. By varying the box size, one can change the cloud cross section and thus the mean free path. Cooling particles this way reduces the random velocities and assists the formation of spiral structure.

Motion of a satellite in a disc potential

B. Zafiropoulos

University of Manchester

1 Equations of the problem

The aim of the present investigation is to obtain analytic expressions for the orbital elements of a satellite moving in a disc potential. The variational equations of the problem of the plane motion will be set up in terms of the rectangular components R, S and W of the disturbing accelerations.

The six elements which specify the orbit of the satellite in space are taken to be the longitude of the ascending node Ω, the inclination ι to the disc, the semi-major axis α, the eccentricity e, the argument of periapse ω and χ the mean anomaly at time $t = 0$. The Gauss-Lagrange planetary equations are used in the following form (*e.g.* Taff 1985, p314)

$$\frac{d\Omega}{dv} = \frac{r^3 \sin u}{\mu p \sin \iota} W, \tag{1}$$

$$\frac{d\iota}{dv} = \frac{r^3 \cos u}{\mu p} W, \tag{2}$$

$$\frac{d\alpha}{dv} = \frac{2\alpha^2}{\mu p} \{ r^2 e \sin v R + r p S \}, \tag{3}$$

$$\frac{de}{dv} = \frac{r^2}{\mu p} \{ p \sin v R + 2r \cos v S + \frac{3er}{2} S + \frac{er}{2} \cos 2v S \}, \tag{4}$$

$$\frac{d\omega}{dv} = \frac{r^2}{\mu e p} \{ -p \cos v R + (p + r) \sin v S \} - \cos \iota \frac{d\Omega}{dv}, \tag{5}$$

$$\frac{d\chi}{dv} = \frac{r^2 \sqrt{1 - e^2}}{\mu e p} \{ (p \cos v - 2er) R - (p + r) \sin v S \}, \tag{6}$$

where u, v are the true anomalies measured from the ascending node and the periapse respectively, p is the semi-latus rectum ($p = r(1 + e \cos v)$), r is the distance from the centre of the disc and μ is a constant.

The gravitational potential of an infinitesimally thin circular disc of radius A, at a distance r ($> A$) from the centre of the disc is (Ramsey 1981, p133)

$$U = -\frac{2\mu}{r} \sum_{\nu=0}^{\infty} \frac{(-1)^{\nu}(2\nu - 1)!!}{(2\nu - 2)!!} \left(\frac{A}{r} \right)^{2\nu} P_{2\nu}(\sin \iota \sin u), \tag{7}$$

where $P_{2\nu}$ denotes the Legendre polynomials and

$$(2\nu - 1)!! = (2\nu - 1)(2\nu - 3)(2\nu - 5) \ldots 2.$$

The odd-numbered Legendre polynomials are absent from (7) due to the symmetry about the plane of the disc.

The rectangular components R, S and W of the disturbing accelerations from the potential (7) have the form:

$$R = -\frac{2\mu}{r^2} \sum_{\nu=1}^{\infty} \left(\frac{A}{r}\right)^{2\nu} \sum_{j=0}^{\nu} \rho_{2\nu,j} \cos(2ju), \qquad (8)$$

$$S = \frac{2\mu f \cos u}{r^2} \sum_{\nu=1}^{\infty} \left(\frac{A}{r}\right)^{2\nu} \sum_{j=0}^{\nu-1} \sigma_{2\nu,j} \sin[(2j+1)u], \qquad (9)$$

$$W = \frac{\mu}{r^2} \sin 2\iota \sum_{\nu=1}^{\infty} \left(\frac{A}{r}\right)^{2\nu} \sum_{j=0}^{\nu-1} \sigma_{2\nu,j} \sin[(2j+1)u], \qquad (10)$$

where $\rho_{2\nu,j}$ and $\sigma_{2\nu,j}$ are functions of $\sin^2 \iota \equiv f$.

2 Solution of equations

The orbital elements of the satellite are presented in the form of summations by means of zero-order Hansen's coefficients. These coefficients are defined by the expression (Kopal 1978, p233)

$$X_0^{n,m}(e) = (1)^m \frac{(n+2)(n+3)\ldots(n+m+1)}{m!} \left(\frac{e}{2}\right)^m$$
$$\times {}_2F_1\left(\frac{m-n-1}{2}, \frac{m-n}{2}; m+1; e^2\right),$$

where ${}_2F_1$ denotes the ordinary hypergeometric series.

The relation which permits us to substitute for the various powers of $(\alpha/r)^l$ in Equations (1) to (6) by means of zero-order Hansen coefficients, has the form

$$\left(\frac{\alpha}{r}\right)^l = \sqrt{1-e^2} \sum_{k=0}^{l} (2-\delta_{0,k}) X_0^{-(l+2),k} \cos(kv),$$

where the symbol $\delta_{0,k}$ is the Kronecker delta.

After substitution for R, S and W from (8) to (10), equations (1) to (6) can be integrated to yield the perturbed elements of the orbit. The results include both secular and periodic terms and are obtained as functions of the true anomaly v, in the form of a terminating convergent series. All elements are subject to periodic perturbations. Secular perturbations do not appear in the semi-major axis, whereas secular terms due to the second spherical harmonic are absent from the elements ι, α and e. The procedure employed can be easily applied to estimate the perturbed elements of a satellite moving in various forms of axisymmetric potentials (ring potentials, spherical potentials *etc.*)

A detailed proof of all the formulae given above as well as the results obtained for the orbital elements will be presented in a fuller account of the present work, to be published shortly.

References

Kopal, Z., 1978. *Dynamics of Close Binary Systems*, Reidel, Dordrecht.
Ramsey, A. S., 1981. *Newtonian Attraction*, Cambridge University Press, Cambridge.
Taff, L. G., 1985. *Celestial Mechanics*, John Wiley and Sons, New York.

Observer's summary

R. J. Cohen

Nuffield Radio Astronomy Laboratories, Jodrell Bank

1 Introduction

The subject of discs is central to most of astrophysics, from the formation and dynamics of planetary systems to the formation of protogalaxies in the early universe. Our meeting this week has been very successful, I feel, in bringing out the connections between disc phenomena of very different types and scale sizes. The planning of sessions has played an important part in bringing this about. I would like to thank the organisers for their careful planning, and for all their efforts in making the meeting a success. It was good that all speakers were given sufficient time explain their ideas.

The Compact Oxford Dictionary defines a disc as a "round flattened part in body, plant *etc.*" In this spirit, and in my capacity as an observer, I will concentrate in my summary on those discs which are actually observed, and which can be assigned a shape. I will start with the smallest discs and work up in size.

2 Planetary rings

Smallest but by no means the least interesting are the planetary disc and ring systems which were reviewed by Jack Lissauer and Nicole Borderies. Because these systems are relatively simple they are ideal to test theories of density waves, bending waves, edge phenomena and gaps. The purely dynamical phenomena can be studied without having to worry about such messy problems as changes of state (star formation), interactions of the disc with central jets or stellar winds, and interactions with unseen halos. A close check with observations is also possible, as the solar system objects can be visited! I found it pleasantly surprising and reassuring that these nearby systems offer such textbook examples of density-waves on our own doorstep. As an outsider to the field I was impressed by the degree of agreement between theory and observation which has already been achieved. Of course our speakers rightly stressed the remaining discrepancies, pointing out that we could not confidently extend the theories to more complicated discs until these simpler structures could be fully explained. I wondered too whether this might not be the place to test numerical codes. After all planetary discs are probably the one type of disc in the universe which most resembles the theoreticians' disc of point masses moving in a known and well-defined potential.

3 High-energy discs

It is a sobering thought that the high-energy discs associated with X-ray binaries and active galactic nuclei are about the same size as planetary discs and the solar system. Andrew King brought us up to date on what X-ray data can tell us about accretion discs in cataclysmic variables and X-ray binaries. In eclipsing systems it is possible to obtain detailed mapping of the temperature and velocity structure of the disc. As a radio astronomer I warmed to this application of the occultation technique, and I recalled that early radio lunar occultation measurements usually turned out to be basically right when compared with modern aperture synthesis maps. As a radio astronomer I could also not help noticing that each X-ray photon collected actually costs some pounds sterling, so perhaps it is a good thing that so much information can now be extracted!

In the field of active galactic nuclei Matt Malkan summarized the mostly indirect evidence for accretion discs, such as the important evidence of ionization cones. He stressed the importance of disc orientation on the observed spectrum, and on the parameters one would derive for the (also unseen) central black hole. The "advantage of having less information" which he mentioned was for me a difficulty, but might well have appealed more to theoreticians than to an observer. Joss Bland introduced us to the HIFI system, a CCD-based spectrophotometer which possibly runs the risk of providing too much information and so eliminating all theories entirely! The new CCD systems are extremely powerful, as shown for example in the maps of $H\alpha$ and [NII] in the Seyfert galaxy NGC 1068, where line-ratio maps distinguish the narrow-line HII regions from the pervasive broad-line regions. The data on NGC 4258 were also very striking in showing a massive concentration of gas in the nucleus, a feature which appears to be crucial to certain types of activity.

4 Protostellar discs

One of the few speakers to actually define a disc was José Torrelles. His criteria were that a protostellar disc should possess elongation, a velocity gradient along the major axis (*i.e.* a rotation curve), and a centre of stellar activity at the kinematic centre. We were shown several fine examples of well-observed discs which meet these strict criteria. José emphasised the usefulness of the ammonia molecule as a probe of the temperature and density distributions in these discs, as well as the kinematics. The special physics of the ammonia molecule make it arguably the most useful molecule of all for probing the physical conditions in molecular clouds (Ho & Townes 1983). There is also the advantage that the VLA is well equipped at the ammonia band!

Molecular line data was only one topic among many covered by Ron Snell in his comprehensive review of the observational evidence for protostellar discs. In the infrared the evidence is still mainly indirect. Infrared excesses from young stars are widespread and provide important constraints on disc masses and accretion rates and pre-main-sequence evolution (Adams *et al.* 1987). Perhaps the most striking infrared evidence is the detection of strong polarization of the radiation from regions of bipolar outflow. The degree of polarization increases with the elongation of the outflow. The direction of the polarization is found to be perpendicular to the outflow axis and parallel to the general interstellar magnetic field (Sato *et al.* 1985, Hodapp 1984). Mike Scarrott reminded us of the difficulties of interpreting such data, and distinguishing the effects of scattering from intrinsic polarization of the central source.

Direct infrared imaging of star-forming discs is difficult for several reasons, including the high extinction to the central parts of the disc, the poor contrast between the disc and the background emission from the parent cloud, and the problem of achieving adequate angular resolution. Lunar occultations of the source M8E give a resolution which is unlikely to be bettered in the near future, and show a compact elongated component only ~ 11 AU in thickness (Simon *et al.* 1985).

Molecular line work offers the most direct evidence on disc structure, and some fine examples were shown by Ron Snell. Unfortunately most molecular "discs" do not have a clear rotational signature. Sometimes there is confusion between rotation and/or expansion, and sometimes there is no systematic velocity pattern discernible. Another problem is to separate the disc emission from that of the rest of the molecular cloud, and this may account for the confused velocity patterns in some cases. The density and temperature structure of the protostellar discs can be probed by observing several different molecular lines which have different excitation requirements.

A fundamental problem in molecular line work is that of surface brightness sensitivity. It is important for the theoreticians to realize that the molecular line maps of discs that we were treated to this week show about as much detail as we are ever likely to see. The angular resolution cannot be improved indefinitely because as your beam gets smaller so you receive less and less flux per beam (from an extended source). Most molecular line emission has a brightness temperature of 10–100 K, and together with the sky background and receiver noise this sets a limit of $\sim 1''$ on the angular resolution which can be employed in order to map a disc in a reasonable time. Interferometry alone cannot overcome this problem, and the limit I have mentioned will be with us for some time, until it becomes possible to build arrays of very large mm-wave telescopes. One way of dealing with the problem is to make better use of the resolution inherent in the measurements. The poster by John Richer and Rachael Padman highlighted one promising technique. They used the maximum entropy method to deconvolve mm line data on several molecular lines, and so were able model the density structure and the kinematics of the protostellar discs. Their technique also worked on longitude-velocity maps, and I am sure it will become increasingly important in the years to come.

Another way around the problem of surface brightness sensitivity is to study the non-thermal emission from interstellar masers. These sample only very high density regions, but because of their high brightness temperatures they can be studied with milli-arcsec precision. One of the most remarkable discs we saw this week was the G35.2-0.7N system described by Genevieve Brebner. Here the masers trace the innermost parts of a massive disc heated from within by a massive star and compact HII region. The molecular disc is almost exactly edge-on, and forms a coherent structure aligned over two orders of magnitude from 0.3 pc down to 0.003 pc (Brebner *et al.* 1987). The high degree of alignment in the molecular gas makes it all the more remarkable that the ionized gas, which also has an elongated and symmetrical distribution, follows a different (*i.e.* misaligned) axis. Precessing jets from the young star are one possible explanation for this puzzle.

5 Milky Way

Moving on we come to the Milky Way, one of the most beautiful discs of all. One very important development which we did not hear much about this week is the disc of Galactic infrared sources detected by IRAS. About 80 000 of these are highly evolved stars with dusty envelopes. A large fraction are believed to be OH maser sources radiating primarily in the 1612 MHz satellite line. Radio searches are now underway at many observatories, and already the first 1000 OH-IR maser sources have been identified. The radio data give the expansion velocity of the envelope and the radial velocity of the star to better than 1 km s^{-1} accuracy, and also allow an estimate of the stellar mass-loss rate. This programme promises to provide a unique database on the galactic distribution and kinematics of a whole population of stars. The poster by Peter te Lintel and Herwig Dejonghe gave us a foretaste of what might be done in the future in understanding the orbits of these stars. There is a problem of another kind looming, however. The 1612 MHz band in which these sources radiate is "shared" with the Russian global navigation system GLONASS. Although the system is not yet complete GLONASS is already causing worldwide disruption to radioastronomical work in the 1612 MHz band. The Russians are doing nothing illegal I must add. The fault is in the Radio Regulations, which do not afford the 1612 MHz line sufficient protection. Satellite transmissions are particularly harmful to radio astronomy because a single transmitter can reach so much of the earth's surface. GLONASS has shown just how easily a radio astronomy band can be destroyed globally.

The IRAS sources trace the galactic disc and central bulge. One somewhat surprising feature is the square shape of the bulge. The bulge has about the same projected size as the massive nuclear disc of molecular clouds and HI. The poster by Mark Bailey and Althea Wilkinson explored the scattering of stars off these clouds into box-type orbits to give a square bulge. We heard a lot this week about edge modes. The nuclear disc is one disc which really does have a sharp edge, and furthermore the neutral gas density represents a significant fraction of the total density in the nucleus. Professor Fridman in his thought-provoking talk drew our attention to the dynamical similarities between rotating gravitational systems and shallow water waves, and then proceeded to describe edge-driven spiral water waves generated in the laboratory. This was analogue modelling at its best, and his photographs showed the clearest spiral patterns I expect to see in a real disc.

6 Galaxies

Astrophysical discs are lively, and nowhere is this better illustrated than in the gaseous discs of spiral galaxies. Renzo Sancisi in his review introduced us to the HI discs of several galaxies, each one an individual. If there was one common feature it was perhaps the very irregular distributions of HI, almost sponge-like in appearance, with many filaments and bubbles. In M101 one large hole in the HI distribution matches up perfectly with high-velocity HI moving at some 150 km s^{-1} with respect to the quiescent gas. Renzo summarized the arguments which suggest that the gas is presently leaving the disc, eventually to fall back as "high-velocity clouds" like those in our Galaxy. In NGC 628 we saw a classical warped HI disc which continues far beyond the optical image of the galaxy. In NGC 3198 on the other hand the hydrogen and optical emission both cut off rather sharply. My ears pricked up when I heard that 100 hrs of VLA time had been needed to establish this.

In the background throughout was the question of galactic halos. The HI rotation curves of many spiral galaxies continue flat to great distances, indicating a steadily increasing mass-to-light ratio. Claude Carignan's poster pointed out that dwarf galaxies can have halos too. With so much matter unseen it is important to find new ways to constrain the possibilities. Linda Sparke has developed the idea that warped HI discs might be precession modes in the potential of a non-spherical halo. Her work shows that an oblate V = constant halo can give warped disc modes, either an up warp or a down warp depending on the halo core size. This seems to be a promising idea to pursue further.

But already I am straying into theoretical matters, which were after all the main concern of the meeting. It is time I handed over to our final speaker.

References

Adams, F. C., Lada, C. J. & Shu, F. H., 1987. *Astrophys. J.,* **312**, 788.

Brebner, G. C., Heaton, B., Cohen, R. J. & Davies, S., 1987. *Mon. Not. R. Astron. Soc.,* **229**, 679.

Ho, P. T. P. & Townes, C., 1983. *Annu. Rev. Astron. Astrophys.,* **21**, 239.

Hodapp, K. W., 1984. *Astron. Astrophys.,* **141**, 255.

Sato, S., Nagata, T., Nakajima, T., Nishida, M., Tanaka, M. & Yamashita, T., 1985. *Astrophys. J.,* **291**, 708.

Simon, M., Peterson, D. M., Longmore, A. J., Storey, J. W. V. & Tokunaga, A. T., 1985. *Astrophys. J.,* **298**, 328.

Common processes and problems in disc dynamics

Scott Tremaine

Canadian Institute for Theoretical Astrophysics, University of Toronto, Canada

1 Introduction

The recognition that astrophysical discs exist was a major intellectual achievement. As Lissauer stressed at this meeting, it was more than 40 years after Galileo discovered peculiar appendages to Saturn ('two servants for the old man, who help him to walk and never leave his side') before Huygens published, as an anagram, the first correct model of the Saturn system ('it is surrounded by a thin flat ring, nowhere touching, and inclined to the ecliptic'). The long delay was due in part to the limited angular resolution of the available telescopes, but also reflects the leap of imagination needed to grasp the true nature of the first known non-spherical celestial body.

Compared with this one example of an astrophysical disc known for over 300 years, the number and variety of discs that have been discovered or inferred in just the last 30 years is remarkable: (1) Saturn's rings have been joined by lesser ring systems around the other three giant planets, all discovered since 1977; (2) there is recent strong evidence that discs are associated with many protostars and young stars (reviewed by Snell), as well as with active galactic nuclei (reviewed by Malkan); (3) it was only in the late 1960's that accretion discs were recognized to be a central ingredient of many close binary star systems, in particular cataclysmic variables and many Galactic X-ray sources; (4) although it has long been known that the solar system formed from a disc, the analysis of realistic models of protoplanetary discs, and direct observations of similar discs (*e.g.* the β Pictoris disc), began only in the last few years; (5) it is likely that discs play a crucial role in collimating the jets discovered in double radio sources, SS433, and bipolar flows from young stars.

One of the themes of this meeting has been that common dynamical processes act in astrophysical discs of various types, and hence that many of the problems confronting astrophysicists dealing with different disc systems can be solved using similar tools. In keeping with this theme, I begin by reviewing a few of the processes that appear to be central to the behaviour of astrophysical discs, and that we are confident we understand. Then I will move on to discuss some of the problems that we do not understand fully as yet, but that we believe are both common in and important to several types of astrophysical disc.

2 Common processes

The fundamental process governing the evolution of astrophysical discs can be stated simply: *energy dissipation makes discs spread*. This process was already understood by Maxwell, in his Adams' Prize essay on Saturn's rings ('as E diminishes, the distribution of the rings must be altered, some of the outer rings moving outwards, while the inner rings move inwards'). Maxwell also estimated the spreading time of the rings assuming that their viscosity was that of water, and commented humorously on the likely fate of material in the inner part of the rings: 'As for the men of Saturn I should recommend them to go by tunnel when they cross the "line" ' (see Brush *et al.* 1983).

The physical reason why disc spreading is associated with lower energy is easy to demonstrate. In the absence of external torques, discs conserve their total angular momentum, and seek to reach the lowest energy state consistent with that angular momentum. If for simplicity we consider a ring that is centrifugally supported in a fixed gravitational potential $\Phi(r)$, the specific energy and angular momentum of a disc element at radius r are given by

$$E(r) = \frac{r}{2}\frac{d\Phi}{dr} + \Phi, \qquad L(r) = \sqrt{r^3\frac{d\Phi}{dr}},$$

and it follows that $dE/dL = \Omega(r)$ where $\Omega = (r^{-1}d\Phi/dr)^{1/2}$ is the angular speed. Thus consider a simple model of disc spreading in which unit mass at radius r_2 moves outward, by acquiring angular momentum dL from a unit mass at $r_1 < r_2$ that moves inward. The resulting net change in energy is $dE = dE_1 + dE_2 = [-(dE/dL)_1 + (dE/dL)_2]dL = [-\Omega(r_1) + \Omega(r_2)]dL$, which is negative since Ω is usually a decreasing function of radius. Thus disc spreading leads to a lower energy state. In general, disc spreading, outward angular momentum flow, and energy dissipation accompany one another in astrophysical discs.

Maxwell considered only dissipation due to molecular viscosity, but we now realize that there are many different mechanisms that a disc can exploit in order to lower its energy. (1) D. N. C. Lin has argued in his talk that turbulent viscosity driven by infall plays a major role in determining the structure of the protoplanetary disc. (2) Magnetic viscosity may be important in accretion discs, although our understanding of the generation and evolution of magnetic fields in discs remains in a primitive state (*cf.* Donner's talk). (3) Lissauer has shown how efficiently density waves transport angular momentum and liberate energy in Saturn's rings – the typical rate for a strong wave turns out to be about a Gigawatt, comparable to Niagara Falls – and Ruden has argued that a strong $m = 1$ density wave may be present in circumstellar discs. (4) Small bodies ('ringmoons') imbedded in or just outside a disc can both transport angular momentum and induce effects such as the formation of sharp edges (Borderies). (5) In addition, a variety of structures within discs, including wakes (Toomre), grooves (Sellwood), bars (Frank), and other non-axisymmetric instabilities (Savonije) can provide effective angular momentum transport. Moreover, in many cases it is energetically favourable for these structures to form precisely because they liberate free energy by redistributing the disc's angular momentum.

One of the most important consequences of this redistribution process is embodied in Lynden-Bell & Pringle's (1974) remarkable formula for the effective temperature $T(r)$ of a steady-state, geometrically thin, optically thick, viscous accretion disc,

$$T(r)^4 = \frac{3GM\dot{M}}{8\pi r^3\sigma}\left(1 - \sqrt{r_\star/r}\right),$$

where M is the central mass, $\dot{M}$ is the accretion rate, σ is the Stefan-Boltzmann constant, and $r_\star$ is the inner boundary of the disc. The surface density of the disc, the equation of state, and the strength and properties of the viscosity do not enter the equation. The simplicity and power of this formula are central to our efforts to understand the spectral properties of unresolved discs in many different contexts (*cf.* talks by King, Malkan and Ruden).

Self-gravity plays an important role in many astrophysical discs. The central parameter governing the effects of self-gravity in discs is Toomre's (1964) Q parameter, usually defined as

$$Q = \frac{\sigma_R \kappa}{3.36 G \Sigma} \text{ for stellar discs} \qquad Q = \frac{c\kappa}{\pi G \Sigma} \text{ for gaseous discs,}$$

where κ is the epicycle frequency, Σ is the surface density, σ_R is the radial velocity dispersion, and c is the sound speed. What might be called 'Toomre's law' states that: *discs with $Q < 1$ are unstable*. Although this result was derived only in a short-wavelength (WKB) approximation, numerical work has shown that the inequality $Q > 1$ provides an accurate necessary condition for stability of a wide range of disc models. In addition, Q has proved to be the most useful single thermometer that we have to measure the importance of self-gravity to disc dynamics.

Toomre's law can be rephrased in terms of other variables. The surface density $\Sigma(r) \approx \rho(r)h(r)$ where ρ is the mean density in the disc and h is its thickness; the epicycle frequency $\kappa \approx \Omega$, where Ω is the angular speed (this holds exactly for a Kepler disc and elsewhere is almost always correct within a factor of two). The angular speed is related to the mass interior to r, M_r, by $\Omega^2 \approx GM_r/r^3$ (this holds exactly for a spherical mass distribution and is approximately correct for a disc). If we define the mean density interior to r, $\rho_m(r) = M_r/(\frac{4}{3}\pi r^3)$, replace the dispersion σ_R or sound speed c by Ωh (approximately correct if the velocity ellipsoid is not very anisotropic and $Q \gtrsim 1$), and drop factors of order unity, then we find the simple result $Q \approx \rho_m(r)/\rho(r)$. In other words, discs are gravitationally unstable at radius r if the mean density in the disc at r exceeds the mean density of the system interior to r. In this form Toomre's law is roughly the inverse of the usual Roche limit, which states that a self-gravitating satellite of density ρ will be disrupted by tidal forces if $\rho_m/\rho \gtrsim 1$.

As this brief survey already shows, work on almost every different type of disc system has contributed to the development of disc dynamics: viscous spreading was first discussed in the context of Saturn's rings, then in models of the protoplanetary disc (Jeffreys 1924, Lüst 1952), but the complete mathematical formulation came from studies of accretion discs (Lynden-Bell & Pringle 1974). Toomre's original (1964) paper discussed stability of self-gravitating galactic discs, but he was to some extent anticipated by Safronov's (1960) study of the stability of the protoplanetary disc. Density and bending waves were investigated by C. C. Lin & Shu (1964) and Hunter & Toomre (1969) as models of spiral arms and warps in galaxies, but waves satisfying the Lin-Shu and Hunter-Toomre dispersion relations have now been exhibited much less ambiguously in Saturn's rings.

3 Common problems

Probably the single most embarrassing aspect of contemporary disc dynamics is that we do not understand why accretion discs accrete. Although accretion is almost certainly the result of energy dissipation and outward angular momentum flow, the nature of the dissipation remains mysterious. This problem was addressed by King in his review, but let me summarize some of the arguments again. Molecular viscosity is certainly negligible in accretion discs, and magnetic viscosity may be present but is not well-understood. Gravitational torques, density waves, and shocks can transport angular momentum in inviscid discs, but so far there is no convincing model for accretion discs based on these processes.

Lastly, but most persistently, it has often been argued that turbulent viscosity must be present in accretion discs, based on the following reasoning. Plane shear flows spontaneously become turbulent if the Reynolds number Re $\approx r^2\Omega/\nu$ (r is the system size, ν is the kinematic viscosity, Ω is the shear in a planar system or the angular speed in a disc) exceeds about 10^3. If other forms of viscosity are negligible, the Reynolds number in an accretion disc would be extremely high if the flow were laminar; hence, it is argued, turbulence must develop, with sufficient vigour to reduce Re to $\lesssim 10^3$.

There are several reasons to be suspicious of this chain of argument. First, inviscid discs are strongly stabilized by angular momentum gradients (*i.e.* the Rayleigh stability criterion $d(r^2\Omega)/dr > 0$ is satisfied, or, in the language of stellar dynamics, the epicycle frequency is real, $\kappa^2 > 0$). Second, linear stability analyses of differentially rotating discs without self-gravity (as described by Glatzel and Papaloizou) generally do not show instability unless there is a reflecting boundary condition at the edge of the disc, and hence imply that only global, not local, instabilities are present. (There is a simple physical argument that suggests why a reflecting boundary induces instability. Waves inside co-rotation have negative energy density and waves outside have positive energy density. The reflecting boundary establishes a cavity between the boundary and the forbidden zone around co-rotation; leakage through the forbidden zone excites waves of the opposite energy density and hence the wave amplitude in the cavity must grow by energy conservation.) A third reason to disbelieve arguments that high Re is always forbidden is that we have a counter-example: Saturn's rings have Re $\approx 10^{14}$! Thus the weight of theoretical evidence strongly suggests that thin, isolated, Keplerian discs without self-gravity are locally stable (although turbulence can develop in discs subjected to a rain of infalling material; see D. N. C. Lin's review). Obviously, alternatives to turbulent viscosity as the angular momentum transport process in accretion discs deserve to be investigated very thoroughly.

It is possible that the question of whether accretion discs are necessarily turbulent will only be answered when the techniques of computational fluid dynamics have improved to the point where we can investigate the high Reynolds number, high Mach number flows characteristic of these systems. However, an intriguing alternative described here by Fridman is to use laboratory studies of shallow water flows, in which the speed of surface waves replaces the sound speed of the gas, to study supersonic flows in disc geometries. I hope that experiments of this kind will be exploited further.

The second common problem that I would like to discuss is the stability of self-gravitating discs. This field of study is almost exactly a quarter-century old: in fact, 1964 and 1965 saw the publication of three seminal papers that together introduced many of the concepts that have proved to be central to the stability of self-gravitating discs. Toomre (1964) invented the Q parameter as a measure of axisymmetric stability; C. C. Lin & Shu (1964) introduced the concept of density waves described by a WKB dispersion relation; and Goldreich & Lynden-Bell (1965) showed that leading density waves can undergo strong transient amplification due to self-gravity as they are sheared into trailing waves by the disc's differential rotation ('swing amplification'). Since that time, perhaps the most important new result has been the discovery that a wide range of discs are subject to violent 'bar-like' ($m = 2$) instabilities that persist even when Q

exceeds unity and the disc is safely stable to axisymmetric disturbances (Hohl 1971, Ostriker & Peebles 1973).

The elucidation of the physics governing the bar instability has proved to be a challenging task. An arsenal of techniques has been brought to bear – including fast Fourier transform and tree-based potential solvers, 'quiet starts' to suppress Poisson noise, and the use of 'softened' gravity to suppress axisymmetric instabilities in cold discs – but it has turned out to be surprisingly difficult to carry out reliable stability studies of self-gravitating discs. One problem is that high spatial resolution is needed to follow the dynamics in the central regions, which have an important influence on disc stability; another problem is that many numerical techniques lead to spurious feedback from trailing to leading waves, which thereafter are sheared and swing amplified into larger trailing waves. Thus it is very encouraging that N-body simulations and linear normal mode calculations now find the same unstable modes in many cases (*e.g.* Sellwood & Athanassoula 1986).

The past few years have seen rapid progress in the sophistication and accuracy of numerical models of self-gravitating discs. These advances have been accompanied by semi-analytic models of the bar instability (Toomre 1981) that are based on the concepts of density waves, swing amplification, and feedback loops and that appear to provide remarkably accurate predictions of the stability properties of many disc models. An excellent summary of our present understanding of the linear stability of self-gravitating discs has been given at this meeting by Papaloizou.

As an outsider, my impression is that with these advances the problem of the stability of self-gravitating discs has largely been solved; while obstacles are still present and new phenomena certainly remain to be investigated (such as the closely related edge instabilities described by Toomre, the groove instabilities described by Sellwood, and the instabilities driven by vortensity gradients described by Papaloizou), what is left to be done would be described in military terms as 'mopping up'. Of course, there are many cases from military history where an announcement of this kind not only demoralized the supposedly victorious troops but also was followed by enemy victory, but I hope that neither of these unfortunate events will occur, and that the stability of self-gravitating discs will soon be regarded as a solved problem.

A closely related subject in disc dynamics is spiral structure theory, which has been intertwined with the study of disc stability ever since C. C. Lin & Shu's (1964) recognition that spiral structure could be treated as a density wave. Their work led to many efforts to demonstrate that theoretical disc models admit long-lived normal modes of spiral form, and to compare the observed shapes and kinematics of spiral arms to these modes, in the hope that ultimately the fits would yield information on the surface density and velocity dispersion in observed discs. It is striking that the presentations at this meeting have contained few direct comparisons of this kind – except, of course, Lissauer's analyses of forced density waves in Saturn's rings – and I think that this is a symptom of fundamental changes in spiral structure theory.

Almost all dynamicists agree that most spiral structure reflects an underlying density wave in the old stars that comprise the bulk of the mass in the galactic disc. However, it now seems likely that in most cases it is an oversimplification to identify the spiral pattern with the most unstable normal mode of the disc, which is the pattern that would develop if a smooth disc evolved in isolation. Discs are subject to a variety of gravitational disturbances – bars (Athanassoula's talk), encounters with other galaxies (Sundin's video paper), and molecular clouds (Toomre's talk), to mention just three – and their response to these disturbances is generally spiral, if there is differential rotation, and often very strong, if Q is not too large. Only in the case where the perturber mainly drives a single mode do we expect the spiral pattern to match the pattern of a discrete mode of the disc.

One consequence of this point of view is a new answer to the old winding problem: why does differential rotation not make the spiral arms wind up? As originally envisaged by many theorists, the answer to this question in density wave theory was that the pattern is composed of one or more modes, and discrete modes are stationary in a rotating frame and hence do not wind up. It now seems more likely that the spiral pattern remains stationary only if the disturbance is persistent (*e.g.* a central bar), and *does* wind up, if it arises from a transient disturbance (*e.g.* an encounter with another galaxy). The detailed agreement of neutral hydrogen kinematics with the predictions of models based on gas flow through a stationary density wave has sometimes been used to argue that spiral patterns must be stationary, but even density waves that are winding up can often produce a kinematic signature in the gas that closely resembles observations.

An important consequence of this change in perspective is that there is now less emphasis on *a priori* predictions of the spiral structure that should be present in a given galactic disc. Instead, we argue that there are many different mechanisms that can excite spiral structure, that we understand in broad outline how spirals originate, but that the prediction of the strength, shape and origin of the spiral pattern in a given galaxy is usually too difficult a task. Instead, it is more fruitful to concentrate, as Sancisi did in his talk, on the effects of the spiral pattern on the disc – on star formation, the energy balance of the interstellar medium, angular momentum transfer, heating the disc stars, and so forth – since these are important for galactic evolution and relatively independent of the detailed mechanism by which the spiral was formed.

A final interesting dynamical puzzle is that many types of astrophysical disc are warped. The only warps whose origin is well-understood are those associated with planetary rings (Lissauer, Borderies), but accretion discs and spiral galaxies also appear to be warped. In her talk, Sparke has advanced the promising idea that galactic warps are discrete bending modes in a disc embedded in a flattened dark halo. The basis for this idea is that the angular momentum vectors of the disc and halo are unlikely to be aligned, since the disc material has a different history from the halo material and hence has been subjected to different torques in the process of galaxy formation (Efstathiou's talk). Since the disc material is dissipative, it settles to a state in which each ring of material precesses at the same rate in the combined field of the halo and the other disc material, and this rate is simply the pattern speed of Sparke's mode. This idea dates back at least to papers by Toomre (1983) and Dekel & Shlosman (1983), and in fact has even earlier roots: the shape of Sparke's mode is simply the invariable surface first discussed by Laplace in studies of solar system dynamics. Sparke has argued that the warps may provide clues to the mass distribution in the halo and disc, but similar optimistic claims have long been made by spiral structure theorists, and there is no strong reason to believe that bending waves are more likely to be useful probes of disc structure than density waves.

Even if galactic warps arise from misalignment of the disc and halo, a number of issues remain to be resolved. It is likely that halos are not only flattened but triaxial, so that the halo not only warps the disc but distorts circular streamlines into ellipses. Unfortunately, the kinematic signature of elliptical streamlines is similar to that of inclined circular streamlines; nevertheless, it is important to investigate whether triaxial halos are compatible with kinematic observations of our own and other disc galaxies. A second issue, already mentioned by Toomre (1983) and Dekel & Shlosman (1983), is that dynamical friction may re-align the disc with the halo in times short compared to the Hubble time.

Accretion discs in binary star systems are also believed to exhibit warps, although here the evidence is less direct. The 35-day modulation of the X-ray flux from Hercules X-1 is believed to arise from occultation of the X-rays by the rim of a warped accretion disc, and the 164-day precession of the jets in SS433 is likely to arise from precession of a warped disc that feeds the jets. The most popular models for these systems require that (1) the spin angular momentum of the companion star that feeds the disc is misaligned with the orbital angular momentum of the binary system; (2) the disc is 'slaved', that is, the viscosity in the disc is so high that the disc orientation rigidly follows the orientation of the precessing companion star (this requires that the viscous diffusion time through the disc is shorter than the precession time). The required viscosity is very high (*e.g.* Katz 1980), which has the advantage that it should be possible to model the discs easily with relatively unsophisticated hydrodynamic codes to see if the model works in detail. I am personally sceptical that slaved discs are the correct explanation for the behaviour of these systems; I hope that some of the theorists at this meeting who have worked on other types of disc system will turn their attention to this problem and seek a radically different solution.

I am grateful to Alar Toomre, both for countless enlightening discussions of disc dynamics over the years and for detailed and constructive comments on this manuscript. I would also like to thank the Berkeley Astronomy Department for their hospitality while this talk was being written up.

References

Brush, S. G., Everitt, C. W. F. & Garber, E., 1983. *Maxwell on Saturn's Rings*, 44, 135, The MIT Press, Cambridge, MA.

Dekel, A. & Shlosman, I., 1983. In *Internal Kinematics and Dynamics of Galaxies*, IAU Symposium 100, p. 187, ed. Athanassoula, E., Reidel, Dordrecht.

Goldreich, P. & Lynden-Bell, D., 1965. *Mon. Not. R. Astron. Soc.*, 130, 125.

Hohl, F., 1971. *Astrophys. J.*, 168, 343.

Hunter, C. & Toomre, A., 1969. *Astrophys. J.*, 155, 747.

Jeffreys, H., 1924. *The Earth*, p. 55, Cambridge University Press, Cambridge.

Katz, J. I., 1980. *Astrophys. J. Lett.*, 20, 135.

Lin, C. C. & Shu, F. S., 1964. *Astrophys. J.*, 140, 646.

Lüst, V. R., 1952. *Z. Naturforsch.*, 7a, 87.

Lynden-Bell, D. & Pringle, J. E., 1974. *Mon. Not. R. Astron. Soc.*, 168, 603.

Ostriker, J. P. & Peebles, P. J. E., 1973. *Astrophys. J.*, 186, 467.

Safronov, V. S., 1960. *Ann. d'Astrophys.*, 23, 979.

Sellwood, J. A. & Athanassoula, E., 1986. *Mon. Not. R. Astron. Soc.*, 221, 195.

Toomre, A., 1964. *Astrophys. J.*, 139, 1217.

Toomre, A., 1981. In *Structure and Evolution of Normal Galaxies*, p. 111, eds. Fall, S. M. & Lynden-Bell, D., Cambridge University Press, Cambridge.

Toomre, A., 1983. In *Internal Kinematics and Dynamics of Galaxies*, IAU Symposium **100**, p. 177, ed. Athanassoula, E. Reidel, Dordrecht.

Citation index

Adams, F. C., Lada, C. J. & Shu, F. H., 1987. *Astrophys. J.*, **312**, 788.   27, 52, 226

Adams, F. C., Lada, C. J. & Shu, F. H., 1988. *Astrophys. J.*, **326**, 865.   52, 53, 119

van Albada, G. D., 1985. *Astron. Astrophys.*, **142**, 491.   145

van Albada, G. D. & Roberts, W. W., 1981. *Astrophys. J.*, **246**, 740.   99

Alexander, A. F., 1962. *The Planet Saturn*, MacMillan Co, New York, NY.   1

Allen, R. J., Atherton, P. D. & Tilanus, R. P. J., 1986. *Nature*, **319**, 296.   129

Allen, R. J. & Goss, W. M., 1979. *Astr. Astrophys., Suppl. Ser.*, **36**, 135.   129

Anderson, C. M., Oliverson, N. A. & Nordsieck, K. H., 1980. *Astrophys. J.*, **242**, 188.   85

Antipov, S. V., et al. 1988. *Physics of Plasmas*, **14**, 1104.   187

Antonov, V. A., 1976. *Uchen. zap. Leningrad Univ. (Sov.)*, **32**, 79.   200

Appenzeller, I., Jankovics, I. & Ostriecher, R., 1984. *Astron. Astrophys.*, **141**, 108.   53

Arnaud, K. A., Branduari-Raymont, G., Culhane, J. L., Fabian, A. C., Hazard, C., McGlynn, T. A., Shafer, R. A., Tennant, A. F. & Ward, M. J., 1985. *Mon. Not. R. Astron. Soc.*, **217**, 105.   82

Artymowicz, P., 1988. *Astrophys. J. Lett.*, **335**, L79.   44, 47

Artymowicz, P., Burrows, C. & Paresce, F., 1989. *Astrophys. J.*, **337**, 494.   43, 44, 45, 48

Athanassoula, E. & Bosma, A., 1985. *Annu. Rev. Astron. Astrophys.*, **23**, 147.   21

Athanassoula, E., Bosma, A., Creze, M. & Schwarz, M. P. 1982. *Astron. Astrophys.*, **107**, 101.   144

Athanassoula, E. & Sellwood, J. A., 1986. *Mon. Not. R. Astron. Soc.*, **221**, 213.   176

Aumann, H. H., 1985. *Publ. Astron. Soc. Pac.*, **97**, 885.   47

Aumann, H. H., 1988. *Astron. J.*, **96**, 1415.   47, 48

Baan, W. A. & Haschick, A. D., 1983. *Astron. J.*, **88**, 1088.   144

Backman, D. E. & Gillett, F. C., 1987. In *Cool Stars, Stellar Systems and The Sun*, p. 340, ed. Linsky, J. L. & Stencel, R. E., Springer-Verlag, New York.   47, 48

Backman, D. E., Gillett, F. C. & Witteborn, F. C., 1989. *Astrophys. J.*, (in press).   45, 47, 48

Bahcall, J. N., 1984. *Astrophys. J.*, **276**, 169.   137

Bahcall, J. N., 1984. *Astrophys. J.*, **287**, 926.   137

Bailey, M. E., 1980. *Mon. Not. R. Astron. Soc.*, **191**, 195.   141

Bailey, M. E., 1982. *Mon. Not. R. Astron. Soc.*, **200**, 247.   141

Bailey, M. E., 1985. In *Cosmical Gas Dynamics*, p. 49, ed. Kahn, F. D., VNU Science Press, Utrecht.   141

Bailey, M. E. & Clube, S. V. M., 1978. *Nature*, **275**, 278.   141

Ball, R., 1986. *Astrophys. J.*, **307**, 453.   132

Bally, J. & Lada, C. J., 1983. *Astrophys. J.*, **265**, 824.   67

Bally, J., Snell, R. L. & Predmore, R., 1983. *Astrophys. J.*, **272**, 154.   56

Bastian, U. & Mundt, R. 1985., *Astron. Astrophys.*, **144**, 57.   53

Batchelor, J., 1973. *An Introduction for Fluid Dynamics*, Mir, Moscow.   190

Becklin, E. E. & Zuckerman, B. 1989. In *Millimetre and Submillimetre Astronomy*, proceedings of the URSI Symposium, ed. Philipps, T., Kluwer Academic Press, Dordrecht, in preparation.   47

Beckwith, S., Zuckerman, B., Skrutskie, M. & Dyck, H., 1984. *Astrophys. J.*, **287**, 793.   54

Begelman, M. C., Blandford, R. D. & Rees, M. J., 1984. *Rev. Mod. Phys.*, **56**, 255.   99

Begelman, M. C., McKee, C. F. & Shields, G. A., 1983. *Astrophys. J.*, **271**, 70.   84

Bell, R., Clarke, C. & Lin, D. N. C., 1989. In preparation.   36

Bell, R., Lin, D. N. C. & Ruden, S. P., 1989. In preparation.   31, 37

Bertin, G. & Lin, C.C., 1988. In *Evolution of Galaxies*, p. 255, (Proceedings of the   199, 200, 204
    10th IAU Regional Astronomy Meeting). Ed. J. Palouš, Publ. Astron. Inst.
    Czech. Acad. Sci. **69**.

Bertin, G., Lin, C. C., Lowe, S. A. & Thurstans, R. P., 1989. *Astrophys. J.*, **338**,   170
    pp. 78 & 104.

Bertin, G. & Romeo, A. B., 1988. *Astron. Astrophys.*, **195**, 105.   209, 210

Bertola, F., Buson, L. M. & Zeilinger, W. W., 1988. *Nature*, **335**, 705.   147

Bertola, F., Galletta, G., Kotanyi, C. & Zeilinger, W. W.,1988. *Mon. Not. R. Astron.*   147
    Soc., **234**, 733.

Bertout, C., Basri, G. & Bouvier, J., 1988. *Astrophys. J.*, **330**, 350.   27, 52

Bieging, J., 1984. *Astrophys. J.*, **286**, 591.   56

Bieging, J., Cohen, M. & Schwartz, P., 1984. *Astrophys. J.*, **282**, 699.   50

Binney, J. & Petrou, M., 1985. *Mon. Not. R. Astron. Soc.*, **214**, 449.   142

Bisnovaty-Kogan, G. S. & Zel'dovich, Ya. B., 1970. *Astrophysica (Sov.)*, **6**, 149.   199

Bland, J. & Tully, R. B., 1989. *Astron. J., in press*.   143

Bodenheimer, P. H. & Pollack, J., 1986. *Icarus*, **67**, 391.   28

Bodenheimer, P., Yorke, H. W., Rozyczka, M. & Tohline, J. E., 1989. In *Formation*   33
    and Evolution of Low-Mass Stars, eds. Dupress, A. & Lago, M. T. V. T.,
    Reidel, Dordrecht, in press.

Borderies, N., 1987. *Bull. Am. Astron. Soc.*, **19**, 891.   21

Borderies, N., 1989. *Icarus, in press*.   21

Borderies, N., Goldreich, P. & Tremaine, S., 1982. *Nature*, **299**, 209.   12, 13, 19, 20, 21

Borderies, N., Goldreich, P. & Tremaine, S., 1983. *Icarus*, **55**, 124.   19, 20

Borderies, N., Goldreich, P. & Tremaine, S., 1983. *Astron. J.*, **88**, 1560.   19, 22

Borderies, N., Goldreich, P. & Tremaine, S., 1983. *Astron. J.*, **88**, 226.   22

Borderies, N., Goldreich, P. & Tremaine, S., 1984. In *Planetary Rings*, p. 713, eds.   21
    Greenberg, R. & Brahic, A., University of Arizona Press, Tuscon, AZ.

Borderies, N., Goldreich, P. & Tremaine, S., 1984. *Astrophys. J.*, **284**, 429.   22

Borderies, N., Goldreich, P. & Tremaine, S., 1985. *Icarus*, **63**, 406.   22

Borderies, N., Goldreich, P. & Tremaine, S., 1986. *Icarus*, **68**, 522.   9, 13, 21

Borderies, N., Goldreich, P. & Tremaine, S., 1989. *Icarus, in press*.   21

Borderies, N., Gresh, D. N., Longaretti, P-Y. & Marouf, E. A., 1988. *Bull. Am.*   22
    Astron. Soc., **20**, 844.

Bosma, A., Goss, W. M. & Allen, R. J., 1981. *Astron. Astrophys.*, **93**, 106.   130

Bothun, G. D. & Dressler, A., 1986. *Astrophys. J.*, **301**, 57.   214

Bottema, R., 1988. *Astron. Astrophys.*, **197**, 105.   177

Brackmann, E. & Scoville, N., 1980. *Astrophys. J.*, **242**, 112.   67

Brebner, G. C., Heaton, B., Cohen, R. J. & Davies, S., 1987. *Mon. Not. R. Astron.*   67, 227
    Soc., **229**, 679.

Bruch, A., 1986. *Astron. Astrophys.*, **167**, 91.   85

Brush, S. G., Everitt, C. W. F. & Garber, E., 1983. *Maxwell on Saturn's Rings*, **44**,   231
    135, The MIT Press, Cambridge, MA.

Burns, J. A., 1976. *Amer. J. Phys.*, **44**, 944.   4

Burns, J. A., Showalter, M. R. & Morfill, G. E., 1984. In *Planetary Rings*, p. 200,   2
    eds. Greenberg, R. & Brahic, A., University of Arizona Press, Tuscon, AZ.

Byrd, G. G., Sundelius, B. & Valtonen, M., 1987. *Astron. Astrophys.*, **171**, 16.   213

Byrd, G. G., Valtonen, M., Sundelius, B. & Valtaoja, L., 1986. *Astron. Astrophys.*,   213
    166, 75.

Cabot, W., Canuto, V. M., Hubickyj, O. & Pallock, J. B., 1987. *Icarus*, **69**, 387.   29

Cabot, W., Canuto, V. M., Hubickyj, O. & Pallock, J. B., 1987. *Icarus*, **69**, 423.   29

Cairns, R. A., 1979. *J. Fluid Mech.*, **92**, 1.   126

Cameron, A. G. W., 1973. *Icarus*, **18**, 407.   34

Dobrovolskis, A. R., 1980. *Icarus*, **43**, 222.   21

Dobrovolskis, A. R., Steiman-Cameron, T. Y. & Borderies, N. J., 1988. *Bull. Am.*   21
    Astron. Soc., **20**, 861.

Dolotin, V. V. & Fridman, A. M., 1989. In *Nonlinear Waves in Physics and Astro-*   72
    physics, eds. Gaponov-Grekhov, A. V. & Rabinovitch, M. I., Springer-Verlag,
    New York.

Dones, L., 1987. *PhD thesis*, University of California, Berkeley.   9, 12, 13, 25

Donner, K., 1979. *PhD thesis*, Cambridge University.   32

Donner, K. J. & Brandenburg, A., 1989. *Geophys. Astrophys. Fluid Dyn.*,   151
    (submitted).

Doyle, L. R., 1987. *PhD thesis*, University of Heidelberg.   25

Draine, B. T. & Lee, H. M., 1984. *Astrophys. J.*, **285**, 89.   47

Dressler, A., 1980. *Astrophys. J.*, **236**, 351.   213

Duschl, W. J., 1983. *Astron. Astrophys.*, **119**, 248.   85

Duschl, W. J., 1988. *Astron. Astrophys.*, **194**, 33.   101

Dyck, H. & Lonsdale, C., 1979. *Astron. J.*, **84**, 1339.   51

Eddington, A. S., 1926. *The Internal Constitution of the Stars*, Cambridge University   73
    Press, Cambridge.

Edelson, R. A. & Malkan, M. A., 1986. *Astrophys. J.*, **308**, 59.   95

Eder, J., Lewis, B. M. & Terzian, Y., 1988. *Astrophys. J.*, **66**, 183.   139

Edwards, S., Cabrit, S., Strom, S. E., Ingeborg, H., Strom, K. & Anderson, E., 1987.   53
    Astrophys. J., **321**, 473.

Elliot, J. L., Dunham, E. W. & Mink, D. J., 1977. *Nature*, **267**, 328.   2

Elmegreen, B. G. & Elmegreen, D. M., 1983. *Astrophys. J.*, **267**, 31.   217

Elmegreen, D. M. & Elmegreen, B. G., 1982. *Mon. Not. R. Astron. Soc.*, **201**, 1021.   217, 219

Elmegreen, D. M. & Elmegreen, B. G., 1987. *Astrophys. J.*, **314**, 3.   217

Elsasser, H. & Staude, H., 1978. *Astron. Astrophys.*, **70**, L3.   50

Erickson, S. A., 1974. *PhD thesis*, MIT.   112

Esposito, L. W., 1986. *Icarus*, **67**, 345.   13

Esposito, L. W., Cuzzi, J. N., Holberg, J. B., Marouf, E. A., Tyler, G. L. & Porco,   11
    C. C., 1984. In *Saturn*, p. 463, eds. Gehrels, T. & Matthews, M. S., University
    of Arizona Press, Tuscon, AZ.

Esposito, L. W., O'Callahan, M. & West, R. A., 1983. *Icarus*, **56**, 439.   11

Evans, N. W., 1989. *Inter. Journ. Comp. Math.*, in press.   184

Evans, N. W. & Lynden-Bell, D., 1989. *Mon. Not. R. Astron. Soc.*, **236**, 801.   183

Faulkner, J., Lin, D. N. C. & Papaloizou, J. C. B., 1983. *Mon. Not. R. Astron. Soc.*,   36
    205, 359.

Ferland, G. J. & Rees, M. J., 1988. *Astrophys. J.*, **332**, 141.   89

Ferrers, N. M., 1877. *Quart. J. Pure Appl. Math.*, **14**, 1.   145

Flynn, B. C. & Cuzzi, J. N., 1989. *Icarus*, in press.   26

Frank, J., King, A. R. & Lasota, J. P., 1986. *Astron. Astrophys.*, **178**, 137.   77

Frank, J., King, A. R. & Raine, D. J., 1985. *Accretion Power in Astrophysics*, Cam-   73
    bridge University Press, Cambridge.

French, R. G., Elliot, J. L., French, L. M., Kangas, J. A., Meech, K. J., Ressler, M.   22
    E., Buie, M. W., Frogel, J. A., Holberg, J. B., Fuensalida, J. J., Joy, M., 1988.
    Icarus, **73**, 349.

Fridman, A. M., 1978. *Usp. Fiz. Nauk*, **125**, 352.   188

Fridman, A. M., 1989. *Pis'ma Astron. Zh.*   185, 192

Fridman, A. M., 1989. *Dokl. Akad. Nauk SSSR.*, in press.   190

Fridman, A. M., Morozov, A. G., Nezlin, M. V. & Snezhkin, E. N. 1985. *Phys.*   186, 194, 195, 196
    Lett. A, **109**, 228.

Fridman, A. M. & Polyachenko, V. L., 1984. *Physics of Gravitating Systems,*   188, 193, 199, 200
     Springer-Verlag, New York.
Friedjung, M. & Muratorio, G., 1987. *Astron. Astrophys.,* **188,** 100.   87
Garcia, M. R., 1986. *Astron. J.,* **91,** 400.   86
Gavazzi, G. & Jaffe, W., 1987. *Astrophys. J.,* **310,** 53.   213
Genzel, R. & Townes, C. H., 1987. *Annu. Rev. Astron. Astrophys.,* 25, 377.   101
Gerin, M., Combes, F. & Athanassoula E., 1989. *Astron. Astrophys.,* Submitted.   219, 220
Geroyannis, V. S., 1988. *Astrophys. J.,* **327,** 273.   127
Gingold, R. A. & Monaghan, J. J., 1982. *J. Comput. Phys.,* **46,** 429.   211
Gisler, G. R., 1980. *Astron. J.,* **85,** 623.   213
Glatzel, W., 1987. *Mon. Not. R. Astron. Soc.,* **228,** 77.   122, 126
Glatzel, W., 1989. *J. Fluid Mech.,* **202,** 515.   123
Glatzel, W., 1989. In preparation.   121
Gledhill, T. & Scarrott, S., 1989. *Mon. Not. R. Astron. Soc.,* **236,** 139.   55
Goguen, J. D., Hammel, H. B. & Brown, R. H., 1989. *Icarus,* **77,** 239.   25
Goldreich, P. & Lynden-Bell, D., 1965. *Mon. Not. R. Astron. Soc.,*   156, 200, 203, 205, 206, 234
     130, 124.
Goldreich, P. & Porco, C. C., 1987. *Astron. J.,* **93,** 730.   22
Goldreich, P. & Tremaine, S., 1977. *Nature,* **277,** 97.   17
Goldreich, P. & Tremaine, S., 1978. *Icarus,* **34,** 227.   1, 7, 20
Goldreich, P. & Tremaine, S., 1978. *Icarus,* **34,** 240.   7, 9, 13, 20, 21
Goldreich, P. & Tremaine, S., 1978. *Astrophys. J.,* **222,** 850.   181
Goldreich, P. & Tremaine, S., 1979. *Astrophys. J.,* **233,** 857.   176
Goldreich, P. & Tremaine, S., 1979. *Astron. J.,* **84,** 1638.   22
Goldreich, P. & Tremaine, S., 1979. *Nature,* **277,** 97.   21
Goldreich, P. & Tremaine, S., 1980. *Astrophys. J.,* **241,** 425.   14
Goldreich, P. & Tremaine, S., 1981. *Astrophys. J.,* **243,** 1062.   22
Goldreich, P. & Tremaine, S., 1982. *Annu. Rev. Astron. Astrophys.,* 20, 249.   14, 20, 38, 103
Goldreich, P., Tremaine, S. & Borderies, N., 1986. *Astron. J.,* **92,** 490.   23
Goldreich, P. & Ward, W. R., 1973. *Astrophys. J.,* **183,** 1051.   35, 37
Goldsmith, P. F. & Arquilla, R., 1985. In *Protostars and Planets II,* p. 137, eds.   33
     Black, D. & Matthews, M. S., University of Arizona Press, Tuscon, AZ.
Gould, A., 1988. preprint.   138
Gradie, J., Hayashi, J., Zuckerman, B., Epps, H. & Howell, R., 1987. In *Proc. 18th*   47
     Lunar and Planetary Conference Vol. I, p. 351, Cambridge University Press
     and the Lunar and Planetary Institute.
Grasdalen, G., Strom, S. E., Strom, K., Capps, R., Thompson, D. & Castelaz, M.,   54
     1984. *Astrophys. J. Lett.,* **283,** L57.
Greenberg, R., Wacker, J. F., Hartmann, W. K. & Chapman, C. R., 1978. *Icarus,*   28
     35, 1.
Greenstein, J. L. & Kraft, R. P., 1959. *Astrophys. J.,* **130,** 99.   74
Gresh, D. L., Rosen, P. A., Tyler, G. L. & Lissauer, J. J., 1986. *Icarus,* **68,** 502.   6, 9, 11
Gusten, R., Chini, R. & Neckel, T., 1984. *Astron. Astrophys.,* **138,** 205.   56
Habe, A. & Ikeuchi, S., 1985. *Astrophys. J.,* **289,** 540.   150
Habe, A. & Ikeuchi, S., 1988. *Astrophys. J.,* **326,** 84.   150
Habing, H. J., 1986. In *Light on Dark Matter,* p. 40, ed. Israel, F. P., Reidel,   139
     Dordrecht.
Habing, H. J., 1987. In *The Galaxy,* p. 173, eds. Gilmore, G. & Carswell, B., Reidel,   142
     Dordrecht.
Hämeen-Anttila, K. A. & Pyykkö, S., 1973. *Astron. Astrophys.,* **19,** 235.   26
Hameury, J. M., King, A. R. & Lasota, J. P., 1986. *Astron. Astrophys.,* **162,** 71.   79
Hapke, B., 1986. *Icarus,* **67,** 264.   26

Kalnajs, A. J., 1976. *Astrophys. J.*, **205**, 745, 751.   200
Kalnajs, A. J., 1977. *Astrophys. J.*, **212**, 637.   175
Kaneko, N., *et al.*, 1989. *Astrophys. J.*, **337**, 691.   144
Kant, I., 1755. *Allgemeine Naturgeschichte und Theorie des Himmels.*   1
Kato, S., 1983. *Publ. Astron. Soc. Jpn*, **35**, 249.   119
Katz, J. I., 1980. *Astrophys. J. Lett.*, **20**, 135.   237
Kawabe, R., Morita, K., Ishiguro, M., Kasuga, T., Chikada, Y., Handa, K., Iwashita, H., Kanzawa, T., Okumura-Kawabe, S., Kobayashi, H., Takahashi, T., Murata, Y. & Hasegawa, T., 1989. preprint.   56
Kawabe, R., Ogawa, H., Fugui, Y., Takano, T., Takaba, H., Fujimoto, Y., Sugitani, K. & Fujimoto, M., 1984. *Astrophys. J. Lett.*, **282**, L73.   56
Keel, W. C., Kennicutt, R. C., Hummel, E. & van der Hulst, J. M., 1985. *Astron. J.*, **90**, 708.   213
Kenyon, S. J. & Hartmann, L., 1987. *Astrophys. J.*, **323**, 714.   27
Kenyon, S. J., Hartmann, L. & Hewett, R., 1988. *Astrophys. J.*, **325**, 231.   35, 36
King, A. R., 1989. In *Classical Novae*, p. 17, eds. Bode, M. F. & Evans, A., Wiley, Chichester.   85
Knapp, G. R., 1987. In *Structure and Dynamics of Elliptical Galaxies*, IAU Symp. 127, p. 145, ed. de Zeeuw, T., Reidel, Dordrecht.   147
Kobayashi, Y., Kawara, K., Maihara, T., Okuda, H., Sato, S. & Noguchi, K., 1978. *Publ. Astron. Soc. Jpn*, **30**, 377.   51
Kopal, Z., 1978. *Dynamics of Close Binary Systems*, Reidel, Dordrecht.   224
Kormendy, J. & Norman, C. A., 1979. *Astrophys. J.*, **233**, 539.   217
Kraft, R. P., 1962. *Astrophys. J.*, **135**, 408.   74
Krumm, N. & Burstein, D., 1984. *Astron. J.*, **89**, 1319.   133
Kuijken, K. & Gilmore, G., 1988a,b,c. *Mon. Not. R. Astron. Soc.*, submitted.   137, 138
Lacey, C. G., 1984. *Mon. Not. R. Astron. Soc.*, **208**, 687.   140, 141
Lada, C. J., 1985. *Annu. Rev. Astron. Astrophys.*, **23**, 267.   50, 67
Lagrange-Henri, A. M., Vidal-Madjar, A. & Ferlet, R., 1988. *Astron. Astrophys.*, **190**, 275.   43
Landau, L. D., 1944. *Dokl. Akad. Nauk SSSR*, **44**, 151.   193, 195
Landau, L. D. & Lifshitz, E. M., 1986. *Hydrodynamics*, Nauka, Moscow.   185, 186, 193
Lane, A. L., *et al.*, 1982. *Science*, **215**, 537.   11
Laplace, P., 1796. *Exposition du systeme du monde*, Paris.   1
Larson, R., 1989. In *The Formation and Evolution of Planetary Systems*, eds Weaver, H. A., Patesce, F. & Danly, L., Cambridge University Press, Cambridge, in press.   32
Lau, Y. Y., Lin, C. C. & Mark, J. W. K., 1976. *Proc. Natl. Acad. Sci. U.S.A.*, **73**, 1379.   112
Lecar, M. & Aarseth, S. J., 1985. *Astrophys. J.*, **305**, 564.   28
Lee, U., 1988. *Mon. Not. R. Astron. Soc.*, **232**, 711.   125
Lee, U. & Saio, H., 1986. *Mon. Not. R. Astron. Soc.*, **221**, 365.   125
Lee, U. & Saio, H., 1987. *Mon. Not. R. Astron. Soc.*, **224**, 513.   125
Lee, U. & Saio, H., 1989. *Mon. Not. R. Astron. Soc.*, in press.   125
Lewis, J. S., 1972. *Icarus*, **16**, 241.   34
Lewis, R. J. & Freeman, K., 1989. Preprint.   139
Lightman, A. P. & White, T. R., 1988. *Astrophys. J.*, **335**, 57.   90
Lin, C. C. & Lau, 1979. *Stud. Appl. Math.*, **60**, 97.   199, 200, 204, 205
Lin, C. C. & Shu, F. H., 1964. *Astrophys. J.*, **140**, 646.   1, 200, 233, 234, 235
Lin, C. C., Yuan, C. & Shu, F. H., 1969, *Astrophys. J.*, **155**, 721.   188
Lin, D. N. C., 1981. *Astrophys. J.*, **242**, 780.   29

Lin, D. N. C., 1989. In *The Formation and Evolution of Planetary Systems*, eds   31
      Weaver, H. A., Patesce, F. & Danly, L., Cambridge University Press, Cam-
      bridge, in press.
Lin, D. N. C. & Papaloizou, J. C. B., 1980. *Mon. Not. R. Astron. Soc.*, **191**, 37.   28, 29
Lin, D. N. C. & Papaloizou, J. C. B., 1985. In *Protostars and Planets II*, p. 981,   28, 29, 32, 34, 35
      eds. Black, D. & Matthews, M. S., University of Arizona Press, Tuscon,
      AZ.
Lin, D. N. C. & Papaloizou, J. C. B., 1986. *Astrophys. J.*, **307**, 395.   39
Lin, D. N. C. & Papaloizou, J. C. B., 1986. *Astrophys. J.*, **309**, 846.   39
Lin, D. N. C., Papaloizou, J. C. B. & Faulkner, J., 1985. *Mon. Not. R. Astron. Soc.*,   36
      212, 105.
Lin, D. N. C., Papaloizou, J. C. B. & Ruden, S. P., 1987. *Mon. Not. R. Astron.*   23
      Soc., **227**, 75.
Lin, D. N. C., Papaloizou, J. C. B. & Savonije, G. J., 1989. In preparation.   32, 38
Lin, D. N. C. & Pringle, J. E., 1987. *Mon. Not. R. Astron. Soc.*, **225**, 607.   32, 33, 103
Lin, D. N. C. & Pringle, J. E., 1989. In preparation.   34
Lin, D. N. C. & Tremaine, S. D., 1983. *Astrophys. J.*, **264**, 364.   141
Lindzen, R. S. & Barker, J. W., 1985. *J. Fluid Mech.*, **151**, 189.   122
te Lintel Hekkert, P., Caswell, J. L., Habing, H. J., Norris, R. P. & Haynes, R. F.,   139
      1989. In preparation.
Lissauer, J. J., 1985. *Nature*, **318**, 544.   22
Lissauer, J. J., 1985. *Icarus*, **62**, 425.   9, 11
Lissauer, J. J. & Cuzzi, J. N., 1982. *Astron. J.*, **87**, 1051.   5, 12
Lissauer, J. J. & Cuzzi, J. N., 1985. In *Protostars and Planets II*, p. 920, eds. Black,   3
      D. C. & Matthews, M. S., University of Arizona Press, Tuscon, AZ.
Lissauer, J. J., Shu, F. H. & Cuzzi, J. N., 1984. In *Planetary Rings*, Proc. IAU   11, 13, 14
      Colloq. **75**, p. 385, ed. Brahic, A., Cepadues, Toulouse.
Lissauer, J. J., Squyres, S. W. & Hartmann, W. K., 1988. *J. Geophys. Res.*, **93**,   14
      13776.
Little, L., Dent, W., Heaton, B., Davies, S. & White, G., 1985. *Mon. Not. R. Astron.*   56
      Soc., **217**, 227.
Lominadze, J. G., Chagelashvily, G. D. & Chanashvili, R. G., 1988. *Letters to Astron.*   205
      J. (Sov.), **14**, 856.
Long, K. N., Helfand, D. J. & Grabelsky, D. A., 1981. *Astrophys. J.*, **248**, 925.   79
Longaretti, P. Y. & Borderies, N., 1986. *Icarus*, **67**, 211.   9, 11, 13, 21
Lumme, K., Irvine, W. M. & Esposito, L. W., 1983. *Icarus*, **53**, 174.   25
Lubow, S. H., Balbus, S. A. & Cowie, L. L., 1986. *Astrophys. J.*, **309**, 494.   211
Lüst, V. R., 1952. *Z. Naturforsch.*, *A*, **7a**, 87.   32, 233
Lynden-Bell, D., 1962. *Mon. Not. R. Astron. Soc.*, **124**, 95.   183
Lynden-Bell, D. & Kalnajs, A. J., 1972. *Mon. Not. R. Astron. Soc.*, **157**, 1.   175
Lynden-Bell, D. & Pringle, J. E., 1974. *Mon. Not. R. Astron. Soc.*, **168**,   27, 28, 32, 53, 232, 233
      603.
MacKenty, J., 1989. *Astrophys. J.*, (in press).   100
Madau, P., 1988. *Astrophys. J.*, **327**, 116.   94
Malkan, M. A., 1983. *Astrophys. J.*, **268**, 582.   90
Malkan, M. A. & Sargent, W. L. W., 1982. *Astrophys. J.*, **254**, 22.   89
Maraschi, L. *et al.*, 1987. Multi-frequency Observations of the Sy 1 Galaxy 3C 120,   95
      In *Evidence of Activity in Galaxies*, IAU Symp. **121**, p215, Byurukan.
Marcaide, J., Torrelles, J. M., Gusten, R., Menten, K., Ho, P. T. P., Moran, J. &   56
      Rodríguez, L. F., 1988. *Astron. Astrophys.*, **197**, 235.
Marochnik, L. S. & Suchkov, A. A., 1984. *Galaxy*, Nauka, Moscow.   186, 195

Mason, K. O., 1986. In *The Physics of Accretion onto Compact Objects*, p. 29, eds. 83
    Mason, K.O., Watson, M. G. & White, N. E., Springer-Verlag, New York.

Massevitch, A. G. & Tutukov, A. V., 1988. *Stellar evolution: theory and observation*, 72
    Nauka, Moscow.

Matsuda, T., Inoue, M., Sawada, K., Shima, E. & Wakamatsu, K., 1987. *Mon. Not.* 99
    R. Astron. Soc., **229**, 295.

May, A., van Albada, T. S. & Norman, C. A., 1985. *Mon. Not. R. Astron. Soc.,* 142
    214, 131.

Maxwell, J. C., 1859. *On the stability of the motions of Saturn's Rings*, Cambridge 1
    and London: MacMillan and Co. Also reprinted in *Maxwell on Saturn's Rings*,
    1983. p. 73, eds. Brush, S. G., Everitt, C. W. F. & Barber, E., The MIT Press,
    Cambridge, MA.

McClintock, J. E. & Remillard, R. A., 1986. *Astrophys. J.,* **308**, 110. 79

Menten, K. Walmsley, C., Krugel, E. & Ungerechts, H., 1984. *Astron. Astrophys.,* 56
    137, 108.

Merritt, D., 1984. *Astrophys. J.,* **276**, 26. 213, 214

Mestel, L., 1963. *Mon. Not. R. Astron. Soc.,* **126**, 553. 136, 155

Meyer, F. & Meyer-Hofmeister, E., 1981. *Astron. Astrophys.,* **104**, L10. 79

Mezger, P., Chini, R., Kreysa, E. & Wink, J., 1987. *Astron. Astrophys.,* **182**, 127. 56

Mineshige, S. & Osaki, Y., 1983. *Publ. Astron. Soc. Jpn,* **35**, 377. 79

Mineshige, S. & Wheeler, J. C., 1989. *Astrophys. J.,* **343**, in press. 80

Monaghan, J. J. & Lattanzio, J. C., 1985. *Astron. Astrophys.,* **149**, 135. 211

Moneti, A., Forrest, W., Pipher, J. & Woodward, C., 1988. *Astrophys. J.,* **327**, 870. 54

Morfill, G. E., Tscharnuter, W. & Völk, H. J., 1985. In *Protostars and Planets II*, 33
    p. 493, eds. Black, D. & Matthews, M. S., University of Arizona Press, Tuscon,
    AZ.

Morfill, G. E. & Völk, H. J., 1984. *Astrophys. J.,* **287**, 371. 28

Morini, M., Scarsi, L., Molteni, D., Salvati, M., Perola, G. C., Piro, L., Simari, G., 82
    Boksenberg, A., Penston, M. V., Snijders, M. A. J., Bromage, G. E., Clavel,
    J., Elvius, A. & Ulrich, M. H., 1986. *Astrophys. J.,* **307**, 486.

Morozov, A. G., 1977. *Pis'ma Astron. Zh.,* **3**, 195. 186, 192, 193, 194, 195, 196

Morozov, A. G., 1979. *Astron. Zh.,* **56**, 498. 186, 192, 193, 194, 195, 196

Morozov, A. G., 1985. *Astron. J. (Sov.),* **62**, 805. 199, 200, 205

Morozov, A. G., Fainshtein, V. G. & Fridman, A. M., 1976. *Dokl. Akad. Nauk SSSR,* 186
    231, 588.

Morozov, A. G., Fainshtein, V. G., Polyachenko, V.L. & Fridman, A. M., 1976. *As-* 186
    tron. Zh., **53**, 946.

Morozov, A. G., Nezlin, M. V., Snezhkin, E. N. & Fridman, A. M., 1984. *Pis'ma* 186, 194, 195, 196
    Zh. Ehksp. Teor. Fiz., **39**, 504.

Morozov, A. G., Nezlin, M. V., Snezhkin, E. N. & Fridman, A. M., 1985. *Usp.* 186, 194, 195, 196
    Fiz. Nauk, **145**, 161.

Morozov, A. G., Polyachenko, V. L. & Shukhman, I. G., 1974. Preprint of SibIZMI- 199
    RAN, 5-74, Irkutsk.

Mundt, R. & Fried, J., 1983. *Astrophys. J. Lett.,* **274**, L83. 53

Mundy, L., Wilking, B. & Myers, S., 1986. *Astrophys. J. Lett.,* **311**, L75. 56

Muratorio, G. & Friedjung, M., 1988. *Astron. Astrophys.,* **190**, 103. 87

Mushotzky, R. F., 1984. *Adv. Space Res.,* **3**, No. 10-13, 157. 90

Myers, P. C., 1983. *Astrophys. J.,* **270**, 105. 72

Nagagawa, Y., Watanabe, S. & Nakazawa, K., 1989. In *The Formation and Evolution* 30, 37
    of Planetary Systems, eds. Weaver, H. A., Patesce, F. & Danly, L., Cambridge
    University Press, Cambridge, in press.

Nagata, T., Sato, S. & Kobayashi, Y., 1983. *Astron. Astrophys.,* **119**, L1. 52

Nezlin, M. V., Rylov, A. Yu., Snezhkin, E. N. & Trubnikov, A. S. 1987. *Zh. Ehksp.*   187
    Teor. Fiz., **92**, 3.
Nicholson, P.D., Cooke, M. L., Matthews, K., Elias, J. H. & Gilmore, G., 1989.   2
    Icarus, , submitted.
Noguchi, M., 1987. *Astron. Astrophys.*, **203**, 259.   214
Noguchi, M., 1987. *Mon. Not. R. Astron. Soc.*, **228**, 635.   177, 219
Norris, R. P., 1984. *Mon. Not. R. Astron. Soc.*, **207**, 127.   67
Obukhov, A. M., 1949. *Izv.AN SSSR, Geography and Geophysics*, **13**, no. 4, 281.   71
Ohashi, T., 1988. In *Physics of Neutron Stars and Black Holes*, p301, ed. Tanaka,   95
    Y., Universal Academy Press, Tokyo.
Osaki, Y., 1974. *Publ. Astron. Soc. Jpn*, **26**, 429.   79
Ostriker, J. P. & Peebles, P. J. E., 1973. *Astrophys. J.*, **186**, 467.   100, 176, 235
Paczynski, B., 1977. *Acta Astron.*, **28**, 91.   103
Padman, R. & Richer, J. S., 1989. In *Sub-millimetre and Millimetre Wave Astronomy*,   65
    ed. Webster, A. S., Kluwer, Dordrecht, in preparation.
Palmer, P. L., Papaloizou, J. C. B. & Allen, A. J., 1989. *Mon. Not. R. Astron. Soc.*,   175, 176
    to appear.
Papaloizou, J. C. B., Faulkner, J. & Lin, D. N. C., 1983. *Mon. Not. R. Astron. Soc.*,   36
    205, 487.
Papaloizou, J. C. B. & Lin, D. N. C., 1985. *Astrophys. J.*, **285**, 818.   39
Papaloizou, J. C. B. & Lin, D. N. C., 1989. *Astrophys. J.*, in press.   31, 113
Papaloizou, J. C. B. & Pringle, J. E., 1984. *Mon. Not. R. Astron. Soc.*, **208**, 721.   103, 110, 121
Papaloizou, J. C. B. & Pringle, J. E., 1985. *Mon. Not. R. Astron. Soc.*, **213**, 799.   103, 110
Papaloizou, J. C. B. & Pringle, J. E., 1987. *Mon. Not. R. Astron. Soc.*, **225**, 267.   103, 110
van Paradijs, J. & Verbunt, F., 1984. In *High Energy Transients in Astrophysics*,   79
    p. 49, ed. Woosley, S. E., AIP Conf. Proc. **115**, New York.
Paresce, F. & Burrows, C., 1987. *Astrophys. J. Lett.*, **319**, L23.   45, 46, 47
Parmar, A. N., White, N. E., Giommi, P. & Gottwald, M., 1986. *Astrophys. J.*, **308**,   84
    199.
Pasha, I. I., 1985. *Sov. Astron. Lett*, **11**, 1.   217
Pasha, I. I. & Fridman, A. M., 1989. *Zh. Ehksp. Teor. Fiz.*, in press.   195
Pedlosky, G., 1984. *Geophysical Hydrodynamics*, vols 1-2, Mir, Moscow.   186
Petrou, M. & Papayannopoulos, T., 1986. *Mon. Not. R. Astron. Soc.*, **219**, 157.   176
Pfenniger, D., 1984. *Astron. Astrophys.*, **134**, 373.   180
Pfenniger, D., 1989. *Astrophys. J.*, in press.   135
Polyachenko, V. L., 1972. Dissertation, Leningrad Univ.   199
Polyachenko, V. L., 1987. *Astron. Circ. (Sov.)*, No.1490.   202
Polyachenko, V. L. & Fridman, A. M., 1976. See: Fridman & Polyachenko 1984.
Polyachenko, V. L. & Fridman, A. M., 1981. *Pis'ma Astron. Zh.*, **7**, 136.   188
Polyachenko, V. L. & Shukhman, I. G., 1972. Preprint of SibIZMIRAN, 1-2, Irkutsk.   199
Polyachenko, V. L. & Shukhman, I. G., 1981. *Astron. J. (Sov.)*, **58**, 933.   199, 200
Polyachenko, V. L. & Strel'nikov, A. V., 1988. *Astron. Circ. (Sov.)*, No.1529.   202, 204
Polyachenko, V. L. & Strel'nikov, A. V., 1989. To be published.   206
Porco, C., Danielson, G. E., Goldreich, P., Holberg, J. B. & Lane, A. L., 1984. *Icarus*,   12
    60, 17.
Porco, C., *et al.*, 1984. *Icarus*, **60**, 1.   12
Porco, C. C. & Goldreich, P., 1987. *Astron. J.*, **93**, 724.   17, 22
Pounds, K. A., 1985. In *Galactic and Extragalactic Compact X-Ray Sources*, p261,   94
    eds. Tanaka, Y. & Lewin, W. H., ISAS, Tokyo.
Prestwich, A., 1985. *MSc thesis*, University of Manchester.   67
Priedhorsky, W. C. & Holt, S. S., 1987. *Space Sci. Rev.*, **34**, 291.   79
Pringle, J. E., 1981. *Annu. Rev. Astron. Astrophys.*, **19**, 137.   73

Sellwood, J. A., 1986. In *The Use of Supercomputers in Stellar Dynamics*, Lecture   157
    Notes in Physics **267**, p. 5, eds. Hut, P. & McMillan, S., Springer-Verlag, New
    York.

Sellwood, J. A., 1989. In *Nonlinear Phenomena in Vlasov Plasmas*, p. 87, ed. Doveil,   155
    F., Éditions de Physique, Orsay.

Sellwood, J. A. & Athanassoula, E., 1986. *Mon. Not. R. Astron. Soc.*, **221**, 195.   235

Sellwood, J. A. & Carlberg, R. G., 1984. *Astrophys. J.*, **282**, 61.   157, 159, 170, 176

Sellwood, J. A. & Kahn, F. D., 1989. In preparation.   162, 167, 168

Sellwood, J. A. & Lin, D. N. C., 1989. *Mon. Not. R. Astron. Soc.*, in press.   31, 32, 168, 169

Shakura, N. I. & Sunyaev, R. A., 1973. *Astron. Astrophys.*, **24**, 337.   28, 33, 81

Shakura, N. I. & Sunyaev, R. A., 1976. *Mon. Not. R. Astron. Soc.*, **175**, 613.   82

Shaviv, G. & Wehrse, R., 1986. *Astron. Astrophys.*, **159**, L5.   76

Shields, G. A., 1978. *Nature*, **272**, 706.   89

Shlosman, I. & Begelman, M. C., 1987. *Nature*, **329**, 810.   99

Shlosman, I., Frank, J. & Begelman, M. C., 1989. *Nature*, **338**, 45.   99

Shostak, G.. & van der Kruit, P. C., 1984. *Astron. Astrophys.*, **132**, 20.   131

Showalter, M. R., Cuzzi, J. N., Marouf, E. A. & Esposito, L. W., 1986. *Icarus*, **66**,   21
    297.

Shu, F. H., 1968. *PhD thesis*, Harvard University.   188

Shu, F. H., 1984. In *Planetary Rings*, p513, eds. Greenberg, R. & Brahic, A.,   3, 4, 8
    University of Arizona Press, Tuscon, AZ.

Shu, F. H., Adams, F. C. & Lizano, S., 1987. *Annu. Rev. Astron. Astrophys.*, **25**,   27, 49
    23.

Shu, F. H., Cuzzi, J. N. & Lissauer, J. J., 1983. *Icarus*, **53**, 185.   2, 5, 6, 7, 9, 13

Shu, F. H., Dones, L., Lissauer, J. J., Yuan, C. & Cuzzi, J. N., 1985. *Astrophys.*   8, 9, 11, 13, 21, 25
    J., **299**, 542.

Shu, F. H., Yuan, C. & Lissauer, J. J., 1985. *Astrophys. J.*, **291**, 356.   9, 21

Siemiginowska, A. & Czerny, B., 1989. In *Theory of Accretion discs*, eds. Duschl, W.   95
    & Meyer, F., NATO Conference Series, in press.

Simkin, S. M., van Gorkom, J., Hibbard, J. & Su, H.-J., 1987. *Science*, **235**, 1367.   132

Simkin, S., Su, H. & Schwarz, M. P., 1980. *Astrophys. J.*, **237**, 404.   100

Simon, M., Peterson, D. M., Longmore, A. J., Storey, J. W. V. & Tokunaga, A. T.,   54, 55, 226
    1985. *Astrophys. J.*, **298**, 328.

Simonson, G. S., 1982. *PhD thesis*, Yale University.   150

Sitko, M. L., 1986. In *Continuum Emission in Active Galactic Nuclei*, p29, N.O.A.O.,   95
    Tucson, AZ.

Sivagnanam, P. & Le Squeren, A. M., 1986. *Astron. Astrophys.*, **168** 374.   139

Skaley, D., 1985.   "Sphärische Dynamos mit differentieller Rotation und   151
    ortsabhängiger Leitfähigkeit", Diploma thesis, University of Freiburg.

Smak, J., 1969. *Acta Astron.*, **19**, 155.   85

Smith, B. A. *et al.*, 1979. *Science*, **204**, 951.   2

Smith, B. A. *et al.*, 1981. *Science*, **212**, 163.   12

Smith, B. A. *et al.*, 1986. *Science*, **233**, 43.   17

Smith, B. A. & Terrile, R. J., 1984. *Science*, **226**, 1421.   45

Smith, B. A. & Terrile, R. J., 1987. *Bull. Am. Astron. Soc.*, **19**, 289.   45, 48

Snell, R. L., 1987, In *Star Forming Regions*, IAU Symposium **115**, p. 213, eds. Pe-   50
    imbert, M. & Jugaku, J., Reidel, Dordrecht.

Snell, R. L., Bally, J., Strom, S. E. & Strom, K., 1985. *Astrophys. J.*, **290**, 587.   50, 51

Snell, R. L., Loren, R. & Plambeck, R., 1980. *Astrophys. J. Lett.*, **239**, L17.   50

Sofue, Y., Fujimoto, M. & Wielebinski, R., 1988. *Annu. Rev. Astron. Astrophys.*,   151
    24, 459.

Sparke, L. S., 1986. *Mon. Not. R. Astron. Soc.*, **219**, 657.   149

Sparke, L. S. & Casertano, S., 1988. *Mon. Not. R. Astron. Soc.*, **234**, 873.   147
Spruit, H. C., 1987. *Astron. Astrophys.*, **184**, 173.   32
Statler, T. S., 1989. *Astrophys. J.*, **344**, in press.   137, 138
Steiman-Cameron, T. Y. & Durisen, R. H., 1988. *Astrophys. J.*, **325**, 26.   149, 150
Steiman-Cameron, T. Y. & Durisen, R. H., 1989. *Astrophys. J.*, submitted.   149
Strom, K., Strom, S. E., Edwards, S., Cabrit, S. & Skrutskie, M., 1989. preprint.   52
Strom, K., Strom, S. E., Wolff, S., Morgan, J. & Wenz, M., 1986. *Astrophys. J. Suppl. Ser.*, **62**, 39.   53
Strom, S. E., Edwards, S. & Strom, K. M., 1989. preprint.   27
Strom, S. E., Strom, K., Grasdalen, G., Capps, R. & Thompson, D., 1985. *Astrophys. J.*, **90**, 2575.   54, 55
Sun, W.-H. & Malkan, M. A., 1987. In *Supermassive Black Holes*, ed. Kafatos, M., Cambridge University Press, Cambridge.   93, 94
Sun, W.-H. & Malkan, M. A., 1989. *Astrophys. J.*, in press.   91
Sundelius, B., Thomasson, M., Valtonen, M. J. & Byrd, G. G., 1987. *Astron. Astrophys.*, **174**, 67.   174, 217
Svitek, T. & Danielson, G. E., 1987. *J. Geophys. Res.*, **92**, 14,979.   26
Tacconi, L. J. & Young, J. S., 1986. *Astrophys. J.*, **308**, 600.   131
Taff, L. G., 1985. *Celestial Mechanics*, John Wiley and Sons, New York.   223
Tagger, M., Sygnet, J. F. & Pellat, R., 1989. *Astrophys. J. Lett.*, **337**, L9.   181
Tagger, M., Sygnet, J. F. & Pellat, R., 1989. In *Proceedings of a course on Plasma Astrophysics in Varenna (Italy)*, p. 335, ESA SP-285, Noordwijk.   181, 182
Tauber, J., Goldsmith, P. & Snell, R. L., 1988. *Astrophys. J.*, **325**, 846.   56
Telesco, C. M., Becklin, E. E., Wolstencroft, R. D. & Decher, R., 1988. *Nature*, **335**, 51.   47, 143
Telesco, C. M., Becklin, E. E., Wynn-Williams, C. G. & Harper, D. A., 1984. *Astrophys. J.*, **282**, 427.   143
Telesco, C. M. & Decher, R., 1988. *Astrophys. J.*, **334**, 573.   144
Thomasson, M., Donner, K. J., Sundelius, B., Byrd, G. G., Huang, T.-Y. & Valtonen, M. J., 1989. *Astron. Astrophys.*, **211**, 25.   217
Thompson, L. A., 1981. *Astrophys. J. Lett.*, **244**, L43.   214
Thompson, W. T., Irvine, W. M., Baum, W. A., Lumme, K. & Esposito, L. W., 1981. *Icarus*, **46**, 187.   25
Tilanus, R. P. J. & Allen, R. J., 1989. *Astrophys. J. Lett.*, **339**, L57.   129
Tilanus, R. P. J., Allen, R. J., van der Hulst, J. M., Crane, P. C. & Kennicutt, R. C., 1988. *Astrophys. J.*, **330**, 667.   129
Tohline, J. E., Simonson G. F. & Caldwell, N., 1982. *Astrophys. J.*, **252**, 92.   150
Toomre, A., 1964. *Astrophys. J.*, **139**, 1217.   31, 108, 119, 199, 233, 234
Toomre, A., 1977. *Annu. Rev. Astron. Astrophys.*, **15**, 437.   103, 156, 173, 176
Toomre, A., 1981. In *Structure and Evolution of Normal Galaxies*, p. 111, eds Fall, S. M. & Lynden-Bell, D., Cambridge University Press, Cambridge.   103, 156, 160, 173, 176, 218, 235
Toomre, A., 1983. In *Internal Kinematics and Dynamics of Galaxies*, IAU Symposium **100**, p. 177, ed. Athanassoula, E. Reidel, Dordrecht.   236, 237
Torrelles, J. M., Cantó, J., Rodríguez, L. F., Ho, P. T. P. & Moran, J., 1985. *Astrophys. J. Lett.*, **294**, L117.   56
Torrelles, J. M., Ho, P. T. P., Moran, J. M., Rodríguez, L. F. & Cantó, J. 1986. *Astrophys. J.*, **307**, 787.   56,61
Torrelles, J. M., Ho, P. T. P., Rodríguez, L. F. & Cantó, J., 1986. *Astrophys. J.*, **305**, 721.   56, 62
Torrelles, J. M., Ho, P. T. P., Rodríguez, L. F. & Cantó, J., 1989. *Astrophys. J.*, submitted.   64

Torrelles, J. M., Ho, P. T. P., Rodríguez, L. F., Cantó, J. & Verdes-Montenegro, L.,   63
    1989. *Astrophys. J.*, in press.
Torrelles, J. M., Rodríguez, L. F., Cantó, J., Carral, P., Marcaide, J., Moran, J. M.   56, 63
    & Ho, P. T. P., 1983. *Astrophys. J.*, **274**, 214.
Tyson, N. D., 1988. *Astrophys. J.*, **329**, L57.   133
Tyson, N. D. & Scalo, J. M., 1988. *Astrophys. J.*, **329**, 618.   133
Vandervoort, P. O., 1970. *Astrophys. J.*, **161**, 67.   188
Verter, F., 1987. *Astrophys. J. Suppl. Ser.*, **65**, 555.   99, 100
Vidal-Madjar, A., Hobbs, L. M., Ferlet, R. & Albert, C. E., 1986. *Astron. Astrophys.*,   43
    167, 325.
Villumsen, J. V., 1985. *Astrophys. J.*, **290**, 75.   140
Vogel, S. N., Kulkarni, S. R. & Scoville, N. Z., 1988. *Nature*, **334**, 402.   129
Wade, R. A., 1984. *Mon. Not. R. Astron. Soc.*, **208**, 381.   93
Wade, R. A., 1988. *Astrophys. J.*, **335**, 394.   76
Walker, H. J. & Wolstencroft, R. D., 1988. *Publ. Astron. Soc. Pac.*, **100**, 1509.   47
Webb, J. & Malkan, M. A., 1986. In *The Physics of Accretion onto Compact Objects*,   93
    p. 15, Lecture notes in Physics, eds. Mason, K. O., Watson, M. G. & White,
    N. E., Springer-Verlag, New York.
Weidenschilling, S. J., 1984. *Icarus*, **60**, 555.   28, 37
Welty, A., Strom, S. E., Hartmann, L. W. & Kenyon, S. J., 1989. preprint.   53
Wetherill G. W. & Stewart, G. R., 1988. *Icarus*, **74**, 543.   28
Wevers, B. M. H. R., 1986. *Astrophys. J. Suppl. Ser.*, **66**, 505.   177
White, N. E., Kaluzienski, J. L. & Swank, J. H., 1984. In *High Energy Transients in*   79, 80
    Astrophysics, p. 31, ed. Woosley, S. E., AIP Conf. Proc. **115**, New York.
White, N. E. & Mason, K. O., 1985. *Space Sci. Rev.*, **40**, 167.   76, 77
Williams, R. E., 1989. *Astron. J.*, in press.   74
Wisdom, J. & Tremaine, S., 1988. *Astron. J.*, **95**, 925.   25
Wolstencroft, R. D., Scarrott, S. M. & Warren-Smith, R. F., 1989. In preparation.   45, 46
Yanke, E., Edme, F. & Lesh, F. 1984. *Special Functions*, Mir, Moscow.   189, 191
Zang, T. A., 1976. *PhD thesis*, MIT.   156
de Zeeuw, P. T., 1985. *Mon. Not. R. Astron. Soc.*, **216**, 273.   184
Zinnecker, H., Chelli, A. & Perrier, C., 1987. In *Star Forming Regions*, IAU Sympo-   54
    sium **115**, p. 71, eds. Peimbert, M. & Jugaku, J., Reidel, Dordrecht.

Index of authors

Co-authors who did not attend the conference are given in italics.

Subject index

Page numbers are given in bold where the subject is the principal topic of the paper.

Printed in the United States
By Bookmasters